Food Technology Disruptions

Food Technology Disruptions

Edited by

Charis M. Galanakis
Galanakis Laboratories, Chania, Greece
Food Waste Recovery Group, Vienna, Austria

ELSEVIER
ACADEMIC PRESS
An imprint of Elsevier

Academic Press is an imprint of Elsevier
125 London Wall, London EC2Y 5AS, United Kingdom
525 B Street, Suite 1650, San Diego, CA 92101, United States
50 Hampshire Street, 5th Floor, Cambridge, MA 02139, United States
The Boulevard, Langford Lane, Kidlington, Oxford OX5 1GB, United Kingdom

Notices

Knowledge and best practice in this field are constantly changing. As new research and experience broaden our understanding, changes in research methods, professional practices, or medical treatment may become necessary.

Practitioners and researchers must always rely on their own experience and knowledge in evaluating and using any information, methods, compounds, or experiments described herein. In using such information or methods they should be mindful of their own safety and the safety of others, including parties for whom they have a professional responsibility.

To the fullest extent of the law, neither the Publisher nor the authors, contributors, or editors, assume any liability for any injury and/or damage to persons or property as a matter of products liability, negligence or otherwise, or from any use or operation of any methods, products, instructions, or ideas contained in the material herein.

Library of Congress Cataloging-in-Publication Data
A catalog record for this book is available from the Library of Congress

British Library Cataloguing-in-Publication Data
A catalogue record for this book is available from the British Library

ISBN: 978-0-12-821470-1

Publisher: Charlotte Cockle
Acquisitions Editor: Nina Rosa de Araujo Bandeira
Editorial Project Manager: Lena sparks
Production Project Manager: Kumar Anbazhagan
Cover Designer: Greg Harris

Typeset by TNQ Technologies

Contents

Contributors

Sena Bakir, Department of Food Engineering, Faculty of Chemical and Metallurgical Engineering, Istanbul Technical University, Maslak, Istanbul, Turkey; Department of Food Engineering, Recep Tayyip Erdogan University, Merkez, Rize, Turkey

Pradip Kumar Bala, Indian Institute of Management Ranchi, Suchana Bhawan, Ranchi, Jharkhand, India

Daniela Braconi, Department of Biotechnology, Chemistry, and Pharmacy, Department of Excellence, University of Siena, Siena, Italy

Maria Rosário Bronze, iBET, Instituto de Biologia Experimental e Tecnológica, Oeiras, Portugal; Instituto de Tecnologia Química e Biológica António Xavier, Universidade Nova de Lisboa, Oeiras, Portugal; iMed. Ulisboa, Faculdade de Farmácia, Universidade de Lisboa, Lisboa, Portugal

Esra Capanoglu, Department of Food Engineering, Faculty of Chemical and Metallurgical Engineering, Istanbul Technical University, Maslak, Istanbul, Turkey

Gizem Catalkaya, Department of Food Engineering, Faculty of Chemical and Metallurgical Engineering, Istanbul Technical University, Maslak, Istanbul, Turkey

Vittoria Cicaloni, Department of Biotechnology, Chemistry, and Pharmacy, Department of Excellence, University of Siena, Siena, Italy; Toscana Life Sciences Foundation, Siena, Italy

Ian R. Cole, Research Centre on Interactive Media, Smart Systems, and Emerging Technologies (RISE), Nicosia, Cyprus; University of Cyprus, Nicosia, Cyprus

Leonor Costa, iBET, Instituto de Biologia Experimental e Tecnológica, Oeiras, Portugal

Maguluri Sree Devi, Kittur Rani Channamma College of Horticulture, Arabhavi, Paramaddi, Karnataka, India; University of Horticultural Sciences, Bagalkot, Karnataka, India

Linh Duong, The National Centre for Food Manufacturing, University of Lincoln, Holbeach, United Kingdom

Farah Bader, Wolfson School of Mechanical, Electrical and Manufacturing Engineering, Loughborough University, Loughborough, United Kingdom

Arnout R.H. Fischer, Wageningen University, Marketing, and Consumer Behaviour Group, Wageningen, The Netherlands

Guillermo Garcia-Garcia, Department of Chemical & Biological Engineering, The University of Sheffield, Sheffield, United Kingdom

Burcu Guldiken, Department of Food and Bioproduct Sciences, University of Saskatchewan, Saskatoon, SK, Canada

Zehra Gulsunoglu, Department of Food Engineering, Faculty of Chemical and Metallurgical Engineering, Istanbul Technical University, Maslak, Istanbul, Turkey

Sandeep Jagtap, Sustainable Manufacturing Systems Centre, School of Aerospace, Transport & Manufacturing, Cranfield University, United Kingdom

Andreas Kamilaris, Research Centre on Interactive Media, Smart Systems, and Emerging Technologies (RISE), Nicosia, Cyprus; Pervasive Systems Group, University of Twente, Enschede, The Netherlands

Svein Halvor Knutsen, Nofima, Norwegian Institute for Food, Fisheries and Aquaculture Research, Aas, Norway

Manjunath Kudari, Kittur Rani Channamma College of Horticulture, Arabhavi, Paramaddi, Karnataka, India; University of Horticultural Sciences, Bagalkot, Karnataka, India

Jie Li, College of Mechanical Engineering, Donghua University, Shanghai, China

Wayne Martindale, The National Centre for Food Manufacturing, University of Lincoln, Holbeach, United Kingdom

Mahantesha B.N. Naika, Kittur Rani Channamma College of Horticulture, Arabhavi, Paramaddi, Karnataka, India; University of Horticultural Sciences, Bagalkot, Karnataka, India

Michael Nickerson, Department of Food and Bioproduct Sciences, University of Saskatchewan, Saskatoon, SK, Canada

Pankaj Pathare, Department of Soils, Water & Agricultural Engineering, College of Agricultural & Marine Sciences, Sultan Qaboos University, Muscat, Oman

Francesc X. Prenafeta-Boldú, Institute of Agri-Food Food Research and Technology (IRTA), Barcelona, Spain

Shahin Rahimifard, Wolfson School of Mechanical, Electrical and Manufacturing Engineering, Loughborough University, Loughborough, United Kingdom

Arghya Ray, Information Technology Area, FORE School of Management, New Delhi, India

Neil J. Rowan, Bioscience Research Institute, Athlone Institute of Technology, Athlone, Ireland

Dhanush Swaroop Sadhu, Kittur Rani Channamma College of Horticulture, Arabhavi, Paramaddi, Karnataka, India; University of Horticultural Sciences, Bagalkot, Karnataka, India

Stefan Sahlstrøm, Nofima, Norwegian Institute for Food, Fisheries and Aquaculture Research, Aas, Norway

Cátia Saldanha do Carmo, Nofima, Norwegian Institute for Food, Fisheries and Aquaculture Research, Aas, Norway

Annalisa Santucci, Department of Biotechnology, Chemistry, and Pharmacy, Department of Excellence, University of Siena, Siena, Italy

Ana Teresa Serra, iBET, Instituto de Biologia Experimental e Tecnológica, Oeiras, Portugal; Instituto de Tecnologia Química e Biológica António Xavier, Universidade Nova de Lisboa, Oeiras, Portugal

George Skouteris, Institute of Fluid Dynamics, Helmholtz-Zentrum Dresden-Rossendorf, Dresden, Germany

Ottavia Spiga, Department of Biotechnology, Chemistry, and Pharmacy, Department of Excellence, University of Siena, Siena, Italy

Suma Sunagar, Kittur Rani Channamma College of Horticulture, Arabhavi, Paramaddi, Karnataka, India; University of Horticultural Sciences, Bagalkot, Karnataka, India

Mark Swainson, The National Centre for Food Manufacturing, University of Lincoln, Holbeach, United Kingdom

Hana Trollman, Wolfson School of Mechanical, Electrical and Manufacturing Engineering, Loughborough University, Loughborough, United Kingdom

Ellen J. Van Loo, Wageningen University, Marketing, and Consumer Behaviour Group, Wageningen, The Netherlands

Preface

One of the keys to any successful firm in the food sector is being able to come up with innovations to keep operations and products, and services, up to date in a fast-changing world. Innovations are essential for all players involved in the food chain as they keep businesses in a privileged position in the market compared to their competitors. Disruption technologies are a revolutionary type of innovation that causes game-changing shifts in established market structures, companies, and institutions. Nowadays, this type of innovation is growing fast in all fields, including food science and technology. Food companies often rely on the sustainability of their established technologies. However, in some cases, these technologies become obsolete, forcing food businesses to risk and produce disruptive innovations in order to become leaders and game-changers in the market.

Food Waste Recovery Group provides insights for the whole food and environment sectors, publishing books on different topics. The books deal with the valorization of food and plant processing by-products (e.g., from grapes, coffee, cereals, olives, and meat) and microalgae, sustainable food systems, food security and nutrition, food waste recovery technologies, biobased products, biobased industries food-saving efforts, innovations in food analysis, traditional foods, and nonthermal processing, as well as innovations strategies in food and environmental science. The group has also prepared books for food toxicology and forensics, shelf-life, food quality, personalized nutrition, wheat and bread-making, gastronomy and food science, nonalcoholic drinks, and textbooks for specific food components such as lipids, glucosinolates, polyphenols, carotenoids, dietary fiber, steviol glycosides, and proteins.

Following the above considerations, the book covers disruptive technologies that cover the whole spectrum of food production (from farm to fork). The main goal is to aspire to the scientific community and professionals active in the field to develop innovative technologies and novel approaches in food production, processing, and delivery.

The book consists of nine chapters. Chapter 1 provides an introduction to food disruptions. Forecasting what constitutes a disruptive technology for the agri-food industry is complicated as the impact is more likely to be measured from a retrospective downstream reflective process. Living in the era of the COVID-19 pandemic and crisis and postlockdown period, modern-day and future food disruption will be influenced by growing demands to produce more safe and nutritious food to meet increasing populations that will respond to dynamic changes in eating habits, personalized nutrition, and consumer attitudes toward alternative protein sources and climate change.

Personalized nutrition should also take into account internal (genetics, microbiome, metabolome interactions) and external (diet and physical activity) factors, keeping in mind that personal, psychological, and social factors are critical to keeping people motivated and engaged. In this context, high-throughput omics approaches have emerged as optimal tools to study the interindividual response to diet and for the possibility to be integrated into a systems biology perspective. To this line, Chapter 2 describes how omics technologies can influence the field of personalized nutrition and the challenges that still lie ahead.

Chapter 3 introduces the key elements in functional food innovations using food processing by-products and emerging ingredients. The role of carbohydrates, proteins, lipids, bioactive compounds, and minerals are discussed in the point of emerging technologies, new ingredients, and the challenges in functional food development. The fortification of food products with functional ingredients to enhance its organoleptic properties, physicochemical properties, preservation, and morphological properties are also discussed.

Chapter 4 introduces alternative protein products as emerging ingredients in the protein industry. It starts by presenting an overview of different alternative protein sources such as those that are land-based (pulses and cereals), microalgae, and insects. It explains how wet and dry extraction methods are applied together with pretreatment processes to recover protein-rich fractions (concentrates and isolates) from these sources. Moreover, a discussion on how these alternative protein products will assist the development of meat analogs is presented.

Chapter 5 focuses on the utilization of the Internet of Things (IoT) technologies in the food supply chain, specifically post farmgate until food reaches the retailer's shelves. It presents relevant applications that have been successfully developed and deployed in food supply chains, as well as advantages and disadvantages of their implementation, and finally, future trends such as elements of Industry 4.0, blockchain, intelligent packaging, and artificial intelligence.

Chapter 6 discusses innovative distribution and delivery of food, before presenting the results of corresponding semistructured interviews and from user-generated data in order to provide an overview of the customer's viewpoints. Food quality, food safety, and sustainability, based on strategic, tactical, and operational planning are also discussed. Finally, modifications of strategies for the effective operations' handling in case of emergency outbreaks are explored.

Chapter 7 focuses on the application of blockchain technology in agriculture and the food supply chain. A wide range of relevant recently finished or ongoing projects and initiatives are introduced, noting relevant barriers, challenges, potential benefits, and opportunities, addressing the question of whether blockchain has matured enough as a technology to be effectively used in real-life applications within the agri-food industry. The use of blockchain as a driver toward a transparent food supply chain is confirmed. However, numerous barriers (e.g., technical, policy, regulatory, and educational issues) still exist, hindering its broader adoption in agri-food systems.

Chapter 8 covers various digital tools that allow delivering agricultural information to farmers. The efficiency of these tools is discussed, taking into account a supporting case study dealing with the impact of digital extension services on farmer's agricultural practices in selected villages of Belagavi district, Karnataka, India.

In Chapter 9, insights about the public's response to disruptive and radical food innovations are presented. After that, the current knowledge of how the public responds to several currently emerging innovations is discussed. These responses can roughly be divided into the search for novel proteins and the convergence of information technology and microelectronics with food production and marketing.

Conclusively, the current book assists food technologists, engineers, scientists, agriculturalists, and chemists working in the food science field, as well as new product developers, researchers, business innovators, policymakers, and commercial entities looking for innovative opportunities in the food supply chain. University Libraries and Institutes could also use it all around the world as ancillary reading in undergraduates and postgraduate level multidiscipline courses dealing with food science and technology, as well as food innovation.

I would like to thank and acknowledge one by one all the authors for their fruitful collaboration and their dedication to editorial guidelines and timeline. I am fortunate to have had the opportunity to collaborate with many international experts from Canada, China, Cyprus, Italy, India, Ireland, Oman, The Netherlands, New Zealand, Spain, Turkey, and the United Kingdom. I would also like to thank the acquisitions editor Nina Bandeira, the book managers Laura Okidi and Lena Sparks, as well as all colleagues from Elsevier's production team, for their assistance during the preparation of this book. Finally, a message for all the readers: those collaborative efforts of hundreds of thousands of words may contain gaps or minor errors. Suggestions, comments, and even criticism are always welcome. In that case, please do not hesitate to contact me to discuss any relevant issues.

Charis M. Galanakis
Research and Innovation Department, Galanakis Laboratories, Chania, Greece
Food Waste Recovery Group, ISEKI Food Association, Vienna, Austria
charismgalanakis@gmail.com

Chapter 1

Introduction to food disruptions

Neil J. Rowan
Bioscience Research Institute, Athlone Institute of Technology, Athlone, Ireland

1.1 Introduction

The agri-food sector is one of the largest manufacturing sectors globally and comprises a dynamic societal-technical innovation ecosystem (Rowan, 2019; Saguy, Roos, & Cohen, 2018). In the EU, this increasingly important sector accounts for €1098 billion turnovers and employs 4.24 million (Saguy et al., 2018). Over the last decade, the food and beverage industry has doubled in size in the USA (Rowan & Galanakis, 2020). The food and drink industry was estimated to be worth £6 trillion in 2015, with packaging comprising almost £1.9 trillion of this value, where digital innovation is rapidly influencing the pace and scale of change (Rowan & Galanakis, 2020). Food manufacturers invested ca. $18 billion in capital expenditures in 2016 (Rowan & Galanakis, 2020). There is an increasing demand for the supply of safe, nutritious food that echoes future projections that support global population growth (Michelini, Principato, & Iseavoli, 2018; (Rowan & Galanakis, 2020). The aforementioned brings challenges and opportunities where diversification of the food supply chain will meet altering diets that respond to increasingly aging, ethnic and cultural populations, diet-related diseases, more personalized products, and the possible emergence of innovations and services to address void created by coronavirus disease (COVID-19) pandemic. The global COVID-19 pandemic has produced a paradigm shift for society, and the food industry is meeting this challenge. However, it will also slow economic recovery or occurrence of the second wave of infection will create additional opportunities for innovators and services such as online retail and deliveries (Rowan & Laffey, 2020). DBEI (2018a, 2018b) projects that there will be a ca. 70% rise in demand for more food products and services over the next 40 years.

Opportunities will be met in part by advances in the digitization of food technologies, processes, and services for a diversity of markets along with commensurate sustaining and disruptive innovation in the adjacent

Food Technology Disruptions. https://doi.org/10.1016/B978-0-12-821470-1.00005-7

manufacturing and materials sectors. In the years ahead, it is envisaged that organic, unprocessed, and healthy food will drive growth in domestic markets (Statista, 2020). For example, the estimated value of shipments of the industry was US$ 795.4 billion in 2019, where. 15.1% of the value of shipments was generated from dairy product manufacturing. Exploiting added-value from premium dairy products has led to the establishment of the dedicated VistaMilk Center in the Republic of Ireland that combines Agri-Food with information communications technology (ICT) research institutes along with prominent companies. VistaMilk Center is funded by Irish Department of Agriculture Food and the Marine (DAFM) and Science Foundation Ireland (SFI) that will forge sustaining and disruptive innovation across animal and human health, integrated and rapid sensing, and communications for intensive sustainability, with an environmental orientation (Science Foundation Ireland, 2020).

The primary purpose of this introductory chapter is to provide a background describing advances in the agri-food sector in the context of articulating what constitutes technology disruption across this domain and potential new emergent disruptors where many activities are evident across many domains, which are addressed in the proceeding linked chapters. The overarching goal is to introduce fundamental, conceptual, and best applied-knowledge underpinning food disruptions, where a more profound and broader appreciation of how these technologies disrupt social and innovation ecosystems are seen to be adaptive and dynamic. Specifically, this also includes a high-level introduction to digital technologies in agriculture; digital disruption in the food industry; personalized nutrition and omics technologies; changes in eating habits; alternative protein sources; artificial meat; innovations in functional foods development, trends in smart packaging; 3D printing; electronic nose for food authentication; the Internet of Things (IoT). Technologies in the food supply chain; innovative distribution and delivery of food; social acceptability of food disruptions; blockchain in agriculture; digital extension service; food-drug interactions; digital technologies and personalized nutrition; food choice, personalized nutrition, and food sustainability; and IoT in the food sector. This book will also inform new education and training programs for the food industry and a variety of other stakeholders globally.

1.1.1 Challenges and opportunities presented by the need to meet food sustainability

Future intensive sustainability of the food sector will also be influenced by pressures applied to supply chain, including uncertainties associated with the impact of global warming on crops that will include more flooding and droughts (O'Neill, Rowan, & Fogarty, 2019). A higher drive to innovate will also lead to commensurate needs to balance the impact on the environment with the emergence of less-energy intensive, eco-friendly processes, products, and services (O'Neill et al., 2019). Fisheries and seafood are viewed as desirable high protein, low carbon-intensive products with the emergence of smart aquaculture processes to meet growing consumer demands (Tahar et al., 2018a); (Tahar et al., 2018b); O'Neill et al., 2019). The role of predictive modeling to inform efficacy for adjacent water security is also becoming more popular (Tahar, Tiedeken, Clifford, Cummins, & Rowan, 2017, Tahar, Tiedeken, & Rowan, 2018c). However, Ruis-Salmón et al. (2020) also reported that the seafood and aquaculture sectors across European countries are embracing opportunities to mitigate key environmental pressure points (depletion of resources and climate change), social needs (changing customer attitudes and preferences) or growth in markets (services and business processes along with enhanced competition and worldwide competitiveness). These pressing challenges are influencing the innovation ecosystem from citizens to policymakers to adopt and foster more sustainable practices. There is a commensurate need to harness and accelerate a diversity of partnerships that traverses geographical boundaries in order to generate more effective and efficient networks across seafood and aquaculture sectors along the entire food supply chain. Ruis-Salmón et al. (2020) stated that such challenges and opportunities would be addressed by "a convergence of thinking" in a connected innovation ecosystem that exploits advances in life cycle assessment and modeling that will enable a sustained unite progression to a circular economy. This timely review highlighted that interfaced between food-energy and water will support the assessment of the life cycle of seafood products and services, which will include tracking trends for regional limitations and strengths. This fact will lead to the sharing of new knowledge for add value across European seafood and aquaculture sector, including innovation in ecolabeling and ecodesign that will have far-reaching and cross-cutting influences to the circular economy. Smart innovations in these areas may lead to disruptive products and businesses.

1.1.2 What are disruptive technologies?

Disruptive technologies or disruptive innovations were initially defined to address market disruption in established markets, where a new product or service (a technology) is introduced (Bower & Christensen, 1995; Christensen, Anthony, & Roth, 2004). Sequentially, over the past 20 years, several researchers have expanded upon the theories of Bower and Christensen (1995) to include low and high-end disruptions to meet convergence of new opportunities from adjacent domains (Govindarajan & Kopalle, 2006; Schuelke-Leech, 2018). DTs arise from a global drive to discover innovations that will lead to greater competitiveness, impact and value to businesses and society (Christensen & Bower, 1996; Geels, 2018; Laurer & Dgostino, 2013; Li, Porter, & Suominen, 2018; Sousa & Rocha, 2018; Yongfu et al., 2017). However, several researchers have espoused to expand upon the classic pattern of disruptive innovation identified by Clayton Christensen in 1997 over these past 2 decades, which may not be aligned with his original thinkings. Christensen recently reaffirmed his definition of disruption that is a theory of competitive response. If one innovates in a certain way, then it is envisaged that the incumbent competitors would be expected to do similarly. If one introduces a sustaining innovation, then the incumbents will typically endeavor to mount a defense with a view to eliminating me. However, if the innovation is disruptive, then one is likely to be ignored, or competitors will flee rather than mounting a fight (Denning, 2016).

(Denning, 2016), in his recent "Christensen Updates Disruption Theory" paper, stated that "this is a theory in which an unobtrusive competitor eats away at the low end of an incumbent's market with a lower quality product. The incumbent is happy to concede the low-value customers and concentrates on adding more features for its base of high-value customers. Next, the disruptor steadily improves quality to move up-market, and then devours the whole market of the incumbent, who often does not perceive the threat until it is too late".

Since the 1990s, researchers have referred to DTs as a whirlwind, ground-breaking, game-changing, earth-quake, and emergent technologies that typically cause a substantial disturbance in established market structure and prominent companies by producing highly efficient products and services that are more competitively priced, less complicated and more accessible than established innovations (Christensen & Bower, 1996; Christensen, 1997; Schuelke-Leech, 2018). Innovations may be viewed as disruptive when they take the place of established or broadly accepted ideas arising from scientific inquiry, or in methodologies or in paradigms that causes disruption in knowledge (Kuhn, 1962). Schuelke-Leech, (2018) also reminded us that disruptions could also be seen in legal and regulatory settings, such as the withdrawal of the United Kingdom as a joint member of the European Union that was commonly referred to as Brexit. DTs are seen as different from sustaining technologies (ST) that offer incremental improvements over products and services already known. However, given the potential impact of DTs on businesses and society, most new technologies are considered as sustaining (Garrison, 2009).

Schuelke-Leech, (2018) recently noted that DTs have historically presented challenges to executive management by way of uncertainty and flux in appreciation and deployment of innovations framed upon a level of prior familiarity, transparency, and experiences in discerning strengths, weaknesses, opportunities, and threats for these technologies. A limiting factor in the uptake of new potential DTs may relate to poor-decision making in adopting and embracing these innovations. However, the challenges for technological forecasters and investors it that DTs are by their nature nascent, meaning that they can only be proven as disruptive in hindsight based upon demonstrating evidence-based impact. This fact infers that technology investors must have an appreciation of what constitutes technology disruption in terms of evaluating candidate innovations that have the potential for paradigm disruptive shift (i.e., DTs), as opposed to an incremental sustaining drift (i.e., STs). Review of the best evidence on this subject highlights that one can only assert "potential" as proof of efficacy for disruption for new product or service as this can only be determined proper on review of the impact on the marketplace by end-users and its utility for application across different domains. Schuelke-Leech, (2018) described that new disruptions might be much more massive if they emerge from disrupting the model of capitalism, organizational structures, or social interactions, which are steps beyond transforming the marketplace or existing technological paradigm. It is evidenced by Beinhocker (2006) that predicting the type of disruption can be challenging in a complex system, such as for innovation that often have multiple and dynamic contributions.

In recent times, definitions of DT focus on broad factors affecting the industry and address the nexus between learning experience arising from substitutable innovations that relate explicitly to competitive pricing and performance (Rowan, 2019). The recent review by Beth Ann Schuelke-Leech, (2018) provides an excellent insight into a diversity of DTs. The author describes disruptive-products that are reduced in size, such as exploiting leading developments in nanotechnology. Those that are more lightweight and efficient, such as exploiting additive manufacturing and material science. Those that are more competitively and affordably priced, such as exploiting resource management and manufacturing, including advances in innovative service and business processes. Where other products are exhibit more excellent dexterity and convenience in design and functionality that includes exploiting researcher creativity blended with artificial intelligence, augmented, and virtual reality that includes future-proofing for needs across various platforms. In addition, products that are more significant performing products and services, such as exploiting Physico-chemical developments combined with the use of robotics and AI for design linked to advances in education and workforce training). For example, this author's research group presented for the first time at Kilmer Sterilization Conference on the combined novel use of educational and immersive (augmented and virtual reality) technologies to inform remote workforce training for adjacent medical technology and terminal sterilization industry; this has disruptive potential, but only over time will this be proven (Murray, Buckley, Seery, & Rowan, 2019). Similarly, this concept may be applied to the food industry for the introduction and training of new technologies across the supply chain from production, distribution, and storage. Developing DTs in the agri-food domain is core to supporting and driving national strategic development plans as these generate the job, add-value, troubleshoot, and enhance quality in changing marketplaces. Schuelke-Leech, (2018) provided an excellent review of the factors that underpin why potentially do localized technological innovations lead to much more significant technological disruptions, which cause a larger society where the more significant longer-term impact occurs. Rowan (2019) recently reviewed the development and potentially disruptive technological potential of pulsed light technology for the food and adjacent industries where he reviewed vital factors and potential magnitude of disruption.

Christensen previously explained that disruptive innovation influenced by three theories. The first is the "Disruption Innovation Theory," where organizations use simple, convenient, cost-effective innovations. The second is the "Resources, Processes, and Values (RPV) theory," where these composite theories frame a company's strengths and weaknesses. The third is the "Value Chain Evolution Theory," where a company requires to control its value chain and solve problems that, if not addressed, would inhibit it from harnessing value from these critical activities) (Christensen et al., 2004). Other scholars have pursued the expansion of Christensen's definition of disruptive technologies, which includes potentially broadening these to include both high and low-end disruptions (Govindarajan & Kopalle, 2006) and potentially distinguishing between disruptive technology and disruptive innovation (Yu & Hang., 2010). Schuelke-Leech, (2018) stated that this presents challenges in defining what constitutes technology disruption as there are several levels of disruption. However, as stated previously, these may not be aligned with the original thinking of Clayton Christensen. (Denning, 2016) reported that Christensen feels that the core concept of the disruption theory has been broadly misunderstood and misapplied due to the success that it has garnered for the past 2 decades, as attested in part by highly effective use of these theories by leading companies and institutions. The theory of disruption, according to Christensen, "is a theory of competitive response. Disruption is a process, not an event, and innovations can only be disruptive relative to something else. Over the last 20 years, little by little, we have realized that we need new theories to account for what is going on." (Denning, 2016). This author reports that Christensen recognizes the existence of multiple patterns of disruptions in today's marketplace. Christensen explained that three types of innovations play different roles in the economy, namely (1) market-creating innovation plays a role in growth (2), sustaining innovations make right products better, and (3), which was not recognized in original thinking of 1997 that the role of efficiency innovations eliminate jobs.

1.1.3 Orders of magnitude for disruptive technologies

In the exciting work of Schuelke-Leech, (2018), Beth Ann described a conceptual model to understand the orders of magnitude of DTs that may disrupt markets, businesses, institutions, and the societal norm, which constitute "the innovation ecosystem." Specifically, such disruptions occur at two different levels. Technologies or innovations that constitute disruption at first order reflects a localized change in a given marketplace or industry sector that aligns with Christensen's conceptualization of disruptive technology. Schuelke-Leech, (2018) gave an example of Keurig K-cup single-serve coffee machine (the single-serve plastic coffee pod) as the first-order disruption where 9.8 billion individual coffee pods where sold in 2014. First order-level disruption is the focus of much the business literature where it considers and addresses disrupters in innovation. Schuelke-Leech, (2018) postulated that if one was to that the underpinning concept of this single-serving machine and combined it with a future point with advances in 3D Printing and ICT to create on-demand meals. Then this would have much more extensive and wider-scale influences, potentially affecting many industries and dramatically altering societal norms or situation, and as such, would represent a second-order disruption. Therefore, second-order are technological disruptions that permeate through society, influencing substantial change. Another example of the first to second-order technological conversion would be the discovery of the enzyme polymerase, and it is used in polymerase chain reaction (PCR) technologies that now have disruptive seismic influences across many sectors, including food that includes diagnostic, quality assurance. PCR transforming our ability to work with sophisticated foodborne viruses and parasites in real-time that are not culturable by conventional methods and has been the go-to method for testing the COVID-19 virus in healthcare for a current pandemic. If one then considers introducing AI, AR/VR, and the Internet of Things, this combined concept could be used for valuable remote workforce training using broad throughput PCR as crucial technology that would affect service, business models, and education, which would be additional second-order disruption. For example, many international educational programs seek to provide valuable mobility for researcher training and professional development; the use of virtual remote training on core technologies would influence other order disruption in knowledge provision that currently relies upon more hands-on in situ demonstrations at a defined location (Rowan, 2019).

Second-order disruptions are broader than first-order disruptions. Described by Schuelke-Leech, (2018), second-order DT (1) are not localized and are dynamic advances on existing technologies that frequently combine several separate innovations that might or might not be causing disruption at the first-order level (i.e., locally seen individually as disrupting technologies); (2) emergent innovations that are broadly applied across many different industries; (3) technologies that disrupt current social, institutional norms and standards, operation, production, trends, not limited to a particular market or industry through restructuring, and reorganizing, and (4) technologies that trigger and develop economy-wide growth similar to Kondratieff's waves. However, different second-order DTs may be combined, resulting in a Kondratieff long wave. Kondratieff identified cyclic patterns in capitalist economies in the 1920s (cited Schuelke-Leech, 2018) where others subsequently developed this concept further noting that cycles are seen as linear and sequential and come to dominate and drive economies (Freeman, Clark, & Soete, 1982; Perez, 2002; Schumpeter, 1939a, 1939b). A single dominant technology is at the center of each long wave, such as Wernher Von Braun liquid-propellant rocket engine invented during WWII that disrupted jet aviation and space exploration. Schukeke-Leech (2018) stated that while second-order disruptions are far-reaching than first-order disruptions, they are still considered smaller in influence than long waves.

Advances in ICTs was provided as an example for driving long wave that came through several technological developments in hardware, telecommunications, the Internet, and networking (Ceruzzi, 1999).

First and second-order DTs are considered revolutionary technologies as opposed to innovations that lead to incremental, evolutionary, or continuous developments (Utterback, 1994). There is a great desire to understand the process whereby one can identify candidate technology disruptors (Nagji & Tuff, 2012). Schuelke-Leech, (2018) noted that factors leading to the creation of DTs arise from meeting a localized opportunity (Christensen, 2003); through creativity and problem solving (Rowan, 2019); through financial investments (such as from self-financing to Venture Capital and Angel investors); (Rowan, 2019); through exploiting appropriate networks (Rowan, 2019); through considering broad applicability for an innovative technology (Schuelke-Leech, 2018); and by providing supporting infrastructure and institutions (such as clustering of human capital, networking to enable the innovative process to occur, Drucker, 1985). Governments play a central role in driving in informing the creation of DTs through investment in education along with providing economic and policy conditions enabling start-ups, innovators, or companies to take on the challenge with high-risk products and services (Streeck, 2011). The reader is encouraged to consult the informative workings of Schuelke-Leech, (2018) for a comprehensive understanding of the role disruptive technologies, factors underpinning the orders of magnitude of disruption, and associated modeling of processes that provide critical insight into these innovations across several domains. Climate change, COVID-19, and other uncertain influences were affecting technology disruption.

The UN Environment Program noted that if we are to achieve the Paris Agreement goal of limiting substantial global warming to 1.5°C, then global carbon emissions must be reduced to 7.6% a year. Failure to achieve this aim by 2030 will result in irreversible impacts, including enhanced extreme weather events along with increased existential threats to humans. For example, in the Republic of Ireland, agriculture accounts for a third of carbon emissions where there is an increased interest in innovation and practices in order to help reduce annual emissions from 40 million tonnes to 19.4 million tonnes. This fact is also set against a global background where the National Centers for Environmental Information in the United States has recently reported that January 2020 was the warmest recorded over the past 141 years (NAOO, 2020). This report is also alarming given that the highest January recorded for land and ocean surface temperature internationally at 2.05°F (1.14°C) above the average reported for the 20th-century. It was noted that the past 4 years also represented the four warmest Januaries recorded in the United States; the 10 warmest have all occurred since 2002. This trend in terms of record climate temperatures is similar, as reported by other continents such as Europe (Mullen et al., 2018). The uncertainly of climate change influences agriculture and food production as increased flooding and droughts disrupt food and water systems. Sustaining water quality and security resources will also be a challenge for future agriculture and food production processes (Tiedeken, Tahar, McHugh, & Rowan, 2017). Rowan (2019) has also described the influence of climate change on linked pollination and ecosystem service management, where pollinators are affected by habitat loss, starvation, and complex diseases. Introduction of nonthermal disruptive technologies such as electron-beam or pulsed UV light treatments may facilitate commercial decontamination of pollen for farmed bumblebees that are now cultured to help with loss of our pollinators globally (Naughton, Tiedeken, Garvey, Stout, & Rowan, 2017). The decline of bee pollinators is an alarming statistic that becomes an even more significant cause for concern when the benefits they provide to humans are weighed up, both from a nutritional and economic point of view. The pollination of crops in the USA alone is estimated to be worth more than $14 billion (cited Rowan, 2019), which is provided by honey bees (*Apis mellifera*). In Europe, 78% of all flowering plants are pollinated by animals, with 84% of crops such as strawberries, plums, cucumbers, and rapeseed oil being carried out by insects, worth 15 billion euro per year (Naughton et al., 2017). Wild bees in Ireland provide the bulk of pollination services to various crops and fruit, making them invaluable to the economy and for food variety.

There is a pressing need for smart solutions for ensuring the intensive sustainability of agriculture and food production processes that respond to the challenges of climate change. Although renewable energy deployment is increasing, fossil fuels remain the primary fuel source for heating and transportation for many countries (Schuelke-Leech, 2018), which is despite the general agreement to reduce global greenhouse gas emissions internationally (Intergovernmental Panel on Climate Change, 2019). O'Neill et al. (2019) also reported a negative correlation for the use of biomonitors of effluent quality on receiving water bodies from aquaculture production processes in the Republic of Ireland. Use of the algae produced opposite than expected ecotoxicological findings arising from a 3-month drought experience in the Republic of Ireland in 2018. Specifically, this novel study used the microalga *Pseudokirchneriella subcapitata* in the form of a new natural whole-organism biosensor to examine the conventional physicochemical parameters monitoring in aquaculture farms (namely suspended solids, oxygen, phosphorus, pH, conductivity, nitrogen, and temperature) on the receiving waters in terms of disturbances to ecosystems and organisms. This fact is quite smart, as use of the microalgae allows for cumulative and sequential stressors along with embracing longitudinal effects of exposure time, where the use of conventional physicochemical measurements are grab samples with a specific window of time, where one may experience variances in differences in representativeness for each parameter depending on nature of the method and frequency of testing. There is significant scope to use ICT for both remote and real-time integration of this approach for smart-aquaculture, such as linked to a new mobile phone app as a next-generation management tool for controlling feed rates, BOD, and effluent quality. Studies revealed that the constructed wetland system was negatively influenced by drought conditions as this natural waste remediation process was not able to treat nitrates and phosphates effectively. Therefore, these advances in using novel algal biomonitors for holistic aquaculture process performance are complementary and potentially superior to as an indicator of standard water quality parameters and will provide an early warning tool for both aquaculture process efficiency and to factor in the influence of climate change. The role of ICT, big data, and automation in smart agriculture will play lead role sustainability moving forward. Advances in bioinformatics and next-generation sequencing will also help with improvements in different micro-algae used for this purpose, as well as the determination of microbial populations in the system, including the emergence of pathogens or problematic microorganisms (Naughton, Kavanagh, Lynch, & Rowan, 2020). This fact is particularly relevant as less than 5% of microorganisms are culturable on conventional agar plates from water samples (Fitzhenry, Rowan, del Rio, Cremillieux, & Clifford, 2019; Rowan, 2011; Rowan, Valdramidis, & Gómez-López, 2015). It is envisaged that there will be continued advances and potential for technology disruption in forestry and horticulture for future environmental-proofing such as the delivery of cocktail of helper microorganisms and bioactive compounds through hydrogels from adjacent manufacturing and materials industry to respond variances in climate change and resilience. This fact is likely to be informed by advances in ICT and digital technologies in agriculture and food production.

At the time of writing, the world is experiencing a coronavirus disease (COVID-19) pandemic (cited in Rowan & Laffey, 2020). Given the necessity for food globally, disruption in products and services is likely to emerge from innovations in the delivery and online retail sector as most people remain at home to prevent infection. This is on top of the pressing needs to develop innovative means to increase food production to meet growing populations internationally informed by digital technologies. This infers a focus on food security, such as blockchain and the Internet of things in the food supply chain, safety, including smart packaging, traceability, and alternative, disruptive approaches to food sources such as protein sources. There will be considerable shot-gun market research occurring to review consumer behavior, such for example, through advances in digital extension services. This is additionally increasing preferences for personalized nutrition. Also, and underappreciated, is the potential role of functional foods and nutraceuticals such as in boosting the immune system and wellbeing for tackling COVID-19 infection (Masterson et al., 2019; Carballo et al., 2019).

The coronavirus COVID-19 pandemic crisis will present both challenges and opportunities for the agri-food sector globally. The theory of antifragility (Rowan & Galanakis, 2020; Taleb, 2012) relates to the observation that some things benefit and thrive from shocks, such as exposure to volatility, disorder, and stressors. Antifragility is the concept of things that get better under the conditions of stress and uncertainty. There are pros and cons to following any given method. In terms of potential global economic recovery plans post-COVID-19, and the emergence of Food DTIFs, the significant value may be placed on such things as the review of antibody testing data where it is hoped that epidemiology will show that many people were infected where this may inform a v-shaped recession with a short sharp recovery. A desirable v-shape economic recovery trend may be more likely due to a wave of online shopping and people working from home. Approximately $6.2 trillion (12.5% of retail total) is spent on food and beverage in the USA: 2015 was the year that more food was brought in than prepared in the home. COVID-19 has shocked that trend, in the US in Q2, 100bn dollars shifted from restaurants to retail space. Migration of how people shop online. Monopolies in food grocery services may arise, where smaller independent stores may struggle. In the UK, 7% of the population shop for groceries online, with 4% in the USA. However, one-third of the US population bought online during the second week of March, and half of them was their first time. Confidence must be provided to ensure continuity in the food supply chain to avoid friction in the food system. Too much friction will lead to price increases, which includes friction in labor with seasonal workers shortages. If there is a shortage, consumers pay more, yet producers get less. This is the wrong message or signal to producers as they will produce less, which will lead to more shortages. In addition to preventing shortages in the food supply chain, there is a commensurate need to addressing food waste and remediation as countries. Thus, the use of ICT for supply chain management and to understand consumer consumption patterns during a pandemic is essential.

It is uncertain as to the state with any degree of confidence what would be the specific impact caused by the global downturn (or potential recession) in the economy as it relates to specific needs and opportunities met by emergent technologies in agri-food (including ingress from adjacent industries). However, Joseph Schumpeter's trampoline theory of rebooting an economy with innovative and growth-hungry companies coming to the fore appears plausible. This also includes the likely consolidation of major industries with strong packaging and capacity for research and innovation that will potentially flourish during and post-COVID-19 when socioeconomic norms are reset, and countries quickly deploy economic recovery plans. The flexibility and adaptability of companies to meet change and adjust business models, including provision for ICT, including online delivery for supply chain, will be better placed for sustain and for potentially cause food disruption practices. Reuters recently polled over 50 economists, with some forecasting that the world economy will shrink as much as 6% in 2020 (Reuters, 2020). However, predicted extremes include 0.7% growth to average 1.2% contraction. The V-shaped economic response is the best case outcome: when a commensurate sharp recovery trend follows a growth plunge. Reuters stated that 'the April–June GDP contraction will likely be on a scale not seen for decades.

Nevertheless, fiscal and monetary stimulus, over $10 trillion, may help with an equally swift rebound (Curran, 2020). However, confidence in V-shaped economic recovery is questionable, as noted by (Bulwark, 2020), as one cannot dismiss a combination of a record-high global debt-to-GDP ratio and deepest worldwide economic recession observed for many decades. (Bulwark, 2020) infer that these risks include an undesirable return of the European sovereign debt crisis, a potentially lower long-run growth path due to a sudden shift in the Chinese economy, and waves of debt defaults in emerging markets.

The occurrences of these combined risks may influence the occurrence of a COVID-19-induced recession. Reuters (2020) described that an alternative U-shaped economic recovery might occur, that takes more than a couple of quarters as economies have suffered a faster and deeper, which Reuters feel may be the likeliest outcome. This reflects thinking that lockdown impact may last for a while after their lifting with a gradual easing of the lockdown where social distancing will continue that will continue to influence the tourist industry negatively. This also reflects the situation where there is still no vaccine in play, and therefore, impossible to role out second waves of infection that add to uncertain (Rowan & Laffey, 2020). Reuters (2020) noted other type recovery trends, including W-shaped Double-dip seen where easing of lockdown restrictions initially, boosts activity but effects of unemployment and corporate bankruptcies then start to filter through, and this may occur if there is a second wave of infection. Also, an L-shaped recovery may occur if growth plunges and does not recover for some time. This reflects continued increasing cases of COVID-19 forcing long-term lockdowns. At the time of writing, it is too early to state with any degree of confidence what will be likely shape of economic recovery, but sustaining development and disruption in agri-food innovation, products, and services would help facilitate this recovery process globally.

Ireland, as a nation, has worked very hard to become a leader in food production, as attested by the premium quality and affordable pricing. However, in response to COVID-19, countries may consider nationalizing their supply chains for greater control to avoid reliance on another country. This has led to export bans. Nevertheless, global trade feeds one-third of the world, and producing locally means buying less and the need for more land. COVID-19 may cause a contraction in the extension of the supply chain, and countries will trade with whom they can trust. The question of relative advantage arises; will countries afford to produce things they are not familiar with or can do, or will this be a necessity arising from potential supply chain shortage issues, for example, Personal and Protection Equipment (Rowan & Laffey, 2020)? There is likely to be an increased demand for ICT, including areas such as robotics, blockchain, algorithms to improve processes, efficiency, and sustain or create more jobs. Everyone is experiencing shock at the same time as a consequence of stay-at-home and social distancing policies, yet some people are more affected than others. The agri-food ecosystem is well-positioned to respond as it is very well defined as a platform. Where next disruptive technology emerges to meet needs of COVID-19 has yet to be determined as this situation is rapidly evolving, but strong candidate DT likely to emerge from ICT and innovations in service and business processes for the food sector such as delivery for the supply chain. Industries will need to adapt in real-time, which is challenging, given very little market data available to underpin critical decisions. Galankis (2020) recently highlighted the important position that ensuring security and safety will play in responding to challenges of COVID-19, including future provision for introducing in industry 4.0 tools to mitigate food waste along with potential opportunities to fortifying foods with ingredients to help boost consumer immunity. DTs may emerge from entrepreneurs, start-ups, SMEs, and large companies where there is growth-hunger to develop new innovations and business models radically to address COVID-19 needs. Trampoline theory of recovery is likely to help reset economies after COVID-19 pandemic that create new opportunities for food disruptive innovations in service and business processes, which may see a drive for consolidation in various industries.

1.1.4 Strategic funding initiatives to identify and accelerate DTs, a case study from the republic of Ireland

Many countries have strategically focused on providing funding initiatives merging academia and industry to identify the next disruptive technology. The Republic of Ireland launched the Enterprise—Ireland Disruptive Technology Innovation Fund (DTIF) in 2018. This is a €500 million DTIF initiative established under Project Ireland 2040 (https://dbei.gov.ie/DTIF) over the 10 years from 2018 to 2027 alongside enterprise cofunding. All questions and responses are available on this host website. These DTIF awards were to align with the Republic of Ireland's refreshed strategic priority areas for research and innovation to 2023 with a view of enhancing job creation. The DTIF Fund is aligned with the Irish Government's Future Jobs Ireland framework with a focus on "Embracing Innovation and Technological Change," where there is an emphasis on creating and advancing technology disruption on a commercial footing. This model will also facilitate uncertainties created by COVID-19 and by the need to respond to climate change in terms of the intensive sustainability of the agri-food sector in Ireland. As such, this presents a fitting example to demonstrate how a developed country strategically uses its resources to support and forge an ecosystem of creativity to facilitate the generation of new disruptive innovation. Therefore, this Irish DTIF initiative potentially represents one of the first funds of its type in the world to support, review, and fund emergent disruptive technologies where the underpinning aim is to advance the Irish knowledge and socioeconomic landscape and to grow employment. It is envisaged that pursuit of these strategic domains, and harnessing the potential of DTIFS emerging from these cross-cutting areas will also support national economic recovery plan. At the time of writing, Ireland's seasonally adjusted unemployment rate has increased to 5.4% due to the lockdown of the nation where only essential services are authorized to operate, which includes food production and supply chain. The unemployment rate in the Republic of Ireland could reach 17% with broad-spectrum imposed closure of many businesses; it is likely that many businesses may not recover and that new opportunities will present for innovators and companies during and post COVID-19.

The DTIF funding initiative in Ireland is resourced to €65m up to 2022 for projects across many thematic domains encompassing emergent preferences for advancing medical devices, ICT, artificial intelligence, blockchain, robotics, nutraceuticals, therapeutics, manufacturing and environmental. Many of the topics embedded in these DTIF funded projects feature strongly in the majority of insightful chapters described in this "Food Disruptive Technology" book. A review of current spend from the Irish Government on this DTIF initiative to date reveals a commitment of €144 million on 43 projects with 159 project partners, with many leads by start-ups and SMEs. These are typically 3-year funding awards. Analysis of the data provided shows that increased funding for this critical initiative in the priority areas Innovations in Services and Business Processing (1 project, €3.9m (2.7%)); Food (3 projects from DTIF 2, €5.2m (3.6%)); Energy, Climate Action and Sustainability (6 projects, €8.3m (5.6%); Manufacturing and Materials – Advanced and Additive Manufacturing (3 projects, €8.7m (6.0%); ICT (11 projects,€31.1m (21.6%)), and Health and Wellbeing, including Medical Devices, Diagnostics, and Therapeutics (21 projects at €86.8m (60.3%) (Figs. 1.1 and 1.2). These domains reflect the Republic of Ireland's priority strategic areas of research and innovation to 2023. Factors underpinning funding successes in different domains across the priority domains are multivariate but aligns best with

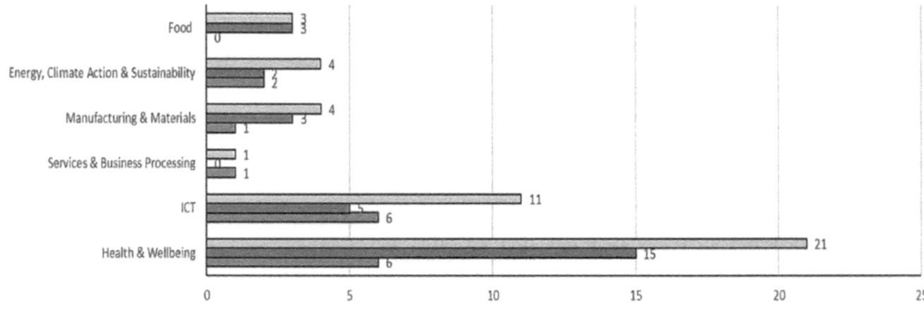

FIGURE 1.1 Number of funded projects awarded per topic in 2018 and 2019 by the Irish Government under the Disruptive Technology Innovation Fund (DTIF).

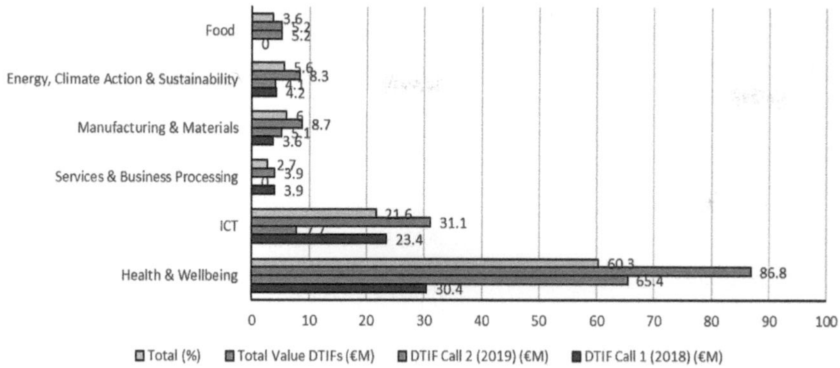

FIGURE 1.2 Amount in Euro of funded projects awarded per topic in 2018 and 2019 by the Irish Government under Disruptive Technology Innovation Fund (DTIF). Proportional representation for each topic in overall DTIF funding is shown as a percentage.

meeting the criteria sought for the award along with matching strong feedback from international peer review. Fig. 1.1 describes the number of DTIF project awards in the various domains, including Food for this Republic of Ireland government initiative, since its launch in 2018. However, it is appreciated that distribution of funding award to date reflects in part the presence of global Medtech and ICT industries in Ireland, in addition to the crucial partnership with leading academic institutions, Science Foundation Ireland-funded Research Centers and Enterprise–Ireland Technology Gateways that all support MNC, SMEs, start-ups, and entrepreneurs in a closely-knit innovation ecosystem. DTIF Call 2 (2019) initiative focused on evaluating how technology-based disruptive innovation was articulated, including collaborations that can alter markets, alter business models, and rationale for developing potentially new disruptive products and services. The initiative paid particular attention to projects of scale with a robust enterprise agenda to harness maximum medium-term economic impact for Ireland. Ideally, it sought enterprise-driven research and development challenges that could demonstrate economic impacts within 3–5 years of project completion.

The three Food DTIF successes (3/43 projects, 3.6%) emerged post review of the second call in 2019 focused on converging technologies across nutraceuticals and traceability (next-generation approaches to advancing sustainability in aquaculture); plant-based proteinaceous ingredients for exploitation as a source of high-quality protein; and beyond food labeling, authentication and certification systems (Table 1.1). However, a top-down review of Ireland's 43 DTIF funded project across all awards reveals the potential for cross-cutting relevance for a number of these platform DTIFs from adjacent domains to potentially increasingly sustain or disrupt agri-food industry, including the three stand-alone food DTIF-specific projects funded under call 2.

TABLE 1.1 Description of awards in the Food domain under Republic of Irelands' Disruptive Technology Innovation Fund (DTIF) initiative.

Title	Project summary	Consortium
HYDRO-fish: Combining targeted nutraceuticals and traceability technology for a smarter and sustainable Irish fish aquaculture industry	HYDRO-fish is a multidisciplinary research program, specifically designed to employ current technologies from other sectors to disrupt and enhance current fish farming practices. The project entails reinforcing the supply chain of Irish salmon production, in particular for organic salmon farming.	National University of Ireland galway, Bio-marine ingredients Ireland, teagasc, marine institute
Optimised commercial-scale cultivation of protein-rich biomass from *Palmaria palmata* for the generation of health-enhancing plant-based proteinaceous ingredients.	This project aims to sustainably generate plant-based proteinaceous ingredients for exploitation as a source of high-quality protein and contribute to meeting the growing global demand for plant-based proteinaceous ingredients for animal and human consumption.	Allihies seafood, carbery, University of limerick
Beyond food labeling	Using massively multiplexed next-generation sequencing to provide a crypto-anchor for food authentication and as a substitute for costly, error-prone labeling and certification systems	IdentiGEN, University College dublin

This initial low number (3/43) of stand-alone Food DTIF projects in Ireland relative to other priority domains, such as health and wellbeing and ICT, reflects in part the significant presence of global multinationals (such as medical devices, ICT) and drive to innovate in these areas that have high-added-value products and services. However, this needs context, as the agriculture-food industry is renowned worldwide for producing and exporting premium high-valued food products along with leading innovation. The agri-food sector in Ireland supports ca. 173,000 jobs that constitute ca. 7.7% of the total employment. Primary production in Ireland is represented by agriculture, fisheries and forestry (includes food, drink, and horticulture), which accounts for 10% of total exports worth €13bn reaching 180 markets worldwide. Central Statistics Office data reveals that Irish food and drink accounts for 21% of all industrial turnover and 23% of all manufacturing industry turnover. Approximately 80% of Ireland's agricultural land is devoted to grasslands, which makes it highly suitable for food production. Key activities include: dairy products and ingredients (34%), meat and livestock (30%), prepared consumer foods (18%), beverages (11%), seafood (2%), and horticulture (2%) (Bord na, 2020). The innovation ecosystem is such that many of the leading food companies such as Diageo, Kerry Group, and Glanbia also are key collaborating partners in national research and innovation centers, enterprise-technology gateways, and benchmarking academic institutions that are focused sustaining and food disruption. For example, the Applied Polymer Technology Center in Athlone Institute of Technology supports ca. 300 projects per year with different start-ups, SMEs, MNCs nationally with cross-cutting links to the agri-food sector, including smart packaging, 3D printing, and nutraceuticals in terms of smart delivery systems. From a review of 143 DTIF awards in Ireland from 2018 to 2019, disruption in the food sector will be strongly influenced by disruptions occurring in other domains that will be considered as converging, and this includes inter alia advances in smart manufacturing (including use of artificial intelligence (athlone institute of technology (AIT)), robotics, augmented and virtual reality (AR/VR)), ICT (including Internet of Things) and innovations in service and business processes. Also, drivers for informing future technology disruption in the agri-food domain will be influenced by needs arising from COVID-10 pandemic along with balancing environmental concerns for more eco-sustainable, climate-friendly products and services.

If one conducts a more in-depth review of the Irelands 43 DTIF projects, it becomes apparent that potentially 20 (46%) at a combined award value of €86.8m (60.2%) have cross-cutting abilities to cause second-order disruption in the food domain (Table 1.2). This would include disruptive training using wearables via wireless communication, use of cytoflow5 for exploiting benefits of nutraceuticals, 3D printing of food, disruptive feed delivery, and future use of nutraceuticals for lung health using aerosol delivery under Health and Wellbeing domain. Disruptive ICT influences on food from these listed

TABLE 1.2 **DTIF Projects funded by Irish Government between 2018 and 2019 – putative relationship with second-order disruption for Food.**

Title of DTIF project	Priority area of award	Potential cross-link to food (potential second-order disruptive technology)	Value of award (€M)
Disruptive gene therapy platform, replacing viruses in the treatment of genetic conditions	Health and wellbeing	Not obvious, as yet	8.4
Holistics - holistic human sensing for health, aging, and wellness	Health and wellbeing	Training, as smart wearables industry value - human-centric intelligent sensors and their wireless communications for products.	7.4
AuriGen solution for persistent atrial fibrillation	Health and wellbeing	Not obvious	5.9
'The future of colorectal cancer diagnosis and treatment: Combining tissue responsive probes, AI and machine learning for medical care	Health and wellbeing	Not obvious	5.7
Therapeutic enzymes as a treatment for sepsis and other immune disorder diseases	Health and wellbeing	Cytoflow5 had the potential for informing new innovation in food – such as nutraceuticals	5
Toward safe and effective off the shelf cellular therapy for cancer	Health and wellbeing	Not obvious.	4.3
Photonics manufacturing pilot Line €4.1 m	ICT-manufacturing	Pilot line hub will develop packaging designs tailored to fast cost-effective packaging processes and equipment and develop and next-gen packaging equipment (including test) with reduced cycle-times.	4.1
Microfluidic gene transfection cell analysis and sorting platform (GTCASP)	Health and wellbeing	Not obvious	3.4

TABLE 1.2 DTIF Projects funded by Irish Government between 2018 and 2019 – putative relationship with second-order disruption for Food.—cont'd

Title of DTIF project	Priority area of award	Potential cross-link to food (potential second-order disruptive technology)	Value of award (€M)
Cooperative energy Trading system (CENTS)	ICT	Consumers and communities will be empowered with the necessary infrastructure to generate their own electricity for artisan food production with lower carbon footprint	3.0
Nex	ICT – Internet of things	Not obvious	3.0
ARDENT II	Health and wellbeing	Not obvious.	2.8
Medical imaging Ireland	Health and wellbeing	Assess impact on new nutraceuticals for lung health (Masterson et al., 2019).	2.2
ArtEngine 2.0 bridging automated, AI-Driven 3D world creation to market	ICT – ai/ar/vr	Food 3D printing using AI as tool creation of 3D models – cost of 3D content creation is prohibitive for small studios and enables codevelopment and adoption of AR/VR.	2.0
BioHealx	Health and wellbeing	Not obvious	1.9
Sustainable Bio-renewable energy from wastewater (S-BREW) for the food and drink wastewater sector that will reduce land-spread waste and produce high-quality renewable energy.	Energy, Climate action and sustainability	Food waste reprocessing	1.8
E-BAMBI - enhanced biocompatibility of additively manufactured Biomedical implants for improved clinical outcomes	Health and wellbeing – Medical devices	Not obvious – but 3D printing focused	1.9

Continued

TABLE 1.2 DTIF Projects funded by Irish Government between 2018 and 2019 – putative relationship with second-order disruption for Food.—cont'd

Title of DTIF project	Priority area of award	Potential cross-link to food (potential second-order disruptive technology)	Value of award (€M)
High throughput microfluidic drug screening platform	Health and wellbeing – Diagnostics/ Therapeutics	Response models for drug testing – may have cross-link to nutraceuticals (GRAS)	1.9
Future software systems architectures	ICT – IoT, AI	Future use to rapidly operationalize new software systems that are slow – with AI	1.6
Irish lasers for the internet of the future (iLife)	ICT – Future networks	Not obvious, as yet	1.6
Connected medical device cybersecurity transparency	ICT – AI, data analytics - blockchain	Possible use of AI, data analytics, blockchain for real-time platform for 2-way communication of safety-critical security information (vulnerability) across food chain	1.5
Creating the bionic many – neural training suit for semimotor impairments	Health and wellbeing	Not obvious	1.5
Advanced environmental decision support system for coastal areas	Energy, climate action and sustainability	Not obvious	1.1
Smart-cardio – a paradigm shift in cardiac arrhythmia treatment	Health and wellbeing – Medical devices	Not obvious	1.1
DEFINE- AM – Disruptive finishing using electrochemical machining for additive manufacturing	Manufacturing and materials	Future link to food for challenges of post processing 3D-printed metallic parts	1.0
Blockchain in technology product supply chain	ICT - blockchain	Food technology product supply chain	1.0
Developing inhaled bioengineered exosome therapeutics	Health and wellbeing	Delivery of smart nutraceuticals via tailored aerosol delivery technology	9.4

TABLE 1.2 DTIF Projects funded by Irish Government between 2018 and 2019 – putative relationship with second-order disruption for Food.—cont'd

Title of DTIF project	Priority area of award	Potential cross-link to food (potential second-order disruptive technology)	Value of award (€M)
Qunatum computing – a software platform for multiple qubit technologies	ICT	Possible role in financial services and logistics supporting food industry	7.3
Point-of-care iron stores/ Ferritin testing for at risk blood donors	Health and wellbeing	Not obvious	7.0
Data-center audio/visual intelligence on-device	ICT	Possible role between in lab and field work for audio and vision-data on devices	6.9
Pharam latch	Health and wellbeing	Not obvious	4.4
Stroke-CIS	Health and wellbeing	Not obvious	4.4
Blockchain and AI-Enabled stratified trial system	Innovations in service, business processes	Food security – ensuring complete (GDPR) trustworthy, control, and ownership of data	3.9
FreeSpace project	ICT	Wireless connectivity with ultra-high capacity wireless laser communication technology for broad food industry – delivers combination of bandwidth, availability and distance	3.6
Transfer print technology for heterogeneous integration of components	Manufacturing and materials	Possibly in food packaging	3.6
EyeVU	Health and wellbeing	Not obvious	3.2
Next-generation heat pump for affordable decarbonization of heating	Energy, climate action, sustainability	Possible role in food distribution and storage as zero carbon-emission, refrigerant-free, heat pump	2.4

Continued

TABLE 1.2 DTIF Projects funded by Irish Government between 2018 and 2019 – putative relationship with second-order disruption for Food.—cont'd

Title of DTIF project	Priority area of award	Potential cross-link to food (potential second-order disruptive technology)	Value of award (€M)
Haemodialysis outcomes and patient empowerment	ICT	Possible role of AI enable software and wearable device for chronic diseases	2.1
Connected enteral feeding healthcare system	Health and wellbeing	New innovative feed delivery device design, connective and apps (possibly COVID-19)	2.0
TRANSPIRE – a trained AI platform for regulation	ICT	Combines human and AI to demystify laws and regulations making it easier to do business while protecting consumers	2
Video intelligent search platform (VISP)	ICT	Not obvious	1.5

projects include the influence of AI, AR/VR and blockchain on distribution, storage of food products and materials, security, training (wearables), and financial services and logistics. In recent times, the immersive experience of using wearables through AR/VR has also been extending to Quality of Experience (QoE) (Murray, Lee, Qiao, & Miro-Muntean, 2017). Also, there is also the emergence of potential disruption on the use of smart audio and visual devices to link between in building (such as laboratory) and the field monitoring of processes. Combined ICT with manufacturing and materials designed DTIF projects include the potential for second-order disruption in smart cost-efficient packaging. Disruption in the area of food security could be potentially achieved through the sole Innovative Services and Business Processes project (Table 1.2). Besides, disruption may be potentially achieved in food distribution and storage through the Energy, Climate Action, and Sustainability project for reveals the use of a new zero-carbon emission, refrigerator-free, heat pump.

Identifying expert technology translators at the interface between research (academia), and enterprise (industry) to push along disruptive products and services along with connecting complementary innovators will be important for evolving the societal-technical ecosystem. Defining optimized business operational structures to harness and sustain disruptive innovation that includes leveraging immediately accessible funds to ensure cash flow and working capital is critical such as for start-ups will also be important. Traditional sources of funding through research is paid post completion of activities and reporting that would present cash flow problems, which may become more challenging to resource post-COVID-19 recovery. This has been exemplified by EU funded instruments, including Rowan & Galanakis, 2020 that includes Future Emerging Technologies (FET) and Interreg programs that focus on regional cohesiveness. Nationally, specific enterprise or state bodies play a pivotal role in supporting innovation and facilitating the emergence of DTIFs. For example, the EU Interreg Atlantic Area Sharebiotech project investigated life science activities for identifying technology core facilities that successfully connect leading academic institutions with SEMs to information technology and business disruption. Many countries now strategically centrally resource through specialist funded centers to support increasingly sustaining and disruptive technologies. In Ireland, this is met jointly by Science Foundation Ireland's funded centers and Enterprise—Ireland's Technology Gateways. Leading scientists and engineers come together through the umbrella of Science Foundation Ireland (SFI) Research Center platforms that harness partnerships with industry and academic institutions, to innovate in order to address complex societal-technical challenges with a view to disruption (https://www.sfi.ie/sfi-research-centres/). For example, the new SFI-funded VistaMilk center combines cutting edge research and innovation for agri-food that partners leading Irish/multinational food and ICT companies. This is represented by converging renowned global expertise of the Tyndall National Institute for biosensors, Ireland's national microelectronics institute, the Insight Center for Data Analytics and the Telecommunications Software and Systems Group (TSSG) at Waterford Institute of Technology that lead have a strong reputation of leading large consortia research in similar topics supported through European funding. VistaMilk will support and accelerate next-generation innovation and decision-support-management tools to optimize efficacy across the production chain for the dairy industry. Whereas, the new Irish APC (Microbiome) SFI-funded Research Center provides another example of convergence for food disruption as it exploits microbial "microbiome" for advancing animal and human health. Key areas, including prevention and treatment of disease through exploring the role of functional food ingredients and novel therapeutics across the lifespan along with disease biomarkers. This APC (Microbiome) also elucidates relevant links between microbes and diet that includes the role of immune-modulation and signaling for wellbeing. Forecasting likely occurrence of next emergent or disruptive technology is desirable as this adds significant competitive advantage and revenue stream to companies. It is not evident, as yet, as to the degree by which investors and researchers consider smart forecast models, but this is likely to accelerate over the coming years. This facility will also help offset in part the void in knowledge created by current shock to markets and businesses caused by COVID-19. An example of a useful tool cited previously by researchers for emergent technologies is the "Gartner Hype Cycle." Lajoie & Bridges (2014) reported that using the Gartner Hype Cycle potentially helps with informed decision-making, and enables an organization to assess risk. Gartner Inc.'s research is available via its analyst webinars and blogs.

1.2 Potential technology, product and business service disruptors in food for 2020 and beyond

1.2.1 Trend toward microbial and plant-based disruptive innovations, next-generation protein sources, and alternative food ingredients

In recent times, there has been increasing interest in the augmented use of microorganisms, such as yeast, microalgae, and bacteria, in the form of protein sources (Fig. 1.3). Microorganisms are commonly used in fermented products that we are very familiar with, such as yogurts and sauerkraut. This offers a more efficient, innovative approaches to producing the same proteins that we are already familiar with (Medical Expo, 2020). The demand for such alternative food ingredients has been pushed by Millennials and Generation X with changes in eating habits (Chapter 5) along with commensurate changes in personalized nutrition. From the disruption of introducing Greek yogurt to the emergence of new functional foods such as seaweeds (Mohamed, Nadia, Hafeedza, & Rahman, 2012), there has been increasing interest in food ingredients. Such things have informed a trend toward personalized nutrition. Considerable development in this space has been the recent partnership of Nestlé with Corbion. This combines exciting expertise of Corbion's microalgae innovation with Nestlé fermentation abilities that is renowned for its smart plant-based products. Another example, include Impossible Foods, who are making soy heme (typically found in soy plants) for plant-based burgers through microbial fermentation. Impossible food burger is made with soy leghemoglobin that mimics the taste of meat. Such innovations in food ingredients may also complement growing consumer demand for eco-sustainable food sources, which also reflects changing eating habits, diets, and the new role of personalized nutrition. StartUS-Insight, (2020) noted that laboratory-cultured meat might provide an alternative or complementary source to actual meat where the latter requires approximately seven tons of water to produce 450 g of beef. Interestingly, the price for producing approximately 140 g of artificial meat dropped to €9,59 Euro in 2017 from a

FIGURE 1.3 Bioreactor culture of microalgae used in freshwater aquaculture industry in Ireland.

non-affordable initial costing base of €274.366. A useful trend to follow for disruption in food production and services is to monitor activities in the USA as more than a third of the world's top food and drink processing companies are headquartered there, including Unilever, Danone, Diageo, Kirin, SAB-Miller, Cadbury Schweppes, Heineken, and Asahi Breweries.

High protein feed for animal and human usage will prove relevant, which has been exemplified by the intensive focus on this product for intensive aquaculture production globally. Aquaculture is rapidly developing worldwide and highlights one of the fastest growth areas for the food industry (Fečkaninová, Koščová, Mudroňová, Popelka, & Toropilová, 2017; Liu, Steele, & Meng, 2017; Tahar, Kennedy, Fitzgerald, Clifford, &&;Rowan, 2018a, Tahar, Kennedy, Fitzgerald, Clifford, & Rowan, 2018b; O'Neill et al., 2019). Aquaculture is recognized to be one of the most affordable and sustainable forms of edible protein (Liu et al., 2017). As described earlier, aquaculture's pace and scale of expansion reflect a substantial increase in our worldwide population and the commensurate demand for more safe and nutritious food (Seoane, Rioboo, Herrero, & Cid, 2014). Between 1983 and 2003, worldwide fisheries production arising from capture increased to 92.6 M tonnes from 71.1M, whereas intensive aquaculture production achieved 70.2 M tonnes from a lower 6 M tonne base (Ottinger, Clauss, & Kuenzer, 2016). In 2014, aquaculture production reached 73.8 M tonnes and now accounts for ~50% of fishery products produced for human consumption (Liu et al., 2017). Fredricks, Jewell, and Survey (2015) suggest that by 2030 aquaculture will provide an estimated 62% of fish for human consumption. Therefore, fish stocks are depleting on the oceans, and there is a countermeasure push to develop sustainable aquaculture processes with a view to enhancing disruption. Hatch-Blue is an example of accelerator SME focused on investing and progressing entrepreneurs to fast-track potential disruptive technologies for the fisheries, seafood, and aquaculture sector (https://www.hatch.blue/) globally. Precisely, hatch-blue constitutes the first accelerator program for sustainable aquaculture that seeks out, develops, and nurtures start-ups for disruptive innovation. The underpinning tenet is that it firmly believes that a sustainable aquaculture industry capable of meeting global demand for food production. Hatch-blue provides an important route to capital and revenue by exploiting considerable networking with industry that includes their worldwide investment community. The author attended this Hatch-Blue program when held in Dublin in 2019 and found it to be excellent with a clear vision in fast-tracking potentially disruptive technologies in aquaculture to market. This 1-week intensive program was strongly supported by Bord Iascaigh Mhara (BIM, Ireland's national seafood development organization), which highlights the importance of national initiatives to capitalize on trending innovation to support job creation framed upon capturing new knowledge for the food sector.

A useful exercise to follow, in terms of mapping potential disruptive technology trends in this space, is to track innovations and approaches in marine and freshwater aquaculture (Tahar et al., 2018a, 2018b). For example, the development of new fish feeds and innovations for remote and real-time use of sensors and technologies to monitor feed rates, physicochemical parameters, and fish health. However, limitations in space that would allow for expansion of existing facilities, challenges with the development of new sites due to licensing, the lack of availability of freshwater, and the ever-growing concerns associated with pollution are thought to be significant hurdles in the further expansion of traditional aquaculture systems (Badiola, Mendiola, & Bostock, 2012; O'Neill et al., 2019). Concerns about the environmental impacts of the rapid expansion of intensive aquaculture systems have also led to increased research interest in integrated multitrophic aquaculture systems or IMTA (Granada, Sousa, Lopes, & Lemos, 2016; O'Neill et al., 2019). BIM undertook a feasibility study to assess the potential use of peatlands (bogs) for sustainable aquaculture diversification. AquaMona is an example of a new concept in integrated multitrophic aquaculture (IMTA) that has the potential to disrupt the production of high-value freshwater fish. This concept, a collaboration between Bord Na Mona and BIM in the Republic of Ireland, uses cutaway peatlands to organically farm Eurasian perch and rainbow trout, which is powered by wind energy. This Aquamona process also exploits algae and duckweed as a natural process for water quality in terms of treating rearing water and waste recycling (O'Neill et al., 2019) (Fig. 1.4). Approximately 5% of Ireland consists of peatlands that are vital for biodiversity, conservation, and maintaining our natural ecosystems. This is set against a growing trend to strategically convert from brown to green innovation, where Bord Na Mona is leading the charge in new sustainable energy-efficient technologies that are exploited to drive their new businesses in the sustainable food production area, such as medicinal plants and herbs. Findings from O'Neill et al., (2019) support the use of peatlands as future locations for integrated aquaculture processes. Bord Na Mona own and manage ca 80,000 ha of peatlands in the Republic of Ireland, where there has been the transition to renewable energy along with exploiting new businesses ventures that includes the production of high-value plants and herbs that can be used for nutraceutical and health benefits linked to workforce training and education.

FIGURE 1.4 Aerial view of 'AquaMona' peatlands freshwater aquaculture RAS process in Irish Midlands. *Picture furnished by Bord na Móna, with permission.*

An example would be tapping of Birch water from trees located across 4 ha of peatlands in the Irish midlands for potential disruption in the health and wellbeing market. Bord na Mona (2020) is developing "birchwater" that is the sap from birch trees as it is rich in natural nutrients and low in sugar with a view to marketing as a new health-promoting beverage. Bord na Mona states that "Birchwater is an electrolyte-replacing beverage, high in antioxidants and similar to coconut or maple water." Bord na Mona has approximately 8000 ha of naturally colonized birch trees on their raised bogs and is also using birchwater as new smart ingredients for cosmetics and personal care products. The use of aquaculture model has also been exploited to investigate potentially new disruptive immune-priming nutraceuticals, such as beta-glucans from yeast, along with microalgae extracts, for fish health that has been reported to have positive implications for the gut microbiome in fish (Carballo et al., 2019).

Aquaculture has progressed toward water, and waste recirculation production models, or disruptive recirculation aquaculture system (RAS) approaches (Tahar et al., 2018a; Tahar et al., 2018b), in order to negate effluent release to environment that is important for the sustainability of the industry (O'Neill et al., 2019). These advanced production systems provide efficient, reliable, and repeatability systems for farmed fish production that includes a trend toward exploiting biomonitoring techniques such as the use of microalgae to detect pollutants or dynamic changes in processes, such as the impact of climate change (O'Neill et al., 2019). Naughton, Kavanagh, Lynch, & Rowan, (2020) has also demonstrated the potential for using flow cytometry in laboratory setting matched with in-field use of AlgaeTorch for monitoring natural microalgae diversity and populations as natural biological means of regulating water quality in aquaculture processes. The AlgaeTorch, produced by bbe Moldaenke, is a potentially disruptive technology as it enables both real-time and in-the-field measurement of microalgae and cyanobacteria in all types of water. AlgaeTorch measures chlorophyll-a of intact cells without sample preparation, where complete measurement needs less than 20 s. No sampling or preparation is necessary. The combined use of Flow-Cytometry and AlgaeTorch may disrupt natural aquaculture processes as there is less reliance on the use of energy-intensive technologies and aeration, where these innovations low-carbon emissions and waste remediation for both food sustainability and bioeconomy. However, time will tell where these approaches are seen as increasing sustaining or disruptive in nature.

1.2.2 Other innovation that will inform food disruption

Blockchain offers an exciting security-proof approach to recording every digital transaction that can inform a broad spectrum of smart innovations from business processes to 5G networks (Sharma & Singh, 2020). In the food disruption context, it has the potential to radically transform and disrupt safety and quality, waste remediation and recycling, security, and authenticity and traceability (Medical Expo, 2020). Medical Expo (2020) noted, by way of example, that a critical driver for this technology in the food space has been the deployment of IBM Food Trust of blockchain for improving food standardization and efficiencies throughout the entire supply chain.

Artificial intelligence (AI) is increasingly used to develop new foods and flavors, such as Coca Cola's research into the Cherry Stripe in 2017 (Medical Expo, 2020). AI will play a prominent role in the personalization of foods and nutrition, exploiting the vast potential of digitalization. The robotics industry is estimated to be work ca. €2.2 billion by 2022 and has to potential to transform the food industry. StartUS-Insight (2020) noted that food and liquid processing might be advanced by exploiting robotics as this innovation can increase output, reduce cost while enhancing the quality of service along the supply chain. Interestingly, safety regulations distinguish food robotics from other automation that will inform efficiency, including using high-quality, sustainable ingredients. For example, StartUS-Insight (2020) reported that Momentum Machines had developed a fully autonomous burger machine that can slice toppings, grill, assemble, and package the finished product without the need for using people. This innovative process also facilities personalized orders, including selecting a variety of sauces and seasonings.

There has been a global push to readdress dependency on single-use plastics with a greater focus on smart packaging, including the emergence of potential for bioplastics. Large companies such as Diago and Nestlé, are leading the way in recycling technologies, such as for multipacks for beer and bottled water, respectively. (Medical Expo, 2020). Food wastes cost the EU approximately €143 billion, where approximately 88 million tonnes are wasted annually (StartUS-Insight, 2020). Several initiatives have progressed to address waste reprocessing, for example, the creation of a dedicated National Bioeconomy Center in the Republic of Ireland to exploit waste streams for the dairy industry. This is also the subject of many transnational research and innovation initiatives such as recently funded by European Commission, Interreg Neptunus project that combines academic expertise with the industry across the Atlantic area to address waste recycling in the fisheries and seafood area, including life cycle assessment, valorization, and eco-labeling (https://neptunus-project.eu/). Also, there is increasing commitment by the takeaway service across many European countries to reduce the amount of edible food thrown away by restaurants (StartUS-Insight, 2020).

There is also increasing interest in the development of 3D printers, also known as additive manufacturing, as a sustaining and potentially disruptive technology for a wide range of possibilities for the food industry. 3D food printers also permit personalized and repeatable nutrition where it is considered to provide the correct amount of nutrients to match different lifestyles, gender, and health requirements. For example, experimental 3D Bioprinters are designed to prints living cells that have the potential to advance food supply chain needs. StartUS-Insight (2020) provided the example of Natural Machines that advance cooking with fresh ingredients. The innovative startup behind this

3D food printer is Foodini (StartUS-Insight, 2020). There is also increasing evidence as to the role of 3D printers for food products. Therefore, 3D printers can help realize the potential to produce intricate food designs that include provision for automation, such as personalized meal preparation. Brunner, Delley, and Denkel (2018) described how 3D printers provide the potential for innovation across the food manufacturing, retail, and catering sectors. However, the role of social marketing and communication to inform behavior changes and to seek feedback on attitudes, perceptions, and barriers for the uptake of this technology will be necessary, particularly as one radically informs new concepts, it is crucial to appreciate the market need for active uptake and to refine products for different opportunities (Brunner et al., 2018).

There has been increasing evidence as to the changing preferences in diets that reflect the lifestyle and emerging therapeutic needs of consumers. This coincides with the dawn of a new era of personalized nutrition, or personalized meals that are steadily informing and transforming the food industry. Tailor-diets are an exciting opportunity. Also, the preference for functional foods or nutraceuticals that boost immune-system and wellbeing for COVID-19 are likely to become famous (example, Masterson et al., 2019), as are innovative products that will help the nutritional and immunomodulatory recovery of patients postcontracting COVID-19 (Masterson et al., 2020). These products may emerge from seaweeds, yeast, algae, plants, and fungi or mushrooms that reduce inflammatory responses that are typically associated with cytokine storm in severe COVID-19 patients. The London-based Nutrifix caters for personalized diets where they recommend meals to cook, buy, or have delivered, tailored to consumers' nutritional needs.

Food delivery companies are beginning to concentrate on exploiting the role of artificial intelligence (AI) for problem-solving matches with automation, such as automated guided vehicles. As an example, slow-moving pavement droids to deliver food have been tested by Just Eat, who has partnered with Starship Technologies for this exciting opportunity. These droids are guided by a GPS signal and cameras to navigate around obstacles (StartUS-Insight (2020). The Internet of Things (IoT) is increasingly becoming relevant for the next-generation food industry, which includes forging innovation in services and business processes. For example, the Internet introduced an innovation suitable for all kitchen devices such as analysis of items for food refrigeration, including taking note of expiration dates with provision for suggesting recipes along with meal preparation.

Additionally, the inbuilt recognition system supports consumers in keeping an eye on their fridge via smartphone or tablet (StartUS-Insight, 2020). Food security is also an essential factor to have to the fore, which includes frequent concerns over contamination of shellfish with Norovirus or the Winter-Vomiting bug. The monitoring of food from field to fork using IoT technologies presents a logical solution to this challenge that also must ensure that such innovative technologies align with food safety standards. StartUS-Insight

(2020) noted that the start-up TellSpec developed a potential solution for cloud-based spectroscopy. The patented AI-based real-time cloud analysis engine helps monitor events of food fraud, as well as of food contamination, where it helps consumers and authorities to make choices to prevent the onset of health issues related to food.

1.3 Summary

Enhanced innovation leading to the creation of new disruptive technologies in the agri-food domain will inform new exciting new products and services that will address challenges and opportunities for the intensive sustainability of the industry, including embracing COVID-19 pandemic crisis. Defining and forecasting what constitutes a disruptive technology is complicated as the impact is more likely to be measured from a retrospective downstream perspective. Disruptive technologies can substantially cause localized change within a market or industry (i.e., first-order disruption) or cause ground-breaking changes across many cross-cutting domains (i.e., second-order disruption) over a relatively short or more extended time period that substantially influences societal norms. Modern-day and future disruptive technologies for the agri-food sector will be influenced by the growing demand to produce more safe, nutritious foods to meet growing populations that reflects dynamic changes in eating habits such as personalized nutrition, alternative protein sources, and attitudes toward climate change. Disruptive technologies may be smaller, lighter, more flexible and convenient products offered at a cheaper price. Exploiting advances in ICT and advanced manufacturing will inform critical areas, including security, standards and quality, and traceability along the entire food supply chain. A review of the recent 43 projects funded by the Irish government under Science Foundation Ireland's Disruptive Technology Initiatives was used to highlight trends in the innovation ecosystem and the potential for both cross-cutting and future ground-breaking disruption in the agri-food sector with global outreach and orientation. Understanding where potential food technology disruptions are likely to occur will be aided by having a holistic perspective and appreciation of the complex socio-technological innovation ecosystem. This timely book provides the best knowledge to meet these needs that will also influence education and workforce training.

References

Badiola, M., Mendiola, D., & Bostock, J. (2012). Recirculating Aquaculture Systems (RAS) analysis: Main issues on management and future challenges. *Aquacultural Engineering, 51*, 26–35.

Beinhocker, E. D. (2006). *The origin of wealth: Evolution, complexity and the radical remaking of economics*. Boston, MA: Harvard Business School Press.

Bord na Mona. (2020). *New innovations for Bord na Mona*. https://www.bordnamona.ie/company/our-businesses/new-business-and-land-development/birchwater/ Accessed 8th August, 2020.

Bower, J. L., & Christensen, C. M. (1995). Disruptive technologies – catching the wave. *Harvard Business Review, 73*(1), 506–520.

Brunner, T. A., Delley, M., & Denkel, C. (2018). Consumers' attitudes and change of attitude toward 3D-printed food. *Food Quality and Preference, 68*, 389–396.

Bulwark. (2020). In *Forget about a V-shaped recovery -massive shocks to the global economy and financial markets portend a slow, difficult recovery from the coronavirus recession.* https://thebulwark.com/forget-about-a-v-shaped-recovery/ (Accessed 8th August, 2020).

Carballo, C., Pinto, P. I. S., Mateus, A. P., la Cláudia, C. B., Guerreiro, C., Martinez-Blanch, J. F., et al. (2019). Yeast β-glucans and microalgal extracts modulate the immune response and gut microbiome in Senegalese sole (Solea senegalensis). *Fish and Shellfish Immunology, 92*(September), 31–39.

Ceruzzi, P. E. (1999). *A history of modern computing.* Cambridge, MA: The MTT Press.

Christensen, C. M. (1997). *The innovator's dilemma. The revolutionary book that will change the way you do business.* Boston, MA: Harvard Business School Press.

Christensen, C. M. (2003). *The innovator's solutions:- creating and sustaining successful growth.* Boston, MA: Harvard Business School Press.

Christensen, C. M., Anthony, S. D., & Roth, E. A. (2004). *Seeing what's next: Using the theories of innovation to predict industry change.* Cambridge, MA: Harvard Business Press.

Christensen, C. M., & Bower, J. L. (1996). Customer power, strategic investment, and the failure of leading firms. *Strategic Management Journal, 17*(3), 197–218.

Curran, E. (2020). *Economics- Morgan Stanley economists double down on V-shape global recovery.* https://www.bloomberg.com/news/articles/2020-06-15/morgan-stanley-economists-double-down-on-v-shape-global-recovery (Accessed 8th August, 2020).

DBEI. (2018a). *Ireland's research priority areas 2018-2023.* https://dbei.gov.ie/en/Publications/Publication-files/Research-Priority-Areas-2018-to-2023.pdf.

DBEI. (2018b). https://dbei.gov.ie/en/Publications/Publicatin-files/Research-Prioity-Areas-2018-2023.pdf. (Accessed 30 April 2020).

Denning, S. (2016). Christensen updates disruption theory. *Strategy and Leadership, 44*(2), 10–16. https://doi.org/10.1108/SL-01-2016-0005.

Drucker, P. F. (1985). *Innovation and entrepreneurship: Practice and principles.* New York, NY: Harper & Row.

DTIF Call 2. (2019). *Disruptive technology innovation fund for the republic of Ireland.* https://dbei.gov.ie/en/Publications/Publication-files/DTIF-Call-2-Guide-for-Applicants.pdf. (Accessed 29 April 2020).

Fečkaninová, A., Koščová, J., Mudroňová, D., Popelka, P., & Toropilová, J. (2017). The use of probiotic bacteria against Aeromonas infections in salmonid aquaculture. *Aquaculture, 469*, 1–8.

Fitzhenry, K., Rowan, N., del Rio, A. V., Cremillieux, A., & Clifford, E. (2019). Inactivation efficiency of Bacillus endospores via modified flow-through PUV treatment with comparison to conventional LPUV treatment. *Journal of water process engineering, 27*, 67–76.

Fredricks, B. K. T., Jewell, S., & Survey, U. S. G. (2015). *Literature review of the potential effects of formalin on nitrogen oxidation efficiency of the biofilters of recirculating aquaculture systems (RAS) for freshwater finfish.*

Freeman, C., Clark, J., & Soete, L. (1982). *Unemployment and technical innovation. A study of long waves and economic development.* Washington, DC: Pinter.

Galankis, C. M. (2020). The food system in the era of the Coronavirus (COVID-19) pandemic crisis. *Foods, 9*, 523. https://doi.org/10.3390/foods.9040535.

Garrison, G. (2009). An assessment of organisational size and sense and response capability on the early-adoption of disruptive technology. *Computers in Human Behavior, 25*, 444–449.

Geels, F. W. (2018). Disruption and low-carbon system transformation; progress and new challenges in socio-technical transitions research and the multi-level perspective. *Energy Research and Social Science, 86*, 2317–2324.

Govindarajan, V., & Kopalle, P. K. (2006). Disruptiveness of innovations: Measurement and an assessment of reliability and validity. *Strategic Management Journal, 27*(2), 189–199.

Granada, L., Sousa, N., Lopes, S., & Lemos, M. F. L. (2016). Is integrated multitrophic aquaculture the solution to the sectors' major challenges? - a review. *Reviews in Aquaculture, 8*, 283–300.

Kuhn, T. S. (1962). *The structure of scientific revolutions*. Chicago, IL: The University of Chicago Press.

Lajoie, E. W., & Bridges, L. (2014). Innovative decisions – using the gartner Hype cycle. *Library Leadership and Management, 28*(4), I. https://doi.org/10.5860/llm.v28i4.7083.

Lauer, M. S., & D'Agostin, R. B. (2013). The randomized registry trial – the next disruptive technology in clinical research. *New England Journal of Medicine, 369*(17), 1579–1581.

Li, M., Porter, A. L., & Suominen, A. (2018). Insights into relationship between disruptive technology/innovation and emerging technology – a bibliometric perspective. *Journal of Food Engineering, 220*, 20–27.

Liu, X., Steele, J. C., & Meng, X. Z. (2017). Usage, residue, and human health risk of antibiotics in Chinese aquaculture: A review. *Environmental Pollution, 223*, 161–169.

Masterson, C. H., Murphy, E. J., Gonzalez, H., Major, I., McCarthy, S. D., O'Toole, D., … Rowan, N. J. (2020). Purified β-glucans from the Shiitake mushroom ameliorates antibiotic-resistant *Klebsiella pneumoniae*-induced pulmonary sepsis. *Letters in Applied Microbiology.* https://doi.org/10.1111/lam.13358.

Masterson, C. H., Murphy, E., Major, I., González, H., O'Toole, D., McCarthy, S., et al. (2019). Purified beta-glucan from the lentinus edodes mushroom Attenuates antibiotic resistant *Klebsiella pneumoniae*-induced pulmonary sepsis. *American Journal of Respiratory and Critical Care Medicine, 199*, A1222. https://doi.org/10.1164/ajrccm-conference.2019.199.1_MeetingAbstracts.A1222, 2019.

Medical Expo. (2020). *Five major trends that will impact the food industry in 2020*. http://emag.medicalexpo.com/five-major-trends-will-impact-the-food-industry-in-2020/.

Michelini, L., Principato, L., & Iseavoli, G. (2018). Understanding food sharing models to tackle sustainability challenges. *Ecological Economics, 145*, 205–217.

Mohamed, S., Nadia, S., Hafeedza, H., & Rahman, A. (2012). Seaweeds: A sustainable functional food for complementary and alternative therapy. *Trends in Food Science and Technology, 23*(3), 83–96.

Mullan, D., Matthews, T., Vandaele, K., Barr, I. D., Swindles, G. T., Meneely, J., et al. (2018). Climate impacts on soil erosion and muddy flooding at 1.5 versus 2°C warming. *Land Degradation and Development, 30*(1), 94–108.

Murray, N., Buckley, J., Seery, N., & Rowan, N. J. (2019). Blending immersive and educational technologies to inform sustainability and diversification of workforce training through machine interface learning using sterilization technologies as model – quo vadis? Kilmer sterilization conference, city west, Dublin, Ireland. https://scholar.google.com/scholar?oi=bibs&cluster=16708275275599861885&btnI=1&hl=en. (Accessed 29 April 2020).

Murray, N., Lee, B., Qiao, Y., & Miro-Muntean, G. (September 2017). The impact of scent type on olfaction-enhanced multimedia quality of experience. *IEEE transactions on systems, man, and cybernetics: Systems, 47*(9), 2503–2515. https://doi.org/10.1109/TSMC.2016.2531654.

Nagji, B., & Tuff, G. (2012). Managing your innovation portfolio. Harv. *The Business Review, 90*(5), 66–74.

NAOO. (2020). https://www.noaa.gov/news/january-2020-was-earth-s-hottest-january-on-record. (Accessed 28 April 2020).

Naughton, J., Tiedeken, E. J., Garvey, M., Stout, J., & Rowan, N. (2017). Pulsed light inactivation of the bumble bee trypanosome parasite Crithidia bombi. *Journal of Apiculture Research, 56*, 144–154.

Ottinger, M., Clauss, K., & Kuenzer, C. (2016). Aquaculture: Relevance, distribution, impacts and spatial assessments - a review. *Ocean and Coastal Management, 119*, 244–266.

Naughton, S., Kavanagh, S., Lynch, M., & Rowan, N. J. (2020). Synchronizing use of sophisticated wet-laboratory and in-field handheld technologies for real-time monitoring of key microalgae, bacteria and physicochemical parameters influencing efficacy of water quality in a freshwater aquaculture recirculation system: A case study from the Republic of Ireland. *Aquaculture*. https://doi.org/10.1016/j.aquaculture.2020.735377.

O'Neill, E. A., Rowan, N. J., & Fogarty, A. M. (2019). Novel use of the alga *Pseudokirchneriella subcapitata*, as an early-warning indicator to identify climate change ambiguity in aquatic environments using freshwater finfish farming as a case study. *The Science of the Total Environment, 692*, 209–218.

Perez, C. (2002). *Technological revolutions and financial capital. The dynamics of bubbles and golden ages*. MA: Edward Elgar Publishing, Inc. Northampton.

Reuters. (2020). *Alphabet soup: How will post-virus economic recovery shape up?*. https://uk.reuters.com/article/uk-health-coronavirus-economy-graphic/alphabet-soup-how-will-post-virus-economic-recovery-shape-up-idUKKCN21R25J?il=0.

Rowan, N. (2011). Defining established and emerging microbial risks in the aquatic environment: Current knowledge, implications, and outlooks. *International Journal of Microbiology*, 15. https://doi.org/10.1155/2011/462832, 2011, Article ID 462832.

Rowan, N. (2019). Pulsed light as an emerging technology to cause disruption for food and adjacent industries – quo vadis? *Trends in Food Science and Technology, 88*, 316–332.

Rowan, N., & Laffey, J. G. (2020). Challenges and solutions for addressing critical shortage of supply chain for personal and protective equipment (PPE) arising from Coronavirus disease (COVID19) pandemic – case study from the Republic of Ireland. *The Science of the Total Environment, 725*. https://doi.org/10.1016/j.scitotenv.2020.138532.

Rowan, N., Valdramidis, V. P., & Gómez-López, V. M. (2015). A review of quantitative methods to describe efficacy of pulsed light generated inactivation data that embraces the occurrence of viable but non culturable state microorganisms. *Trends in Food Science and Technology, 44*(1), 79–92.

Rowan, N. J., & Galanakis, C. M. (2020). Unlocking challenges and opportunities presented by COVID-19 pandemic for cross-cutting disruption in agri-food and green deal innovations: Quo Vadis? *Science of the Total Environment*. https://doi.org/10.1016/j.scitotenv.2020.141362.

Ruiz-Salmón, I., Margallo, M., Laso, J., Villanueva-Rey, P., Mariño, D., Quinteiro, P., et al. (2020). Addressing challenges and opportunities of the European seafood sector under a circular economy framework. *Current Opinion in Environmental Science and Health, 13*, 101–106.

Saguy, I. S., Roos, Y. H., & Cohen, E. (2018). Food engineering and food science and technology: Forward looking journey to future horizons. *Innovative Food Science and Emerging Technologies, 47*, 326–334.

Schuelke-Leech, B. A. (2018). A model for understanding the orders of magnitude of disruptive technologies. *Technological Forecasting and Social Change, 129*, 261–274. https://doi.org/10.1016/j.techfore.2017.09.033.

Schumpeter, J. A. (1939a). *Business cycles: A theoretical, historical, and statistical analysis of the capitalist process* (Vol. 1). New York, NY: McGraw-Hill-Book Company, Inc.

Schumpeter, J. A. (1939b). *Business cycles: A theoretical, historical, and statistical analysis of the capitalist process* (Vol. 2). New York, NY: McGraw-Hill Book Company, Inc.

Science Foundation Ireland. (2020). *SFI research centres*. https://www.sfi.ie/sfi-research-centres/. (Accessed 28 April 2020).

Seoane, M., Rioboo, C., Herrero, C., & Cid, Á. (2014). Toxicity induced by three antibiotics commonly used in aquaculture on the marine microalga *Tetraselmis suecica* (Kylin) Butch. *Marine Environmental Research, 101*, 1–7.

Sharma, S. K., & Singh, V. (2020). In *Applications of blockchain technology in the food industry*. New Food Magazine. https://www.newfoodmagazine.com/article/110116/blockchain/ (Accessed 8th August, 2020).

Sousa, M. J., & Rocha, A. (2018). Skills for the disruptive digital business. *Journal of Business Research, 94*, 257–263. https://doi.org/10.1016/j.busres.2017.12.051

StartUS-Insight. (2020). *Disrupting the food industry: A breakdown on startup driven innovation*. https://www.startus-insights.com/innovators-guide/disrupting-food-industry-breakdown-startup-driven-innovation/ (Accessed 8th August, 2020).

Statista. (2020). *Industry report on food manufacturing*. www.statista.com/study/15804/industry-report–food-manufacturing/. (Accessed 23 April 2020).

Streeck, W. (2011). Taking capitalism seriously: towards an institutionalist approach to contemporary political economy. *Soc. Econ. Rev., 9*(1), 137–167.

Tahar, A., Kennedy, A., Fitzgerald, R., Clifford, E., & Rowan, N. (2018a). Full water quality monitoring of a traditional flow-through rainbow trout farm. *Fishes, 3*, 28.

Tahar, A., Kennedy, A. M., Fitzgerald, R. D., Clifford, E., & Rowan, N. (2018b). Longitudinal evaluation of the impact of traditional rainbow trout farming on receiving water quality in Ireland. *PeerJ, 6*, 1–22.

Tahar, A., Tiedeken, E. J., Clifford, E., Cummins, E., & Rowan, N. J. (2017). Development of a semi-quantitative risk assessment model for evaluating environmental threat posed by the three first EU watch-list pharmaceuticals to urban wastewater treatment plants – a Irish case study. *The Science of the Total Environment, 603*, 627–638.

Tahar, A., Tiedeken, E. J., & Rowan, N. J. (2018c). Occurrence and geodatabase mapping of three contaminants of emerging concern in receiving water and at effluent from waste water treatment plants – a first overview of the situation in the Republic of Ireland. *The Science of the Total Environment, 616*, 187–197.

Taleb, N. N. (2012). *Antifragile – the things that gain from disorder*. Random House Publishers, ISBN 978-1-4000-6782-4.

The Intergovernmental Panel on Climate Change. (2019). https://www.ipcc.ch/. (Accessed 29 April 2020).

Tiedeken, E. J., Tahar, A., McHugh, B., & Rowan, N. J. (2017). Monitoring, sources, receptors and control measures for three European Union watch list substances of emerging concern in receiving waters – a 20 year systematic review. *The Science of the Total Environment, 574*, 1140–1163.

Utterback, J. M. (1994). *Mastering the dynamics of innovation*. Boston MA: Harvard Business School Press.

Yongfu, S., Liheng, W., Zontan, S., Kunsheng, W., Liangyan, J. H., Jian, C., et al. (2017). *Connotation and selection of disruptive technologies that lead industrial change*. Strategic Study of CAR. https:/doi.org/10.153302/J-SSCAE-2017.05.002.

Yu, D., & Hang., C. C. (2010). A reflective review of disruptive innovation theory. *International Journal of Management Reviews, 12*(4), 435–452. https://doi.org/10.1111/j.1468-2370.2009.00272.x.

Further reading

Businesswire. (2020). *2020 global food manufacturing market, size, share, outlook and growth opportunities*. Forecast to 2026 https://www.businesswire.com/news/home/20200220005671/en/2020-Global-Food-Manufacturing-Market-Size-Share. (Accessed 29 April 2020).

Department of Agriculture Food and the Marine. (2015a). *National strategic plan for sustainable aquaculture development. Dublin.*

Department of Agriculture Food and the Marine. (2015b). *Local roots global reach: Food Wise 2025*. Dublin: A 10-year vision for the Irish agri-food industry.

Irish Peatland Conservation Council. (2019). *Raised bogs in Ireland* [WWW Document]. A to Z Peatlands. URL http://www.ipcc.ie/a-to-z-peatlands/raised-bogs/. (Accessed 29 July 2019).

Zhang, M., Wang, Z., Xu, J., Liu, Y., Ni, L., Cao, T., et al. (2011). *Ammonium, microcystins, and hypoxia of blooms in eutrophic water cause oxidative stress and C-N imbalance in submersed and floating-leaved aquatic plants in Lake Taihu, China.*

Živić, I., Marković, Z., Filipović-Rojka, Z., & Živić, M. (2009). Influence of a trout farm on water quality and macrozoobenthos communities of the receiving stream (trešnjica river, Serbia). *International Review of Hydrobiology, 94*, 673–687.

Chapter 2

Personalized nutrition and omics technologies: current status and perspectives

Daniela Braconi[1], Vittoria Cicaloni[1,2], Ottavia Spiga[1], Annalisa Santucci[1]
[1]*Department of Biotechnology, Chemistry, and Pharmacy, Department of Excellence, University of Siena, Siena, Italy;* [2]*Toscana Life Sciences Foundation, Siena, Italy*

2.1 Introduction

2.1.1 Personalized nutrition: historical background, significance, and strategies

The modern era has witnessed a revolution in the way we think of our diet. In essence, the concept of food being poor nutrition is now surpassed (Sikalidis, 2019), because the diet is often shaped in terms of individual needs (such as gender, physiological or pathological conditions, physical activity, presence of diseases), social relationships, economic factors, and traditional and cultural background (Kohlmeier et al., 2016). The concept of adapting the diet according to specific needs dates back to the discovery of inborn errors of metabolism made by Garrod more than one century ago. In his pioneering studies, Garrod suggested that biological individuality is reflected in the chemistry of body fluids, and that "… *just as no two individuals of a species are absolutely identical in bodily structure, neither are their chemical processes carried out on exactly the same lines …* " (Garrod, 1902). Once Garrod had shown that genes could influence metabolic processes, research on model organisms (mainly bacteria and their mutants) delivered a better view on the relationships between genotype and nutrition, as described by Roper in 1960. Later on, the Human Genome Project provided scientists with unprecedented knowledge to address the study of the interactions between human genes and diet. It is thanks to such achievements that recently, the disciplines known as "nutrigenetics" and "nutrigenomics" emerged, which are both aimed at clarifying the molecular mechanisms through which dietary components might affect cellular homeostasis and overall health outcomes, defining the modern nutrition field (Peregrin, 2001).

Food Technology Disruptions. https://doi.org/10.1016/B978-0-12-821470-1.00007-0

In light of these considerations, it appears straightforward that standardized and generalized recommendations motivating people to follow healthy lifestyles (diet included) may not be efficacious. People are far from being standardized entities. An absolute healthy individual does not exist; instead, several population groups might have specific requirements, as in the case of children, pregnant women, elderly, or diseased subjects. In turn, it can be stated that "one-size-fits-all" diet strategies are not useful: they cannot manage the biological variability of individuals and often lead to modest improvements in food intake behaviors. However, compelling evidence is already available, demonstrating that a proper lifestyle may be more effective than drugs to prevent diseases or regain/maintain health (Knowler et al., 2002).

Nevertheless, there is still a pandemic rise in noncommunicable diseases accounting for the vast majority of deaths worldwide and accounting for significant economic and societal burdens (World Health Organization, 2016). A possible solution might come from the personalization of diet. In essence, the customization of specific nutritional advice, products, and services could successfully motivate people and assist them in achieving or maintaining effective behavioral changes with beneficial impacts on their health status (Gibney, Walsh, & Goosens, 2016; Ordovas, Ferguson, Tai, & Mathers, 2018). Furthermore, personalization might also assist in reaching other goals related to physical appearance, weight control, endurance, or sports activity, taking into account personal preferences, the need to control specific biochemical parameters, or even include achieving a well-being state, ameliorated attention and mood.

Overall, personalized nutrition may be viewed as a strategy in which the collection of individual information at genetic, phenotypic, medical, and nutritional levels could drive the setting up of customized dietary guidelines. Either healthy or diseased/susceptible subjects may be targeted by using this strategy. Nevertheless, despite an easy conceptualization, still, there is no consensus definition of personalized nutrition. Tentative definitions have been provided, taking into account clinical and biological aspects of nutrition practice (van Ommen et al., 2017) with a focus on managing diseases and maintaining health (Guest, Horne, Vanderhout, & El-Sohemy, 2019; Laddu & Hauser, 2019; Michel & Burbidge, 2019; Ordovas et al., 2018; van Ommen et al., 2017; Wang & Hu, 2018; Westerman et al., 2018), sometimes extending the framework beyond clinical interventions to include advice, products, and services (Ordovas et al., 2018). Furthermore, slightly different descriptors may be used, as in the case of "stratified" or "tailored" interventions, which may imply subtle differences (Ordovas et al., 2018). In an attempt to fully exploit the potential of personalized nutrition toward health and wellness, increased acceptance, application, and expansion of this approach, Bush et al. proposed a definition of personalized nutrition recently as "*a field that leverages human individuality to drive nutrition strategies that prevent, manage, and treat disease and optimize health*" (Bush et al., 2020).

Originally developed to be an individually tailored, gene-centered approach (Ordovas et al., 2018), the concept of personalized nutrition was later revised as a level-based approach (Fig. 2.1). It was nicely exemplified by the recent Food4me pan-European study, the largest randomized controlled trial investigating the efficacy of personalized nutrition (Celis-Morales et al., 2017). The Food4me study enrolled nearly 1600 participants that were randomly assigned for 6 months to three different intervention groups receiving non-personalized general food advice (*level 0*, control group) or personalized advice based on dietary intake only (*level 1*), dietary intake and phenotype (*level 2*) or dietary intake, genotype and phenotype (*level 3*). A novel tendency later emerged, stemming from the same Food4me study, shifting from single individuals to groups of people sharing a typical metabolic profile that is supposed to respond differently to dietary interventions (Fig. 2.1). This approach is known as *metabotyping* (Gibney & Walsh, 2013; O'Donovan et al., 2017) and relies on clustering strategies (e.g., K-nearest neighbors and K-means) to group subjects. The major strength of metabotyping is its ability to encompass several biological processes, since genetic background, environmental factors, and gut microbiota are all reflected in the final metabotype (Brennan, 2017). Within Food4me, for instance, the study

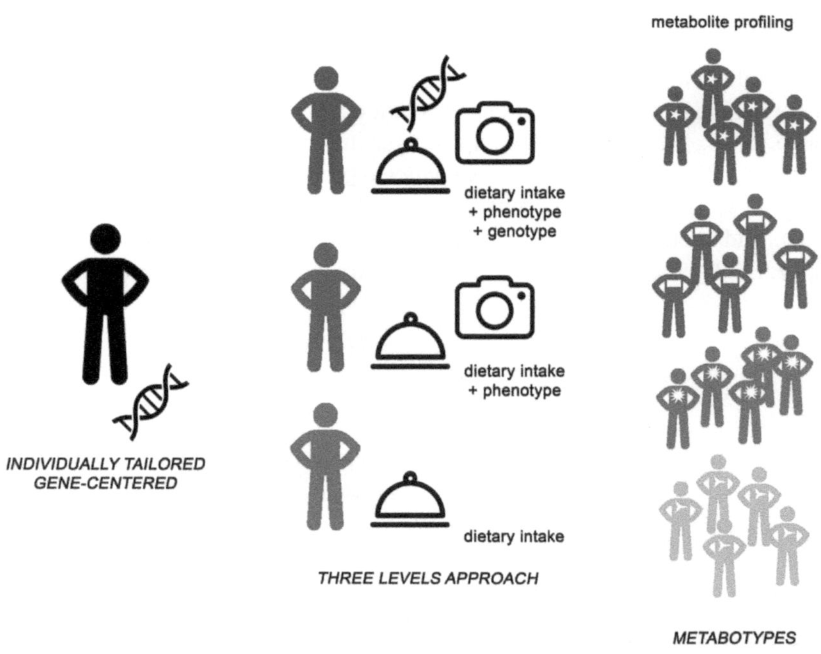

FIGURE 2.1 Personalized nutrition. From the original, individually tailored, gene-centered approach (left) through the three levels approach (middle) up to clustering into metabotypes according to metabolite profiling (right).

participants were grouped according to different metabotypes that proved to be related to nutrient intake, supporting their use; furthermore, dietary recommendations provided through a decision-tree approach were in good agreement with professional advice by nutritionists (O'Donovan et al., 2017).

2.2 From omics to systems biology

The postgenomics era offers unprecedented opportunities to get a holistic view of the effects of food and personalized dietary interventions on complex biological systems thanks to the combination of cutting-edge high-throughput analytical platforms and tools for data analysis (Bland, Minich, & Eck, 2017; Braconi, Bernardini, Millucci, & Santucci, 2018; Badimon, Vilahur, & Padro, 2017). In this context, omics technologies might help to (Fig. 2.2) (Braconi et al., 2018; Bush et al., 2020):

i. Capture complex individual's data
ii. Analyze data to get relevant information, highlighting specific biomarkers
iii. Design the dietary intervention to be undertaken
iv. Monitor the efficacy of the intervention through dedicated biomarkers.

On the other hand, omics technologies might also assist in evaluating the presence of bioactive compounds in food, thus providing relevant knowledge to discover mechanistic insights (Braconi et al., 2018) (Fig. 2.2).

The power of omics technologies lies on the fact that they can be applied to many biological samples like blood, serum, saliva, and stool (Hasin, Seldin, & Lusis, 2017) in order to detect changes that occur in DNA (*nutrigenetics*) (Di Renzo et al., 2014), including its epigenetic modifications (*epigenomics*) (Feinberg, 2001), mRNA (*nutrigenomics*), proteins (*proteomics* and *nutriproteomics*) (Kussmann, Panchaud, & Affolter, 2010; Özdemir & Kolker, 2016), and metabolites (*metabolomics* and *nutrimetabolomics*) (Bordoni & Capozzi, 2014; Sébédio, 2017; Özdemir & Kolker, 2016). Thus, moving from

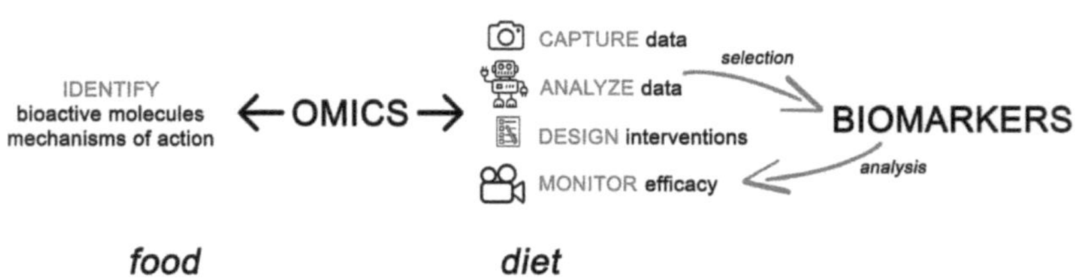

FIGURE 2.2 The contribution of omics technologies for personalized nutrition approaches. Omics technologies are valuable tools to identify bioactive molecules in food and help to understand their mechanisms of action (left panel). For personalized nutrition interventions, omics technologies can act at different levels: capture the data and analyze them to highlight relevant biomarkers, help the design of dietary interventions and assess their efficacy through monitoring of specific biomarkers (right panel) (Braconi et al., 2018; Bush et al., 2020).

genomics, assessing the stable repertoire of genome, downstream omics repertoires up to integrated systems biology approaches can offer natural synergies and a final integrated view of biological variation under the environmental context, which is fundamental knowledge to develop more robust and validate personalization of nutritional strategies.

Three key elements have been identified in personalized nutrition, requiring a multidisciplinary framework: research, education, and practice. Omics disciplines and technologies can provide relevant information to feed all of these three areas and inform all the involved recipients about critical molecules (genes, proteins, and metabolites, either from the host or its microbiota) (Bush et al., 2020) allowing to get a more comprehensive view of an individual's status following specific nutritional and environmental inputs (Bland et al., 2017; Di Renzo et al., 2019; Nilsson, Newsome, Santos, & Schiller, 2019). They are valuable tools to assess, monitor, and map complex molecular networks; however, although holding undoubted potential, today, our ability to translate omics data into practical interventions lags, and efforts are needed to fill this gap (Bland et al., 2017), as discussed later on.

2.2.1 Nutrigenetics and nutrigenomics

The human genome can provide valuable information on how specific gene variants can affect the nutrition-related process, although depicting the underlying molecular processes might be more complicated. Positional cloning through genome-wide linkage scans and candidate gene association studies were the two most frequently adopted strategies until the advent of genome-wide association studies (GWAS) and DNA chips, which allowed the analysis of several hundred single nucleotide polymorphisms (SNPs) simultaneously (Lango & Weedon, 2008). Then, fine-mapping arrays, exome-sequencing, and whole-genome sequencing could help the definition of regions of genetic associations and identify a reliable set of variants; however, causal variant(s) often remained hidden (Smith, Deloukas, & Munroe, 2018). In order to fill this gap, in recent years, genome-editing technology was undertaken to evaluate the functionality of GWAS-associated variants for complex traits (Smith et al., 2018). The relevance of tools such as zinc-finger nucleases, transcription activator-like effector nucleases, and clustered regularly interspaced short palindromic repeats with Cas9 nuclease (CRISPR-Cas9) is now increasingly established to assess the effects of single variants in vivo. However, extreme caution is recommended on commercially available nutrigenomics kits, as a meta-analysis by Pavlidis et al. focusing on 38 genes failed to highlight any significant association with several diet-related diseases, implying that at present such tests cannot be recommended although holding some potential (Pavlidis, Lanara, et al., 2015; Pavlidis, Patrinos, & Katsila, 2015).

Nutrition and diet are potent environmental stimuli, and the impact of the human diet on the genome can likely play a critical role in the development of metabolic diseases. Therefore, two disciplines known as "nutrigenetics," and "nutrigenomics" emerged to assess gene-environment interactions occurring in our dietary landscape, and their primary goal is to help the design of personalized dietary recommendations (Kang, 2012). Nutrigenetics can be defined as the field of nutritional genomics focusing on how the individual genetic background can modify nutrient requirement, susceptibility to developing certain diseases and response to food; *viceversa*, nutrigenomics deals with the study of how food can impact transcription of genes, expression of proteins, and metabolic processes (Kussmann, Raymond, & Affolter, 2006; Peregrin, 2001).

Many variants within the human genome are involved in nutrition-related processes (even if sometimes with poorly understood mechanisms) (van Ommen et al., 2017), finding that corroborates the idea that "one-size-fits-all" recommendations might not be efficacious (Kohlmeier et al., 2016; Stover, 2004). Different genetic polymorphisms might (El-Sohemy, 2007):

i. Affect concrete metabolic pathways, for which precise dietary adjustments may be needed.
ii. Alter the absorption, transport, digestion, or excretion of nutrients or bioactive food molecules.
iii. Induce altered response to nutrients or dietary regimens.

Getting a better knowledge of polymorphisms associated with diseases could help to design and deliver personalized dietary interventions (Kohlmeier et al., 2016). This is well exemplified by monogenic disorders such as the rare condition phenylketonuria, in which an enzyme deficiency within the phenylalanine/tyrosine pathway requires dietary restriction in these two amino acids. Similarly, in galactosemia, a deficiency in enzymes of galactose metabolism requires galactose restriction. A lactose-free diet is needed to manage the decrease in lactase biosynthesis observed with aging, for which a genetic component has been indicated, as well (Deng, Misselwitz, Dai, & Fox, 2015). Carriers of the C677T gene polymorphism in the methylenetetrahydrofolate reductase (*MTHFR*) gene may require supplementation with folate (substrate) and riboflavin (cofactor) due to a defective MTHFR enzyme, a condition that is linked to neural tube defects and hyperhomocysteinemia (a risk factor for cardiovascular diseases) (Hiraoka & Kagawa, 2017).

Despite the inherent complexity, significant advancements have been obtained in the knowledge of multifactorial diseases, too, since specific genetic variants in the human genome have been associated to the development of noncommunicable complex diseases such as type 2 diabetes, obesity, and cancer (Lango & Weedon, 2008; Smith et al., 2018; Kang, 2012; Young, Graff, Fernandez-Rhodes, & North, 2018; Skrypnik et al., 2017). For instance, the *FTO* (fat mass and obesity-associated) gene has been linked both to obesity

(Peeters et al., 2008; Villalobos-Comparán et al., 2008; Tönjes et al., 2010; Hunt et al., 2008; Hotta et al., 2008; Hinney et al., 2007; Dina et al., 2007; Vasan et al., 2014; Frayling et al., 2007) and type 2 diabetes (Bego et al., 2019; Ghafarian-Alipour et al., 2018; Khoshi, Bajestani, Shakeri, Goodarzi, & Azizi, 2019; Lin, Wang, Zhang, & Jin, 2018; Naaz, Kumar, & Choudhury, 2019; Sabarneh et al., 2018; Younus, Algenabi, Abdul-Zhara, & Hussein, 2017), suggesting the need for personalized dietary interventions. The same *MTHFR* gene mentioned previously has been linked to obesity as well.

For such complex diseases, delivering tailored, personalized interventions has proven much more complicated, and so far, with limited value (Di Renzo et al., 2014; Fan et al., 2015; Bashiardes, Godneva, Elinav, & Segal, 2018). However, several studies assess the combined effect of genetic polymorphisms and dietary interventions in complex diseases. In this framework, Table 2.1 provides some example studies on the effects of the Mediterranean diet in subjects genetically susceptible to develop type 2 diabetes, obesity, and cancer. Transferring these findings among different populations has to be avoided since the observed effects might often be due to a population-specific gene make-up. As an example, observations in the traditional Inuit population show a low prevalence of cardiovascular diseases despite a diet rich in polyunsaturated fatty acids. This phenomenon is linked to the presence of different variants of the fatty acids desaturase (*FADS*) gene that is rarely found among Europeans; such variants can lower the production of $n = 3$ and $n = 6$ fatty acids counteracting the intake of harmful polyunsaturated fatty acids (Fumagalli et al., 2015). Similarly, a variant of the fructosamine three kinase-related (*FN3KRP*) gene seems to be associated with increased protection from oxidative stress, which might be related to PUFA in the Inuit population (Mathieson et al., 2015).

2.2.2 Epigenetics

Epigenetics studies those mechanisms not directly related to the DNA sequence itself (Bird, 2007 4477143 2007) that modulate and regulate gene expression through specific "marks" altering the spatial conformation of chromatin: either compacting it (preventing transcription) or opening it (allowing transcription, usually upregulating cell processes) (Tiffon, 2018). DNA methylation (usually acting as silencing), histone acetylation (usually associated with chromatin opening, and hence, transcription activity), noncoding RNAs (functional RNA molecules transcribed from DNA but not translated into proteins) and variants of histone molecules are the epigenetic marks so far known (Tiffon, 2018). These active or repressive marks also depend on environmental factors and are addressed by environmental epigenetics studies (Reamon-Buettner, Mutschler, & Borlak, 2008). In this context, nutrition is one of the most studied and better understood environmental epigenetic factors, acting from gestation to death and mostly impacting human health.

TABLE 2.1 Example studies assessing the effects of the Mediterranean diet on type 2 diabetes, obesity, and cancer, and associated gene polymorphisms.

Condition	Gene (polymorphism)	Main findings	Ref.
Type 2 diabetes	*CLOCK* (rs4580704)	Association between CLOCK rs4580704 and incidence of type 2 diabetes and cardiovascular diseases in type 2 diabetes subjects PREDIMED study examined the modulation by the Mediterranean diet (high in MUFA) on type 2 diabetes and stroke, showing a decreased risk in the intervention group Relevant strengths of the study are the large population ($n = 7018$) and the long follow-up period (4.8 years)	Corella et al. (2016)
	GCKR (rs780094)	Potential protective role of the Mediterranean diet (assessed by a score) on markers of cardiometabolic risk, confirming the association of genetic variation in GCKR with lipid profile Separate and joint associations of diet and genotype (but not their interaction) found with metabolic profile	Sotos-Prieto, Luben, Khaw, Wareham, and Forouhi (2014)
	NLRP3 (rs4612666, rs10733113)	CORDIOPREV study examined the effects of the Mediterranean diet or low-fat diet ($n = 1002$ subjects, 3 years) according to genetic variants of the *NLRP3* inflammasome gene. The study showed beneficial effects on glucose homeostasis associated with a Mediterranean diet in nondiabetic subjects (prevention strategy)	Roncero-Ramos et al. (2018)
	49 established loci	The large EPIC-InterAct study (eight European countries, 16,154 individuals) showed that: (i) a genetic risk score based on 49 established loci for type 2 diabetes is strongly associated with the risk of developing the disease (ii) this relative genetic risk is most significant in those who are younger and leaner at baseline (iii) modifiable factors, mainly obesity dominate the absolute risk of type 2 diabetes (iv) no significant interaction of the genetic risk score with dietary habits (through a Mediterranean diet score)	(Langenberg et al., 2014)
	FTO (rs9939609) *MC4R* (rs17782313)	Neither of the polymorphisms was associated with type 2 diabetes in the whole population; nevertheless, significant gene-diet interactions following a Mediterranean diet were found for both the polymorphisms and their aggregate score, suggesting that the diet could counteract genetic predisposition Preliminary data on a statistically significant interaction between folate intake and fasting plasma glucose concentrations in nondiabetic subjects.	Ortega-Azorín et al. (2012)

Obesity	*FTO* (rs9939609)	Four-week study on 188 Italian subjects showing that the Mediterranean diet could reduce body fat mass, but further studies are needed to confirm the role of *FTO* polymorphism	Di Renzo et al. (2018)
	FTO (rs1421085, rs1121980, rs17817449, rs8050136, rs9939973, and rs3751812)	Study on 1000 subjects on 12 months frequency questionnaire data More strict adherence to the Mediterranean diet (assessed through a score) was associated with lower obesity risk in subjects genetically prone to develop obesity compared to those with lower adherence to the diet or lower obesity genetic risk score	Hosseini-Esfahani et al. (2017)
	ADIPOQ (rs1501299)	82 obese subjects on a 3 months Mediterranean hypocaloric diet The study indicated that T allele of ADIPOQ (rs1501299) could be a predictor of the lack of response of a homeostasis model assessment for insulin resistance, insulin, fasting glucose, and LDL cholesterol	de Luis et al. (2019)
	ADIPOQ (rs266729)	83 obese subjects on a 12 months Mediterranean hypocaloric diet The study showed that ADIPOQ gene variant (rs266729) was associated with increased adiponectin levels and decreased low-density lipoprotein cholesterol, insulin and homeostasis model assessment for insulin resistance after weight loss	de Luis et al. (2019)
Cancer	*SMAD7* (rs4939827)	The study involved 1087 diseased subjects and 2409 controls from the MCC-Spain study and showed a decreased risk for colorectal cancer in CC compared to the TT genotype of the rs4939827 SNP, and an association with the Mediterranean diet pattern as a protective factor	Alonso-Molero et al. (2017)
	OGG1 (rs1052133)	Subjects of the PREDIMED study ($n = 7170$) at high risk for cardiovascular diseases The tested polymorphism seemed to be mainly related to all-cause mortality, mainly due to cardiovascular disease rather than cancer. No statistically significant interactions for total or cardiovascular disease mortality were found for the Mediterranean diet intervention, although significant protective interactions were highlighted for vegetable intake	Corella et al. (2018)
	GSTP1 (rs1695) *NAT2* (rs1799930)	Increased adherence to the Mediterranean diet reduced breast cancer risk in 1109 women with at least one GSTP1 or one NAT2 590G allele	Kakkoura et al. (2017)

Evidence is available linking prenatal nutrition with postnatal health and increased risk to develop diseases: the extreme starvation observed during the Dutch famine has been used to investigate a link with the possible later onset of type 2 diabetes, cardiovascular diseases, metabolic disorders, and cognitive impairment (Davis & Ross, 2007; Tiffon, 2018). Nutritional changes in early life might be associated with long-term changes in DNA driving the development of age-related disorders during the whole life (Chol, Friso, & Choi, 2009; Lillycrop, Hoile, Grenfell, & Burdge, 2014), modifying the expression of critical genes that can impact overall health and longevity (reviewed in (Choi & Friso, 2010; Davis & Ross, 2007)). In particular, maternal folate availability can impact DNA methylation, with consequences for fetal development and children's disease susceptibility (Choi & Friso, 2010). On the contrary, moderate caloric restriction seems to be associated with increased lifespan also through epigenetic modifications (Gensous et al., 2019).

Nutritional epigenetics is still in its infancy, and the knowledge of the precise mechanisms linked to specific food (bioactive) components or complex dietary regimens is so far quite limited. Deciphering their epigenetic signatures would surely help to pave the way for personalized nutrition interventions (Davis & Ross, 2007; Tiffon, 2018), as demonstrated by recent studies on:

- Fruit and fruit juice, correlated with global epigenetic variations showing largely independent signatures related to different immune cell populations and different aspects of immune function (Nicodemus-Johnson & Sinnott, 2017).
- Saturated fatty acids and polyunsaturated fatty acids that induce different methylation signatures in a large number of genes from human subcutaneous adipose tissue, possibly predicting weight gain (Perfilyev et al., 2017).
- The Mediterranean diet that was associated with changes in the methylation status of eight genes (involved in inflammation and immunocompetence) from peripheral blood cells in a group of 36 individuals enrolled in the PREDIMED study (Arpón et al., 2016).

These examples clearly show that understanding the molecular mechanisms and signaling pathways involved in nutriepigenetics can have significant consequences for both public health and targeted interventions to reduce economic and societal impact of human diseases (Tiffon, 2018).

2.2.3 Proteomics and nutriproteomics

Proteomics aims to define the whole set of proteins expressed in a particular organism, tissue, or organ at a given moment. Proteomics technologies represent promising strategies because they can be coupled to the high-throughput capacity of mass spectrometry to achieve fast, robust, and sensitive protein and peptide characterization, detection, and quantification, providing information on protein structure, function, abundance, modification, localization, and interactions. *Qualitative* proteomics is aimed at the characterization of proteins, including information on post-translational modifications that may ensue following environmental inputs. *Quantitative* (or differential) proteomics compares the abundance of specific proteins under different conditions, this variability being the focus of comparative proteomics studies (Özdemir & Kolker, 2016). Differently from the static genome, the proteome is intrinsically variable and context-dependent. *Functional* proteomics is the study of functional interactions between proteins and other macromolecules (Ortea, O'Connor, & Maquet, 2016). *Bottom-up* and *top-down* proteomics strategies can be used. In bottom-up approaches, proteins have to be digested before mass spectrometry, and a fractionation step can be performed before or after the digestion step to reduce sample complexity. Fractionation can be achieved through gel-based two-dimensional electrophoresis (orthogonal separation of proteins in a gel matrix according to their isoelectric point and molecular size) or chromatography techniques. In the case of top-down approaches, fragments dissociating from intact proteins within the mass spectrometer are analyzed without any prior digestion.

Two main investigation areas are well suited to proteomics in nutrition research: discovery of health/disease biomarkers (e.g., related to diabetes and obesity (Kratchmarova et al., 2002; Moulder, Schvartz, Goodlett, & Dayon, 2018; Sundsten & Ortsäter, 2009; Zhao, Barrere-Cain, & Yang, 2015), or celiac disease (Ferretti, Bacchetti, Masciangelo, & Saturni, 2012; KhalKhal et al., 2019; Lexhaller, Ludwig, & Scherf, 2019; Valitutti & Fasano, 2019)), and identification of food bioactive molecules and their mechanisms of action (e.g., from milk and dairy (Basilicata et al., 2018; Chatterton, Nguyen, Bering, & Sangild, 2013; Dallas & Nielsen, 2018; Giacometti & Buretić-Tomljanović, 2017; Lu et al., 2018; Pessione & Cirrincione, 2016; Zhao, Du, Gao, Zhan, & Mao, 2019), or soybean (Erdmann, Cheung, & Schoder, 2008; Fuchs et al., 2005; Fuchs, Dirscherl, Schroot, Daniel, & Wenzel, 2006; Gallagher et al., 2000; Kussmann et al., 2010; Mamone, Picariello, Caira, Addeo, & Ferranti, 2009; Sun, Biela, Hamilton, & Reardon, 2012)). In the study of bioactive food components, proteomics seems to be even more useful than nutrigenetics and nutrigenomics, as these molecules have usually a limited impact on the genome but more significant effects on the proteome (Picó, Serra, Rodriguez, Keijer, & Palou, 2019). Informative biomarker proteins may be searched in several, sometimes easily accessible, body fluids since most human diseases encompass changes in the proteome or induce protein modifications perturbing physiology; the same holds after dietary changes (Picó et al., 2019).

Despite its potential and constant progress, the use of proteomics in nutritional research (and particularly in personalized nutrition) has not met the early expectations yet, although exciting applications are available (Kussmann et al., 2006). Data from a mouse model showed that proteomics could shed light on the mechanisms by which dietary interventions with different sources of fatty acids can affect lipid metabolism and related pathways, with consequences for lipid and glucose metabolism (de Roos et al., 2005). Similarly, a proteomics study was carried out in a mouse model of diet-induced hepatic lipid accumulation (Krahmer et al., 2018). Proteomics can also be exploited to monitor weight-loss dietary interventions in metabolic syndrome (Bruderer et al., 2019; Geyer et al., 2016) or the effects of the Mediterranean diet (Djuric, 2011; Richard et al., 2014). Another application could be the elucidation of the molecular mechanisms responsible for phenotypic changes observed on healthy diets (e.g., rich in vegetables) promoting metabolic health (Ayoub, McDonald, Sullivan, Tsao, & Meckling, 2018). From a broader perspective, proteomics has been suggested as an optimal tool to inform physicians about possible nutritional interventions to be recommended to cancer patients (Schroll & Hummon, 2018). Not only the identity of possible protein biomarkers but also their cellular locations, functions, along with their modifications, holds excellent potential defining novel hypotheses and discover molecular mechanisms that may be beneficial to support personalized nutrition strategies. These future challenges will need to be paralleled by technological advancements in analytical platforms and bioinformatics tools, as discussed later on.

2.2.4 Metabolomics and nutrimetabolomics

Metabolomics aims to provide a holistic and systematic description of all the metabolites (low molecular weight compounds) in a biological system. Metabolomics, embracing a wide range of applications, is probably one of the most active fields of investigation in the food domain. In particular, in nutrition studies, metabolomics can offer its intrinsic ability to connect the metabolism with clinical interventions, optimally fulfilling the need to deliver clinically relevant personalized nutritional interventions (Tebani & Bekri, 2019).

Two complementary approaches are available in metabolomics: *metabolite profiling* (a hypothesis-driven approach analyzing a small subset of metabolites only) and *metabolic fingerprinting* (a high-throughput, hypothesis-free approach trying to capture the overall metabolic "fingerprint" of a system, including all of its metabolites). Under a technological point of view, mass spectrometry, nuclear magnetic resonance spectroscopy, and chromatographic separations are available, each of them providing different levels of sensibility, speed, resolving power and accuracy, thus offering the possibility of integrated studies. Various chemometric tools are available for data analysis and interpretation.

By definition, metabolomics is the elective technique to define metabotypes (metabolic phenotypes, driving individual differences for nutritional requirements and responses to diet and medication) for stratification purposes (Holmes, Wilson, & Nicholson, 2008; O'Donovan, Walsh, Gibney, Gibney, & Brennan, 2016). This approach was undertaken to identify responders to vitamin D supplementation thanks to the analysis of biochemical markers of the metabolic syndrome (O'Sullivan, Gibney, Connor, et al., 2011), and to segregate individuals according to the metabolic response recorder after an oral glucose tolerance test (Morris et al., 2013) identifying those at higher risk presenting with a more severe metabolic dysfunction. Similarly, representative lipoproteins were used to identify subjects with positive lipid profiles upon fenofibrate therapy and with different underlying disturbances in lipoprotein metabolism (van Bochove et al., 2012). One notable example where metabotyping was coupled with personalized dietary advice is available, showing the feasibility of this approach and its potential translation into clinical settings. This study involved a large cohort of Irish subjects who underwent routine biochemical assessments on markers of metabolic health (including lipid- and glucose-related biomarkers) for which a clustering analysis and a decision tree delivering personalized dietary recommendations were developed based on the above biomarkers and additional anthropometry and clinical data. These recommendations were found to be in good agreement with manually compiled individualized advice (O'Donovan et al., 2015). More recently, polyphenol-related metabotypes were suggested as biomarkers of a specific type of microbiota with a distinctive impact on health, suggesting that they could represent useful tools to predict the functionality/status of gut microbiota and to understand the effects of bioactive compounds and the large interindividual variability observed in response to them (Cortés-Martín, Selma, Tomás-Barberán, González-Sarrías, & Espín, 2020). The use of metabolomics to define useful markers for cardiometabolic disease-related metabotypes has been discussed, as well (Palmnäs et al., 2020).

Since the metabolome is the repertoire that is closest to the final phenotype, one can easily assume that metabolomics is the optimal tool to investigate the impact of food and diet on individuals' health (Tebani & Bekri, 2019). Three main areas can be identified through which metabolomics can enhance our understanding of the interactions between food and health/diseases (Tebani & Bekri, 2019), as follows:

- The identification of dietary intake biomarkers (Atkinson, Downer, Lever, Chambers, & George, 2007; Cross, Major, & Sinha, 2011; Gibbons et al., 2015; Heinzmann, Holmes, Kochhar, Nicholson, Schmitt-Kopplin, 2015; Jacobs et al., 2012; Van Velzen et al., 2009). These biomarkers, however, often lack independent large-scale validation studies, and most of them have limited relevance because they have been detected in urine (thus, they are most acute food intake biomarkers); hence, long-term biomarkers are needed here (Tebani & Bekri, 2019).
- The analysis of diet-related diseases in cohort studies (Garcia-Aloy et al., 2014; Garcia-Aloy et al., 2014; Lloyd, Beckmann, Haldar, Seal, Brandt, & Draper, 2013; Wittenbecher et al., 2015). In this area, diet-related data are often collected using traditional self-reports (which are notoriously subject

to bias), and hence, used to group individuals according to the frequency of consumption of specific food or food group. The metabolomics profiles in these groups are then compared. It has to be noted that these studies are frequently based on associations only, as they do not infer any causal relationship (Praticò et al., 2018). Hence, there is a need here for additional intervention studies to validate the discovered biomarkers (Tebani & Bekri, 2019). Population stratification might also be achieved by metabolomics (Allam-Ndoul et al., 2016; Bakker et al., 2010; Paquette et al., 2017).

A worth mentioning example, representing an essential step toward personalized nutrition, can be cited in this field. In fact, serum metabolomics analysis could indicate that biomarkers related to red meat intake were associated with an increased risk of type 2 diabetes (Wittenbecher et al., 2015). Unveiling similar links and validating them in intervention studies is mandatory to understand and confirm the role of metabolic pathways as potential disease drivers (Tebani & Bekri, 2019).

- The analysis of how nutritional interventions can modify metabolic patterns (Andersen et al., 2014; Bouchard-Mercier, Rudkowska, Lemieux, Couture, & Vohl, 2013; Krishnan, Newman, Hembrooke, & Keim, 2012; Li et al., 2016; Moazzami, Shrestha, Morrison, Poutanen, & Mykkanen, 2014; Wang, Edwards, & Clevidence, 2013). This concept is based on the pioneering work by O'Sullivan et al. combining dietary and metabolic patterns to identify metabolites linked to the intake of specific food groups (O'Sullivan, Gibney, & Brennan, 2011), soon followed by other groups (Bouchard-Mercier et al., 2013; Floegel et al., 2013; Menni et al., 2013; O'Gorman et al., 2014; Peré-Trepat et al., 2010). Potential applications include not only the possibility to check compliance to diet, but also and more importantly, to assess critical relationships between diet and diseases (Houston, 2018; Riedl, Gieger, Hauner, Daniel, & Linseisen, 2017; Sébédio, 2017).

2.2.5 Meta-omics approaches

One key player in response to diets and nutrients is undoubtedly the human commensal gut microbiota, which turns humans into "superorganisms" (Greer, Dong, Morgun, & Shulzhenko, 2016) and has implications in either physiological (Moco, Martin, & Rezzi, 2012; Guven-Maiorov, Tsai, & Nussinov, 2017) or pathological processes (Larroya-García, Navas-Carrillo, & Orenes-Pinero, 2019; Brahe, Astrup, & Larsen, 2016; Nishida et al., 2018; Goto, Kurashima, & Kiyono, 2015; Maguire & Maguire, 2017; Harmsen & de Goffau, 2016). The gut microbiota can modulate our response to diet, and at the same time, be modulated by the diet (Moco et al., 2012; Zhang, Ju, & Zuo, 2018; Zinöcker & Lindseth, 2018; Hadrich, 2018; David et al., 2014;

Faith et al., 2013; Wu et al., 2011). Hence, a two-fold approach can be postulated to design personalized nutritional interventions: either modulate the richness and diversity of the gut microbiota (Kang, 2013) also thanks to prebiotics, probiotics, and symbiotic molecules (Hibberd et al., 2019; Lazar et al., 2019), or identify the most suitable diets/nutrients according to the existing microbiota (Rohrmann et al., 2013). Knowledge of the microbiome (the collection of the genomes of the microorganisms composing the microbiota) can also have predictive value to assess the response to dietary interventions, as exemplified recently by Korem et al. In this study, the authors highlighted through a randomized crossover trial that parameters related to the microbiome could predict a priori the response to dietary interventions. In particular, the glycemic response was found to be specific according to bread type and individuals' microbiome, highlighting the importance of personalized nutrition approaches (Korem et al., 2017).

In such a scenario, omics technologies offer the possibility to capture data on genes, transcripts, proteins, or metabolites of a community of organisms (i.e., humans and their gut microbiota) (Rowland et al., 2018). Such analyses, often carried out in an untargeted fashion, are referred to as meta-omics approaches. Integrated meta-omics datasets linking diseases to gut microbiota are being collected (Putignani & Dallapiccola, 2016), taking into account information from metagenomics, metaproteomics, and phenomics (link to clinical conditions). A significant focus has been so far the discrimination between healthy and diseased subjects through their metabolic pathways (Santoru et al., 2017). On the contrary, the effects onto gut microbiota have been investigated only for a few food components, leaving room for novel studies. The potential of such studies is well exemplified by a crossover analysis using meta-omics to get mechanistic knowledge on the effects of resistant starch on the microbiome, metaproteome, and metabolome (Maier et al., 2017). Similarly, a randomized crossover trial assessed the effects of two different orange juices on fecal microbiota and metabolome using an integrated meta-omics approach (Brasili et al., 2019).

2.2.6 Big data and machine learning

The evolution of personalized nutrition through omics is a relevant research task from the scientific and technological point of view to be addressed in the era of *big data* (Ferguson, 2016; Pavlidis et al., 2015, 2016). To this end, machine learning algorithms can assist data mining and create data-driven models transforming information into valuable knowledge through supervised, unsupervised, and reinforced learning systems. Concerted efforts combining omics, bioinformatics, machine learning, and big data are needed to exploit the full potential of personalized nutrition (Bashiardes et al., 2018; Maher, Pooler, Kaput, & Kussmann, 2016; McDonald, Glusman, & Price, 2016; Pranavchand & Reddy, 2016) and translate its applicability at the population level. There are several examples of machine learning approaches in health studies in metabolic diseases. For instance, in the management of diabetes (Kavakiotis et al., 2017), machine learning could help the biomarker discovery process, the assessment of biomarker accuracy, the prediction of disease, and the selection of effective therapeutic regimens. In this context, Zeevi et al. (2015) highlighted the substantial interindividual variability in postprandial blood glucose levels of 800 diabetic subjects after identical meals. Then they devised a machine-learning algorithm based on several parameters (blood, diet, anthropometrics, activity, gut microbiota) that could predict accurately postprandial glycemic response and validated the results in a small cohort of subjects. In the end, they also carried out a blind, randomized controlled trial that demonstrated the efficacy of the developed algorithm in predicting postprandial glucose levels after personalized nutrition intervention (Zeevi et al., 2015). Albers et al. generated a computation machinery approach for the management of diabetes using data assimilation forecasts, which compared well against postprandial glucose measurements and showed superiority when compared to predictions made by trained experts basing on historical nutritional records and glucose levels (Albers et al., 2017). Both these innovative prediction algorithms suggest that they can be further fed by omics data, possibly providing sophisticated tools with exciting applications in the field of personalized nutrition (de Toro-Martín, Arsenault, Despres, & Vohl, 2017), including the evaluation of efficacy or the adherence to the personalized dietary interventions (Sevilla-Villanueva, Gibert, Sanchez-Marre, Fito, Covas, 2017; Spanakis, Weiss, Boh, Lemmens, & Roefs, 2017).

In a future perspective, personalized nutrition and omics can also benefit from what demonstrated by the new Geometric Framework for Nutrition, a tool to transform nutrient and diet-related information into relevant knowledge to understand health and disease (Simpson et al., 2017). This approach proved useful for chronic diseases related to aging and obesity (Gosby et al., 2011; Le Couteur, Solon-Biet, Cogger, et al., 2016; Le Couteur, Solon-Biet, Wahl, et al., 2016; Raubenheimer & Simpson, 2016; Simpson & Raubenheimer, 2005), and a first proof-of-concept was obtained in mice, showing a new link with metabolomics data (Solon-Biet et al., 2014; Wang et al., 2011). The next steps in its implementation might be obtained by feeding the system with data from other omics repertoires (e.g., microbiome, and proteome) or even from multiomics profiling studies (Simpson et al., 2017).

2.3 Conclusions and future perspectives

The last years have documented significant advancements in personalized nutrition, yet the early high expectations are to be met. Concerns about overpromising stimulating hype have been expressed (Chatelan, Bochud, & Frohlich, 2019; Ferguson et al., 2016; Kohlmeier et al., 2016; Stenne,

Hurlimann, & Godard, 2012, 2013), underscoring the need for additional research in the following areas (Camp & Trujillo, 2014; Schork & Goetz, 2017):

(a) Proving definite links between diet and specific individual profile
(b) Deciphering the molecular and physiological processes underlying the above links
(c) Developing non-trivial, original approaches to assess the efficacy of interventions with clear inference about the individual's response.

Nutrigenetics studies available so far: (i) are based on associations only; (ii) are limited to a candidate gene locus or a specific genotype; (iii) lack reproducibility or validation by other groups; (iv) have limited statistical power; (v) do not incorporate gender differences (Corella et al., 2018; de Roos & Brennan, 2017), which represent critical limits. Notable exceptions to this generalization are represented by the FINGEN study (Caslake et al., 2008) and the study on blood pressure related to *MTHFR* polymorphisms (Wilson et al., 2012). This fact implies that conclusive evidence on the effects of dietary changes onto metabolic pathways and disease risk is missing, although the field is rapidly expanding (Hesketh, 2013).

In order to exploit the potential of personalized nutrition, several challenges need to be faced shortly, also within the field of omics (de Toro-Martín et al., 2017; Ohlhorst et al., 2013; O'Sullivan et al., 2018). These challenges include the need for working guidelines to be delivered jointly by omics and nutrition experts (Murgia & Adamski, 2017), turning current knowledge into practical recommendations. Furthermore, there is the need to identify novel omic biomarkers able to capture the impact of nutrition on health appropriately, to monitor disease progression and patient's response to nutritional interventions. The translation of omics data into action for personalized nutrition is in progress and facing challenges, with only a small number of clinically meaningful biomarkers among the thousands discovered (Poste, 2011). To fill this gap, more robust databases, bioinformatics tools, novel statistical methodologies, and machine learning algorithms are mandatory. The example provided by the Food4me study also suggests the need for suitable decision trees to be built based on accurate molecular and phenotype information to deliver customized dietary guidance (Murgia & Adamski, 2017; van Ommen et al., 2017).

The field of omics repertoire is expanding rapidly, and the more classic omics (e.g., transcriptomics, proteomics, metabolomics, epigenomics, interactomics) are now paralleled by emerging omics such as exposomics, lipidomics, regulomics, exosomics that offer novel tools also for personalized nutrition. Including these new repertoires in future research may help the discovery of additional biomarkers to understand bioactivity and bioefficacy of food components, or to understand metabolic interindividual variability (Badimon et al., 2017; Rubio-Aliaga, Kochhar, & Silva-Zolezzi, 2012) better.

Similarly, additional knowledge might come from an enhanced understanding of food-related post-translational modification of protein biomarkers.

One of the most significant challenges will be the quantification and modeling of robust individual omics signatures obtained integrating different omics technologies and multiple biomarkers toward integrative omic profiling (Hesketh, 2013). Such an achievement is expected to increase the efficacy of interventions through better follow-up and monitoring. It could also help to overcome the current limits of biomarkers as indicators or predictors of the impact of diet or single nutrients since the single nutritional biomarkers currently in use (e.g., serum folate, ferritin, zinc, retinol) lack the intrinsic relevance of a comprehensive set of biomarkers representing the complexity of the nutritional and metabolic phenotype (Rubio-Aliaga et al., 2012). The complexity of human diseases and the period they need to fully develop (they can start very early in life (Mensink et al., 2003; Riccardi et al., 2004; Vasan, 2006)), together with the fact that subtle dietary modifications may not be captured by biomarkers (de Roos & Brennan, 2017), are the primary limits toward this goal.

The identification of complex biomarker signatures adopting a systems biology perspective able to phenotype individuals through nutrigenomics, nutriproteomics, and nutrimetabolomics require several analytical challenges. Nevertheless, unraveling multiple omics repertoires at once would provide a robust framework for a deeper understanding of the interactions between complex organisms (such as humans with their microbiota) in a continuously changing environment. The Integrated Personalized Omics Profiling (iPOP), a multiomics strategy proposed for improved management of diseases (Li-Pook-Than & Snyder, 2013), has been paralleled by several reports underscoring the potential of multiomics approaches in physiological and pathological conditions, paving the way toward precision medicine (Altay, Nielsen, Uhlen, Boren, & Mardinoglu, 2019; Bland et al., 2017; Chen et al., 2020; Ghaemi et al., 2019; Karczewski & Snyder, 2018; Olivier, Asmis, Hawkins, Howard, & Cox., 2019). Potential applications are now suggested also in the field of nutrition, as exemplified in the pioneering study of caloric restriction-mediated changes in insulin sensitivity (Dao et al., 2019). In this work, deep-biological phenotyping of overweight and obese subjects using genomics, metagenomics, and metabolomics, coupled to the integration of big data, provided novel insights into the relationships among insulin sensitivity and adipose tissue genes, metabolites, lifestyle factors, and microbiota during caloric restriction.

Microbiota-based nutrition is getting into action now to predict variable clinical phenotypes or to guide personalized therapies in metabolic syndrome or gastrointestinal disorders. Despite promising early studies (Albers et al., 2017; Zeevi et al., 2015), designing personalized nutrition strategies based on the microbiome is still challenging. In essence, this is due to the correlative nature of most of the studies linking physiology, food, and microbiome, together with the fact the underlying mechanisms are only partly understood or derived from sub-optimal animal models (Beura et al., 2016; Kolodziejczyk, Zheng, & Elinav, 2019). Humans, gut microbiota, and food are all very complex systems, and their characterization through omics faces intrinsic and technical limitations reflected in the lack of standardized lab protocols (including for storage conditions), suitable quality controls, or shared data analysis strategies. Furthermore, we know the function of only a fraction of the genes encoded in the microbiome and predicting the function of the remaining ones is based on sequence similarity (Kolodziejczyk et al., 2019), which may be subjected to bias.

Multi-omics approaches pose significant challenges also for methodological issues related to data management and analysis (van Ommen et al., 2017). In this scenario, bioinformatics is expected to face the "big data" challenge and deliver novel tools to manage, integrate, and analyze large datasets. The integration of omics big data and bioinformatics in a systems biology perspective is a demanding task per se, but it holds enormous potential to get a better understanding of diseases and for translating personalized nutrition into clinical practice. The complexity of the information gathered thus increases significantly, and computational tools designed to manage and interpret it are still lagging and facing logistical challenges mainly related to integrating and comprehensively analyzing heterogeneous data (Cambiaghi, Ferrario, & Masseroli, 2017; O'Sullivan et al., 2018). To fill this gap, suitable big data and machine learning tools are needed to get an enhanced understanding of the mechanistic underpinnings of personalized diets along with simplification of the approach to enable its scaled-up utilization by large populations (Kolodziejczyk et al., 2019; O'Sullivan et al., 2018).

Finally, scientific validation is mandatory for personalized nutrition approaches, which often lacks specific standards. This fact has been addressed recently for nutrigenetics with a draft framework setting up criteria to establish the validity of genes and diet interactions, and to determine the consistency and reproducibility of predicted outcomes. However, this preliminary attempt needs to be regularly revised in order to develop a solid ground for scientifically sound and transparent personalized advice (Grimaldi et al., 2017).

In conclusion, even if challenges are significant, efforts will pay off since personalized nutrition can change lives (McDonald et al., 2016).

2.4 Fundings

This work was supported by Regione Toscana [grant OPENRICCIO (PSR 2014–20) and grant BEERBONE (Nutraceutica 2015)] and by the Ministry for Environment, Land and Sea Protection of Italy (MATTM, CUP B68D19000040001). The Ministry of Education, University and Research (MIUR) has granted the Department of Biotechnology, Chemistry, and Pharmacy as "Department of Excellence 2018–22."

References

Albers, D. J., Levine, M., Gluckman, B., Ginsberg, H., Hripcsak, G., & Mamykina, L. (2017). Personalized glucose forecasting for type 2 diabetes using data assimilation. *PLoS Computational Biology, 13*. https://doi.org/10.1371/journal.pcbi.1005232. e1005232.

Allam-Ndoul, B., Guénard, F., Garneau, V., Cormier, H., Barbier, O., Pérusse, L., et al. (2016). Association between metabolite profiles, metabolic syndrome and obesity status. *Nutrients, 8*, 324. https://doi.org/10.3390/nu8060324.

Alonso-Molero, J., González-Donquiles, C., Palazuelos, C., Fernández-Villa, T., Ramos, E., Pollán, M., et al. (2017). The RS4939827 polymorphism in the SMAD7 GENE and its association with Mediterranean diet in colorectal carcinogenesis. *BMC Medical Genetics, 18*. https://doi.org/10.1186/s12881-017-0485-5.

Altay, O., Nielsen, J., Uhlen, M., Boren, J., & Mardinoglu, A. (2019). Systems biology perspective for studying the gut microbiota in human physiology and liver diseases. *EBioMedicine, 49*, 364–373. https://doi.org/10.1016/j.ebiom.2019.09.057.

Andersen, M. B. S., Kristensen, M., Manach, C., Pujos-Guillot, E., Poulsen, S. K., Larsen, T. M., et al. (2014). Discovery and validation of urinary exposure markers for different plant foods by untargeted metabolomics. *Analytical and Bioanalytical Chemistry, 406*, 1829–1844. https://doi.org/10.1007/s00216-013-7498-5.

Arpón, A., Riezu-Boj, J. I., Milagro, F. I., Marti, A., Razquin, C., Martínez-González, M. A., et al. (2016). Adherence to Mediterranean diet is associated with methylation changes in inflammation-related genes in peripheral blood cells. *Journal of Physiology and Biochemistry, 73*, 445–455. https://doi.org/10.1007/s13105-017-0552-6.

Atkinson, W., Downer, P., Lever, M., Chambers, S. T., & George, P. M. (2007). Effects of orange juice and proline betaine on glycine betaine and homocysteine in healthy male subjects. *European Journal of Nutrition, 46*, 446–452. https://doi.org/10.1007/s00394-007-0684-5.

Ayoub, H., McDonald, M., Sullivan, J., Tsao, R., & Meckling, K. (2018). Proteomic profiles of adipose and liver tissues from an animal model of metabolic syndrome fed purple vegetables. *Nutrients, 10*, 456. https://doi.org/10.3390/nu10040456.

Badimon, L., Vilahur, G., & Padro, T. (2017). Systems biology approaches to understand the effects of nutrition and promote health. *British Journal of Clinical Pharmacology, 83*, 38–45. https://doi.org/10.1111/bcp.12965.

Bakker, G. C., van Erk, M. J., Pellis, L., Wopereis, S., Rubingh, C. M., Cnubben, N. H., et al. (2010). An antiinflammatory dietary mix modulates inflammation and oxidative and metabolic stress in overweight men: A nutrigenomics approach. *American Journal of Clinical Nutrition, 91*, 1044–1059. https://doi.org/10.3945/ajcn.2009.28822.

Bashiardes, S., Godneva, A., Elinav, E., & Segal, E. (2018). Towards utilization of the human genome and microbiome for personalized nutrition. *Current Opinion in Biotechnology, 51*, 57–63. https://doi.org/10.1016/j.copbio.2017.11.013.

Basilicata, M. G., Pepe, G., Sommella, E., Ostacolo, C., Manfra, M., Sosto, G., et al. (2018). Peptidome profiles and bioactivity elucidation of buffalo-milk dairy products after gastrointestinal digestion. *Food Research International, 105*, 1003–1010. https://doi.org/10.1016/j.foodres.2017.12.038.

Bego, T., Čaušević, A., Dujić, T., Malenica, M., Velija-Asimi, Z., Prnjavorac, B., et al. (2019). Association of FTO gene variant (rs8050136) with type 2 diabetes and markers of obesity, glycaemic control and inflammation. *Journal of Medical Biochemistry, 38*, 153–163. https://doi.org/10.2478/jomb-2018-0023.

Beura, L. K., Hamilton, S. E., Bi, K., Schenkel, J. M., Odumade, O. A., Casey, K. A., et al. (2016). Normalizing the environment recapitulates adult human immune traits in laboratory mice. *Nature, 532*, 512–516. https://doi.org/10.1038/nature17655.

Bird, A. (2007). Perceptions of epigenetics. *Nature (London), 447*, 396–398. https://doi.org/10.1038/nature05913.

Bland, J. S., Minich, D. M., & Eck, B. M. (2017). A systems medicine approach: Translating emerging science into individualized wellness. *Advances In Metals In Medicine, 2017*, 1–5. https://doi.org/10.1155/2017/1718957.

van Bochove, K., van Schalkwijk, D. B., Parnell, L. D., Lai, C. Q., Ordovás, J. M., de Graaf, A. A., et al. (2012). Clustering by Plasma Lipoprotein profile reveals two distinct subgroups with positive lipid response to Fenofibrate therapy. *PLoS One, 7*. https://doi.org/10.1371/journal.-pone.0038072. e38072.

Bordoni, A., & Capozzi, F. (2014). Foodomics for healthy nutrition. *Current Opinion in Clinical Nutrition and Metabolic Care, 17*, 418–424. https://doi.org/10.1097/MCO.0000000000000089.

Bouchard-Mercier, A., Rudkowska, I., Lemieux, S., Couture, P., & Vohl, M. C. (2013). The metabolic signature associated with the western dietary pattern: A cross-sectional study. *Nutrition Journal, 12*, 158. https://doi.org/10.1186/1475-2891-12-158.

Braconi, D., Bernardini, G., Millucci, L., & Santucci, A. (2018). Foodomics for human health: Current status and perspectives. *Expert Review of Proteomics, 15*, 153–164. https://doi.org/10.1080/14789450.2018.1421072.

Brahe, L. K., Astrup, A., & Larsen, L. H. (2016). Can we prevent obesity-related metabolic diseases by dietary modulation of the gut microbiota? *Advances in Nutrition, 7*, 90–101. https://doi.org/10.3945/an.115.010587.

Brasili, E., Hassimotto, N. M. A., Del Chierico, F., Marini, F., Quagliariello, A., Sciubba, F., et al. (2019). Daily consumption of orange juice from Citrus sinensis L. Osbeck cv. Cara Cara and cv. Bahia differently affects gut microbiota profiling as unveiled by an integrated meta-omics approach. *Journal of Agricultural and Food Chemistry, 67*, 1381–1391. https://doi.org/10.1021/acs.jafc.8b05408.

Brennan, L. (2017). Use of metabotyping for optimal nutrition. *Current Opinion in Biotechnology, 44*, 35–38. https://doi.org/10.1016/j.copbio.2016.10.008.

Bruderer, R., Muntel, J., Müller, S., Bernhardt, O. M., Gandhi, T., Cominetti, O., et al. (2019). Analysis of 1508 plasma samples by capillary-flow data-independent acquisition profiles proteomics of weight loss and maintenance. *Molecular and Cellular Proteomics, 18*, 1242–1254. https://doi.org/10.1074/mcp.RA118.001288.

Bush, C. L., Blumberg, J. B., El-Sohemy, A., Minich, D. M., Ordovás, J. M., Reed, D. G., et al. (2020). Toward the definition of personalized nutrition: A proposal by the American nutrition association. *Journal of the American College of Nutrition, 39*, 5–15. https://doi.org/10.1080/07315724.2019.1685332.

Cambiaghi, A., Ferrario, M., & Masseroli, M. (2017). Analysis of metabolomic data: Tools, current strategies and future challenges for omics data integration. *Briefings in Bioinformatics, 18*, 498–510. https://doi.org/10.1093/bib/bbw031.

Camp, K. M., & Trujillo, E. (2014). Position of the academy of nutrition and dietetics: Nutritional genomics. *Journal of the Academy of Nutrition and Dietetics, 114*, 299–312. https://doi.org/10.1016/j.jand.2013.12.001.

Caslake, M. J., Miles, E. A., Kofler, B. M., Lietz, G., Curtis, P., Armah, C. K., et al. (2008). Effect of sex and genotype on cardiovascular biomarker response to fish oils: The FINGEN study 1-3. *American Journal of Clinical Nutrition, 88*, 618–629.

Celis-Morales, C., Livingstone, K. M., Marsaux, C. F., Macready, A. L., Fallaize, R., O'Donovan, C. B., et al. (2017). Effect of personalized nutrition on health-related behaviour change: Evidence from the Food4Me European randomized controlled trial. *International Journal of Epidemiology, 46*, 578–588. https://doi.org/10.1093/ije/dyw186.

Chatelan, A., Bochud, M., & Frohlich, K. L. (2019). Precision nutrition: Hype or hope for public health interventions to reduce obesity? *International Journal of Epidemiology, 48*, 332–342. https://doi.org/10.1093/ije/dyy274.

Chatterton, D. E. W., Nguyen, D. N., Bering, S. B., & Sangild, P. T. (2013). Anti-inflammatory mechanisms of bioactive milk proteins in the intestine of newborns. *The International Journal of Biochemistry and Cell Biology, 45*, 1730–1747. https://doi.org/10.1016/j.biocel.2013.04.028.

Chen, D., Zhao, X., Sui, Z., Niu, H., Chen, L., Hu, C., et al. (2020). A multi-omics investigation of the molecular characteristics and classification of six metabolic syndrome relevant diseases. *Theranostics, 10*, 2029–2046. https://doi.org/10.7150/thno.41106.

Choi, S.-W., & Friso, S. (2010). Epigenetics: A new bridge between nutrition and health. *Advances in Nutrition, 1*, 8–16. https://doi.org/10.3945/an.110.1004.

Chol, K. K., Friso, S., & Choi, S. W. (2009). DNA methylation, an epigenetic mechanism connecting folate to healthy embryonic development and aging. *The Journal of Nutritional Biochemistry, 20*, 917–926. https://doi.org/10.1016/j.jnutbio.2009.06.008.

Corella, D., Asensio, E. M., Coltell, O., Sorlí, J. V., Estruch, R., Martínez-González, M.Á., et al. (2016). CLOCK gene variation is associated with incidence of type-2 diabetes and cardiovascular diseases in type-2 diabetic subjects: Dietary modulation in the PREDIMED randomized trial. *Cardiovascular Diabetology, 15*, 1–12. https://doi.org/10.1186/s12933-015-0327-8.

Corella, D., Coltell, O., Portolés, O., Sotos-Prieto, M., Fernández-Carrión, R., Ramirez-Sabio, J., et al. (2018). A guide to applying the sex-gender perspective to nutritional genomics. *Nutrients, 11*, 4. https://doi.org/10.3390/nu11010004.

Corella, D., Ramírez-Sabio, J. B., Coltell, O., Ortega-Azorín, C., Estruch, R., Martínez-González, M. A., et al. (2018). Effects of the Ser326Cys polymorphism in the DNA repair OGG1 gene on cancer, cardiovascular, and all-cause mortality in the PREDIMED study: Modulation by diet. *Journal of the Academy of Nutrition and Dietetics, 118*, 589–605. https://doi.org/10.1016/j.jand.2017.09.025.

Cortés-Martín, A., Selma, M. V., Tomás-Barberán, F. A., González-Sarrías, A., & Espín, J. C. (2020). Where to look into the puzzle of polyphenols and health? The postbiotics and the gut microbiota associated with human metabotypes. *Molecular Nutrition and Food Research, 64*, 1900952. https://doi.org/10.1002/mnfr.201900952. 32196920.

Cross, A. J., Major, J. M., & Sinha, R. (2011). Urinary biomarkers of meat consumption. *Cancer Epidemiology Biomarkers and Prevention, 20*, 1107–1111. https://doi.org/10.1158/1055-9965.EPI-11-0048.

Dallas, D., & Nielsen, S. D. (2018). Milk peptidomics to identify functional peptides and for quality control of dairy products. *Methods in Molecular Biology, 1719*, 223–240. https://doi.org/10.1007/978-1-4939-7537-2_15.

Dao, M. C., Sokolovska, N., Brazeilles, R., Affeldt, S., Pelloux, V., Prifti, E., et al. (2019). A data integration multi-omics approach to study calorie restriction-induced changes in insulin sensitivity. *Frontiers in Physiology, 10*, 1958. https://doi.org/10.3389/fphys.2018.01958.

David, L. A., Maurice, C. F., Carmody, R. N., Gootenberg, D. B., Button, J. E., Wolfe, B. E., et al. (2014). Diet rapidly and reproducibly alters the human gut microbiome. *Nature, 505*, 559–563. https://doi.org/10.1038/nature12820.

Davis, C. D., & Ross, S. A. (2007). Dietary components impact histone modifications and cancer risk. *Nutrition Reviews, 65*, 88–94. https://doi.org/10.1111/j.1753-4887.2007.tb00285.x.

Deng, Y., Misselwitz, B., Dai, N., & Fox, M. (2015). Lactose intolerance in adults: Biological mechanism and dietary management. *Nutrients, 7*, 8020–8035. https://doi.org/10.3390/nu7095380.

Di Renzo, L., Cioccoloni, G., Falco, S., Abenavoli, L., Moia, A., Sinibaldi Salimei, P., et al. (2018). Influence of FTO rs9939609 and mediterranean diet on body composition and weight loss: A randomized clinical trial NCT01890070 NCT. *Journal of Translational Medicine, 16*, 308. https://doi.org/10.1186/s12967-018-1680-7.

Di Renzo, L., Gualtieri, P., Romano, L., Marrone, G., Noce, A., Pujia, A., et al. (2019). Role of personalized nutrition in chronic-degenerative diseases. *Nutrients, 11*, 1707. https://doi.org/10.3390/nu11081707.

Di Renzo, L., Marsella, L. T., Sarlo, F., Soldati, L., Gratteri, S., Abenavoli, L., et al. (2014). C677T gene polymorphism of MTHFR and metabolic syndrome: Response to dietary intervention. *Journal of Translational Medicine, 12*, 329. https://doi.org/10.1186/s12967-014-0329-4.

Dina, C., Meyre, D., Gallina, S., Durand, E., Körner, A., Jacobson, P., et al. (2007). Variation in FTO contributes to childhood obesity and severe adult obesity. *Nature Genetics, 39*, 724–726. https://doi.org/10.1038/ng2048.

Djuric, Z. (2011). The Mediterranean diet: Effects on proteins that mediate fatty acid metabolism in the colon. *Nutrition Reviews, 69*, 730–744. https://doi.org/10.1111/j.1753-4887.2011.00439.x.

El-Sohemy, A. (2007). Nutrigenetics. *Nutrigenomics Opportunities in Asia, 60*, 25–30. https://doi.org/10.1159/000107064. KARGER.

Erdmann, K., Cheung, B. W. Y., & Schröder, H. (2008). The possible roles of food-derived bioactive peptides in reducing the risk of cardiovascular disease. *The Journal of Nutritional Biochemistry, 19*, 643–654. https://doi.org/10.1016/j.jnutbio.2007.11.010.

Faith, J. J., Guruge, J. L., Charbonneau, M., Subramanian, S., Seedorf, H., Goodman, A. L., et al. (2013). The long-term stability of the human gut microbiota. *Science (80)*, 341. https://doi.org/10.1126/science.1237439.

Fan, S. J., Yang, B. Y., Zhi, X. Y., He, M., Wang, D., Wang, Y. X., et al. (2015). Are MTHFR C677T and MTRR A66G polymorphisms associated with overweight/obesity risk? From a case-control to a meta-analysis of 30,327 subjects. *International Journal of Molecular Sciences, 16*, 11849–11863. https://doi.org/10.3390/ijms160611849.

Feinberg, A. P. (2001). Methylation meets genomics. *Nature Genetics, 27*, 9–10. https://doi.org/10.1038/83825.

Ferguson, L. R. (2016). The value of nutrigenomics science. *OMICS: A Journal of Integrative Biology, 20*, 122. https://doi.org/10.1089/omi.2015.0164.

Ferguson, L. R., De Caterina, R., Görman, U., Allayee, H., Kohlmeier, M., Prasad, C., et al. (2016). Guide and position of the international society of nutrigenetics/nutrigenomics on personalised nutrition: Part 1 - fields of precision nutrition. *Journal of Nutrigenetics and Nutrigenomics, 9*, 12–27. https://doi.org/10.1159/000445350.

Ferretti, G., Bacchetti, T., Masciangelo, S., & Saturni, L. (2012). Celiac disease, inflammation and oxidative damage: A nutrigenetic approach. *Nutrients, 4*, 243–257. https://doi.org/10.3390/nu4040243.

Floegel, A., Von Ruesten, A., Drogan, D., Schulze, M. B., Prehn, C., Adamski, J., et al. (2013). Variation of serum metabolites related to habitual diet: A targeted metabolomic approach in EPIC-potsdam. *European Journal of Clinical Nutrition, 67*, 1100–1108. https://doi.org/10.1038/ejcn.2013.147.

Frayling, T. M., Timpson, N. J., Weedon, M. N., Zeggini, E., Freathy, R. M., Lindgren, C. M., et al. (2007). A common variant in the FTO gene is associated with body mass index and predisposes to childhood and adult obesity. *Science (80), 316*, 889–894. https://doi.org/10.1126/science.1141634.

Fuchs, D., De Pascual-Teresa, S., Rimbach, G., Virgili, F., Ambra, R., Turner, R., et al. (2005). Proteome analysis for identification of target proteins of genistein in primary human endothelial cells stressed with oxidized LDL or homocysteine. *European Journal of Nutrition, 44*, 95–104. https://doi.org/10.1007/s00394-004-0499-6.

Fuchs, D., Dirscherl, B., Schroot, J. H., Daniel, H., & Wenzel, U. (2006). Soy extract has different effects compared with the isolated isoflavones on the proteome of homocysteine-stressed endothelial cells. *Molecular Nutrition and Food Research, 50*, 58–69. https://doi.org/10.1002/mnfr.200500133.

Fumagalli, M., Moltke, I., Grarup, N., Racimo, F., Bjerregaard, P., Jørgensen, M. E., et al. (2015). Greenlandic Inuit show genetic signatures of diet and climate adaptation. *Science (80), 349*, 1343–1347. https://doi.org/10.1126/science.aab2319.

Gallagher, D., Heymsfield, S. B., Heo, M., Jebb, S. A., Murgatroyd, P. R., & Sakamoto, Y. (2000). Major dietary patterns are related to plasma concentrations ofmarkers of inflammation and endothelial dysfunction. *American Journal of Clinical Nutrition, 72*, 694–701. https://doi.org/10.1093/AJCN.

Garcia-Aloy, M., Llorach, R., Urpi-Sarda, M., Tulipani, S., Estruch, R., Martínez-González, M. A., et al. (2014). Novel multimetabolite prediction of walnut consumption by a urinary biomarker model in a free-living population: The predimed study. *Journal of Proteome Research, 13*, 3476–3483. https://doi.org/10.1021/pr500425r.

Garcia-Aloy, M., Llorach, R., Urpi-Sarda, M., Tulipani, S., Salas-Salvadó, J., Martínez-González, M. A., et al. (2014). Nutrimetabolomics fingerprinting to identify biomarkers of bread exposure in a free-living population from the PREDIMED study cohort. *Metabolomics, 11*, 155–165. https://doi.org/10.1007/s11306-014-0682-6.

Garrod, A. E. (1902). The incidence of alkaptonuria: A study in chemical individuality. *Lancet, 160*, 1616–1620. https://doi.org/10.1016/S0140-6736(01)41972-6.

Gensous, N., Franceschi, C., Santoro, A., Milazzo, M., Garagnani, P., & Bacalini, M. G. (2019). The impact of caloric restriction on the epigenetic signatures of aging. *International Journal of Molecular Sciences, 20*. https://doi.org/10.3390/ijms20082022.

Geyer, P. E., Wewer Albrechtsen, N. J., Tyanova, S., Grassl, N., Iepsen, E. W., Lundgren, J., et al. (2016). Proteomics reveals the effects of sustained weight loss on the human plasma proteome. *Molecular Systems Biology, 12*, 901. https://doi.org/10.15252/msb.20167357.

Ghaemi, M. S., Digiulio, D. B., Vin Contrepois, K., Callahan, B., Ngo, T. T. M., Lee-Mcmullen, B., et al. (2019). Multiomics modeling of the immunome, transcriptome, microbiome, proteome and metabolome adaptations during human pregnancy. *Bioinformatics, 35*, 95–103. https://doi.org/10.1093/bioinformatics/bty537.

Ghafarian-Alipour, F., Ziaee, S., Ashoori, M. R., Zakeri, M. S., Boroumand, M. A., Aghamohammadzadeh, N., et al. (2018). Association between FTO gene polymorphisms and type 2 diabetes mellitus, serum levels of apelin and androgen hormones among Iranian obese women. *Gene, 641*, 361–366. https://doi.org/10.1016/j.gene.2017.10.082.

Giacometti, J., & Buretić-Tomljanović, A. (2017). Peptidomics as a tool for characterizing bioactive milk peptides. *Food Chemistry, 230*, 91–98. https://doi.org/10.1016/j.foodchem.2017.03.016.

Gibbons, H., McNulty, B. A., Nugent, A. P., Walton, J., Flynn, A., Gibney, M. J., et al. (2015). A metabolomics approach to the identification of biomarkers of sugar-sweetened beverage intake. *American Journal of Clinical Nutrition, 101*, 471–477. https://doi.org/10.3945/ajcn.114.095604.

Gibney, M. J., & Walsh, M. C. (2013). The future direction of personalised nutrition: My diet, my phenotype, my genes. *Proceedings of the Nutrition Society, 72*, 219–225. https://doi.org/10.1017/S0029665112003436.

Gibney, M., Walsh, M., & Goosens, J. (2016). Personalized Nutrition: Paving the way to better population health. In K. Kraemer, J. B. Cordaro, J. Fanzo, M. Gibney, E. Kennedy, A. B. Labrique, et al. (Eds.), *Good Nutr. Perspect. 21st Century* (pp. 235–248). Karger. https://doi.org/10.1159/000452389.

Gosby, A. K., Conigrave, A. D., Lau, N. S., Iglesias, M. A., Hall, R. M., Jebb, S. A., et al. (2011). Testing protein leverage in lean humans: A randomised controlled experimental study. *PLoS One, 6*. https://doi.org/10.1371/journal.pone.0025929. e25929.

Goto, Y., Kurashima, Y., & Kiyono, H. (2015). The gut microbiota and inflammatory bowel disease. *Current Opinion in Rheumatology, 27*, 388–396. https://doi.org/10.1097/BOR.0000000000000192.

Greer, R., Dong, X., Morgun, A., & Shulzhenko, N. (2016). Investigating a holobiont: Microbiota perturbations and transkingdom networks. *Gut Microbes, 7*, 126–135. https://doi.org/10.1080/19490976.2015.1128625.

Grimaldi, K. A., van Ommen, B., Ordovas, J. M., Parnell, L. D., Mathers, J. C., Bendik, I., et al. (2017). Proposed guidelines to evaluate scientific validity and evidence for genotype-based dietary advice. *Genes and Nutrition, 12*, 1–12. https://doi.org/10.1186/s12263-017-0584-0.

Guest, N. S., Horne, J., Vanderhout, S. M., & El-Sohemy, A. (2019). Sport nutrigenomics: Personalized nutrition for athletic performance. *Frontiers in Nutrition, 6*, 8. https://doi.org/10.3389/fnut.2019.00008.

Guven-Maiorov, E., Tsai, C. J., & Nussinov, R. (2017). Structural host-microbiota interaction networks. *PLoS Computational Biology, 13*. https://doi.org/10.1371/journal.pcbi.1005579. e1005579.

Hadrich, D. (2018). Microbiome research is becoming the key to better understanding health and nutrition. *Frontiers in Genetics, 9*, 1–10. https://doi.org/10.3389/fgene.2018.00212.

Harmsen, H. J. M., & de Goffau, M. C. (2016). The human gut microbiota. *Advances in Experimental Medicine and Biology, 902*, 95–108. https://doi.org/10.1007/978-3-319-31248-4_7.

Hasin, Y., Seldin, M., & Lusis, A. (2017). Multi-omics approaches to disease. *Genome Biology, 18*, 83. https://doi.org/10.1186/s13059-017-1215-1.

Heinzmann, S. S., Holmes, E., Kochhar, S., Nicholson, J. K., & Schmitt-Kopplin, P. (2015). 2-Furoylglycine as a candidate biomarker of coffee consumption. *Journal of Agricultural and Food Chemistry, 63*, 8615–8621. https://doi.org/10.1021/acs.jafc.5b03040.

Hesketh, J. (2013). Personalised nutrition: How far has nutrigenomics progressed? *European Journal of Clinical Nutrition, 67*, 430–435. https://doi.org/10.1038/ejcn.2012.145.

Hibberd, A. A., Yde, C. C., Ziegler, M. L., Honoré, A. H., Saarinen, M. T., Lahtinen, S., et al. (2019). Probiotic or synbiotic alters the gut microbiota and metabolism in a randomised controlled trial of weight management in overweight adults. *Beneficial Microbes, 10*, 121–135. https://doi.org/10.3920/BM2018.0028.

Hinney, A., Nguyen, T. T., Scherag, A., Friedel, S., Brönner, G., Müller, T. D., et al. (2007). Genome Wide Association (GWA) study for early onset extreme obesity supports the role of fat mass and obesity associated gene (FTO) variants. *PLoS One, 2*. https://doi.org/10.1371/journal.pone.0001361.

Hiraoka, M., & Kagawa, Y. (2017). Genetic polymorphisms and folate status. *Congenital Anomalies, 57*, 142–149. https://doi.org/10.1111/cga.12232.

Holmes, E., Wilson, I. D., & Nicholson, J. K. (2008). Metabolic phenotyping in health and disease. *Cell, 134*, 714–717. https://doi.org/10.1016/j.cell.2008.08.026.

Hosseini-Esfahani, F., Koochakpoor, G., Daneshpour, M. S., Sedaghati-Khayat, B., Mirmiran, P., & Azizi, F. (2017). Mediterranean dietary pattern adherence modify the association between FTO genetic variations and obesity phenotypes. *Nutrients, 9*. https://doi.org/10.3390/nu9101064.

Hotta, K., Nakata, Y., Matsuo, T., Kamohara, S., Kotani, K., Komatsu, R., et al. (2008). Variations in the FTO gene are associated with severe obesity in the Japanese. *Journal of Human Genetics, 53*, 546–553. https://doi.org/10.1007/s10038-008-0283-1.

Houston, M. (2018). The role of noninvasive cardiovascular testing, applied clinical nutrition and nutritional supplements in the prevention and treatment of coronary heart disease. *Therapeutic Advances in Cardiovascular Disease, 12*, 85–108. https://doi.org/10.1177/1753944717743920.

Hunt, S. C., Stone, S., Xin, Y., Scherer, C. A., Magness, C. L., Iadonato, S. P., et al. (2008). Association of the FTO gene with BMI. *Obesity, 16*, 902–904. https://doi.org/10.1038/oby.2007.126.

Jacobs, D. M., Fuhrmann, J. C., Van Dorsten, F. A., Rein, D., Peters, S., Van Velzen, E. J. J., et al. (2012). Impact of short-term intake of red wine and grape polyphenol extract on the human metabolome. *Journal of Agricultural and Food Chemistry, 60*, 3078–3085. https://doi.org/10.1021/jf2044247.

Kakkoura, M. G., Loizidou, M. A., Demetriou, C. A., Loucaides, G., Daniel, M., Kyriacou, K., et al. (2017). The synergistic effect between the Mediterranean diet and GSTP1 or NAT2 SNPs decreases breast cancer risk in Greek-Cypriot women. *European Journal of Nutrition, 56*, 545–555. https://doi.org/10.1007/s00394-015-1099-3.

Kang, J. X. (2012). The coming of age of nutrigenetics and nutrigenomics. *Journal of Nutrigenetics and Nutrigenomics, 5*, I–II. https://doi.org/10.1159/000339375.

Kang, J. X. (2013). Gut microbiota and personalized nutrition. *Journal of Nutrigenetics and Nutrigenomics, 6*, I–II. https://doi.org/10.1159/000353144.

Karczewski, K. J., & Snyder, M. P. (2018). Integrative omics for health and disease. *Nature Reviews Genetics, 19*, 299–310. https://doi.org/10.1038/nrg.2018.4.

Kavakiotis, I., Tsave, O., Salifoglou, A., Maglaveras, N., Vlahavas, I., & Chouvarda, I. (2017). Machine learning and data mining methods in diabetes research. *Computational and Structural Biotechnology Journal, 15*, 104–116. https://doi.org/10.1016/j.csbj.2016.12.005.

KhalKhal, E., Rezaei-Tavirani, M., Razzaghi, M., Rezaei-Tavirani, S., Zali, H., & Rostamii-Nejad, M. (2019). The critical role of dysregulation of antioxidant activity and carbohydrate metabolism in celiac disease. *Gastroenterology and Hepatology from Bed to Bench, 12*, 340–347.

Khoshi, A., Bajestani, M. K., Shakeri, H., Goodarzi, G., & Azizi, F. (2019). Association of Omentin rs2274907 and FTO rs9939609 gene polymorphisms with insulin resistance in Iranian individuals with newly diagnosed type 2 diabetes. *Lipids in Health and Disease, 18*. https://doi.org/10.1186/s12944-019-1085-5.

Knowler, W. C., Barrett-Connor, E., Fowler, S. E., Hamman, R. F., Lachin, J. M., Walker, E. A., et al. (2002). Reduction in the incidence of type 2 diabetes with lifestyle intervention or metformin. *New England Journal of Medicine, 346*, 393–403. https://doi.org/10.1056/NEJMoa012512.

Kohlmeier, M., De Caterina, R., Ferguson, L. R., Görman, U., Allayee, H., Prasad, C., et al. (2016). Guide and position of the international society of nutrigenetics/nutrigenomics on personalized nutrition: Part 2 - ethics, challenges and endeavors of precision nutrition. *Journal of Nutrigenetics and Nutrigenomics, 9*, 28–46. https://doi.org/10.1159/000446347.

Kolodziejczyk, A. A., Zheng, D., & Elinav, E. (2019). Diet–microbiota interactions and personalized nutrition. *Nature Reviews Microbiology, 17*, 742–753. https://doi.org/10.1038/s41579-019-0256-8.

Korem, T., Zeevi, D., Zmora, N., Weissbrod, O., Bar, N., Lotan-Pompan, M., et al. (2017). Bread affects clinical parameters and induces gut microbiome-associated personal glycemic responses. *Cell Metabolism, 25*, 1243–1253.e5. https://doi.org/10.1016/j.cmet.2017.05.002.

Krahmer, N., Najafi, B., Schueder, F., Quagliarini, F., Steger, M., Seitz, S., et al. (2018). Organellar proteomics and phospho-proteomics reveal subcellular reorganization in diet-induced hepatic steatosis. *Developmental Cell, 47*, 205–221.e7. https://doi.org/10.1016/j.devcel.2018.09.017.

Kratchmarova, I., Kalume, D. E., Blagoev, B., Scherer, P. E., Podtelejnikov, A. V., Molina, H., et al. (2002). A proteomic approach for identification of secreted proteins during the differentiation of 3T3-L1 preadipocytes to adipocytes. *Molecular and Cellular Proteomics, 1*, 213–222. https://doi.org/10.1074/mcp.M200006-MCP200.

Krishnan, S., Newman, J. W., Hembrooke, T. A., & Keim, N. L. (2012). Variation in metabolic responses to meal challenges differing in glycemic index in healthy women: Is it meaningful? *Nutrition and Metabolism, 9*, 26. https://doi.org/10.1186/1743-7075-9-26.

Kussmann, M., Panchaud, A., & Affolter, M. (2010). Proteomics in nutrition: Status quo and outlook for biomarkers and bioactives. *Journal of Proteome Research, 9*, 4876–4887. https://doi.org/10.1021/pr1004339.

Kussmann, M., Raymond, F., & Affolter, M. (2006). OMICS-driven biomarker discovery in nutrition and health. *Journal of Biotechnology, 124*, 758–787. https://doi.org/10.1016/j.jbiotec.2006.02.014.

Laddu, D., & Hauser, M. (2019). Addressing the nutritional phenotype through personalized nutrition for chronic disease prevention and management. *Progress in Cardiovascular Diseases, 62*, 9–14. https://doi.org/10.1016/j.pcad.2018.12.004.

Langenberg, C., Sharp, S. J., Franks, P. W., Scott, R. A., Deloukas, P., Forouhi, N. G., et al. (2014). Gene-lifestyle interaction and type 2 diabetes: The EPIC InterAct case-cohort study. *PLoS Medicine, 11*. https://doi.org/10.1371/journal.pmed.1001647. e1001647.

Lango, H., & Weedon, M. N. (2008). What will whole genome searches for susceptibility genes for common complex disease offer to clinical practice? *Journal of Internal Medicine, 263*, 16–27. https://doi.org/10.1111/j.1365-2796.2007.01895.x.

Larroya-García, A., Navas-Carrillo, D., & Orenes-Piñero, E. (2019). Impact of gut microbiota on neurological diseases: Diet composition and novel treatments. *Critical Reviews in Food Science and Nutrition, 59*, 3102–3116. https://doi.org/10.1080/10408398.2018.1484340.

Lazar, V., Ditu, L. M., Pircalabioru, G. G., Picu, A., Petcu, L., Cucu, N., et al. (2019). Gut microbiota, host organism, and diet trialogue in diabetes and obesity. *Frontiers in Nutrition, 6*, 21. https://doi.org/10.3389/fnut.2019.00021.

Le Couteur, D. G., Solon-Biet, S., Cogger, V. C., Mitchell, S. J., Senior, A., De Cabo, R., et al. (2016). The impact of low-protein high-carbohydrate diets on aging and lifespan. *Cellular and Molecular Life Sciences, 73*, 1237–1252. https://doi.org/10.1007/s00018-015-2120-y.

Le Couteur, D. G., Solon-Biet, S., Wahl, D., Cogger, V. C., Willcox, B. J., Willcox, D. C., et al. (2016). New horizons: Dietary protein, ageing and the Okinawan ratio. *Age and Ageing, 45*, 443–447. https://doi.org/10.1093/ageing/afw069.

Lexhaller, B., Ludwig, C., & Scherf, K. A. (2019). Comprehensive detection of isopeptides between human tissue transglutaminase and gluten peptides. *Nutrients, 11*. https://doi.org/10.3390/nu11102263.

Li-Pook-Than, J., & Snyder, M. (2013). IPOP goes the world: Integrated personalized omics profiling and the road toward improved health care. *Chemistry & Biology, 20*, 660–666. https://doi.org/10.1016/j.chembiol.2013.05.001.

Li, K., Brennan, L., McNulty, B. A., Bloomfield, J. F., Duff, D. J., Devlin, N. F. C., et al. (2016). Plasma fatty acid patterns reflect dietary habits and metabolic health: A cross-sectional study. *Molecular Nutrition and Food Research, 60*, 2043–2052. https://doi.org/10.1002/mnfr.201500711.

Lillycrop, K. A., Hoile, S. P., Grenfell, L., & Burdge, G. C. (2014). DNA methylation, ageing and the influence of early life nutrition. *Proceedings of the Nutrition Society, 73*, 413–421. https://doi.org/10.1017/S0029665114000081.

Lin, Z., Wang, Y., Zhang, B., & Jin, Z. (2018). Association of type 2 diabetes susceptible genes GCKR, SLC30A8, and FTO polymorphisms with gestational diabetes mellitus risk: A meta-analysis. *Endocrine, 62*, 34–45. https://doi.org/10.1007/s12020-018-1651-z.

Lloyd, A. J., Beckmann, M., Haldar, S., Seal, C., Brandt, K., & Draper, J. (2013). Data-driven strategy for the discovery of potential urinary biomarkers of habitual dietary exposure. *American Journal of Clinical Nutrition, 97*, 377–389. https://doi.org/10.3945/ajcn.112.048033.

de Luis, D. A., Izaola, O., Primo, D., Gómez-Hoyos, E., Ortola, A., López-Gómez, J. J., et al. (2019). Role of rs1501299 variant in the adiponectin gene on total adiponectin levels, insulin resistance and weight loss after a Mediterranean hypocaloric diet. *Diabetes Research and Clinical Practice, 148*, 262–267. https://doi.org/10.1016/j.diabres.2017.11.007.

Lu, J., Zhang, S., Liu, L., Pang, X., Ma, C., Jiang, S., et al. (2018). Comparative proteomics analysis of human and ruminant milk serum reveals variation in protection and nutrition. *Food Chemistry, 261*, 274–282. https://doi.org/10.1016/j.foodchem.2018.04.065.

Maguire, M., & Maguire, G. (2017). The role of microbiota, and probiotics and prebiotics in skin health. *Archives of Dermatological Research, 309*, 411–421. https://doi.org/10.1007/s00403-017-1750-3.

Maher, M., Pooler, A. M., Kaput, J., & Kussmann, M. (2016). A systems approach to personalised nutrition: Report on the keystone symposium "human nutrition, environment and health.". *Applied and Translational Genomics, 10*, 16–18. https://doi.org/10.1016/j.atg.2016.08.001.

Maier, T. V., Lucio, M., Lee, L. H., Verberkmoes, N. C., Brislawn, C. J., Bernhardt, J., et al. (2017). Impact of dietary resistant starch on the human gut Microbiome, Metaproteome, and Metabolome. *mBio, 8*. https://doi.org/10.1128/mBio.01343-17.

Mamone, G., Picariello, G., Caira, S., Addeo, F., & Ferranti, P. (2009). Analysis of food proteins and peptides by mass spectrometry-based techniques. *Journal of Chromatography A, 1216*, 7130–7142. https://doi.org/10.1016/j.chroma.2009.07.052.

Mathieson, I., Lazaridis, I., Rohland, N., Mallick, S., Patterson, N., Roodenberg, S. A., et al. (2015). Genome-wide patterns of selection in 230 ancient Eurasians. *Nature, 528*, 499–503. https://doi.org/10.1038/nature16152.

McDonald, D., Glusman, G., & Price, N. D. (2016). Personalized nutrition through big data. *Nature Biotechnology, 34*, 152–154. https://doi.org/10.1038/nbt.3476.

Menni, C., Zhai, G., MacGregor, A., Prehn, C., Römisch-Margl, W., Suhre, K., et al. (2013). Targeted metabolomics profiles are strongly correlated with nutritional patterns in women. *Metabolomics, 9*, 506–514. https://doi.org/10.1007/s11306-012-0469-6.

Mensink, R. P., Aro, A., Hond, E. D., German, J. B., Griffin, B. A., Ter Meer, H. U., et al. (2003). PASSCLAIM - diet-related cardiovascular disease. *European Journal of Nutrition, 42*, 6–27. https://doi.org/10.1007/s00394-003-1102-2. i.

Michel, M., & Burbidge, A. (2019). Nutrition in the digital age - how digital tools can help to solve the personalized nutrition conundrum. *Trends in Food Science and Technology, 90*, 194–200. https://doi.org/10.1016/j.tifs.2019.02.018.

Moazzami, A. A., Shrestha, A., Morrison, D. A., Poutanen, K., & Mykkänen, H. (2014). Metabolomics reveals differences in postprandial responses to breads and fasting metabolic characteristics associated with postprandial insulin demand in postmenopausal women. *Journal of Nutrition, 144*, 807–814. https://doi.org/10.3945/jn.113.188912.

Moco, S., Martin, F. P. J., & Rezzi, S. (2012). Metabolomics view on gut microbiome modulation by polyphenol-rich foods. *Journal of Proteome Research, 11*, 4781–4790. https://doi.org/10.1021/pr300581s.

Morris, C., O'Grada, C., Ryan, M., Roche, H. M., Gibney, M. J., Gibney, E. R., et al. (2013). Identification of differential responses to an oral glucose tolerance test in healthy adults. *PLoS One, 8*. https://doi.org/10.1371/journal.pone.0072890. e72890.

Moulder, R., Schvartz, D., Goodlett, D. R., & Dayon, L. (2018). Proteomics of diabetes, obesity, and related disorders. *Proteomics - Clinical Applications, 12*. https://doi.org/10.1002/prca.201600134.

Murgia, C., & Adamski, M. M. (2017). Translation of nutritional genomics into nutrition practice: The next step. *Nutrients, 9*, 366. https://doi.org/10.3390/nu9040366.

Naaz, K., Kumar, A., & Choudhury, I. (2019). Assessment of FTO gene polymorphism and its association with type 2 diabetes mellitus in North Indian populations. *Indian Journal of Clinical Biochemistry, 34*, 479–484. https://doi.org/10.1007/s12291-018-0778-2.

Nicodemus-Johnson, J., & Sinnott, R. A. (2017). Fruit and juice epigenetic signatures are associated with independent immunoregulatory pathways. *Nutrients, 9*, 752. https://doi.org/10.3390/nu9070752.

Nilsson, P. D., Newsome, J. M., Santos, H. M., & Schiller, M. R. (2019). Prioritization of variants for investigation of genotype-directed nutrition in human superpopulations. *International Journal of Molecular Sciences, 20*, 3516. https://doi.org/10.3390/ijms20143516.

Nishida, A., Inoue, R., Inatomi, O., Bamba, S., Naito, Y., & Andoh, A. (2018). Gut microbiota in the pathogenesis of inflammatory bowel disease. *Journal of Clinical Gastroenterology, 11*. https://doi.org/10.1007/s12328-017-0813-5.

Ohlhorst, S. D., Russell, R., Bier, D., Klurfeld, D. M., Li, Z., Mein, J. R., et al. (2013). Nutrition research to affect food and a healthy life span. *American Journal of Clinical Nutrition, 143*, 1349–1354. https://doi.org/10.3945/ajcn.114.093914.

Olivier, M., Asmis, R., Hawkins, G. A., Howard, T. D., & Cox, L. A. (2019). The need for multi-omics biomarker signatures in precision medicine. *International Journal of Molecular Sciences, 20*. https://doi.org/10.3390/ijms20194781.

van Ommen, B., van den Broek, T., de Hoogh, I., van Erk, M., van Someren, E., Rouhani-Rankouhi, T., et al. (2017). Systems biology of personalized nutrition. *Nutrition Reviews, 75*, 579–599. https://doi.org/10.1093/nutrit/nux029.

Ordovas, J. M., Ferguson, L. R., Tai, E. S., & Mathers, J. C. (2018). Personalised nutrition and health. *BMJ, 361*. https://doi.org/10.1136/bmj.k2173.

Ortea, I., O'Connor, G., & Maquet, A. (2016). Review on proteomics for food authentication. *Journal of Proteomics, 147*, 212–225. https://doi.org/10.1016/j.jprot.2016.06.033.

Ortega-Azorín, C., Sorlí, J. V., Asensio, E. M., Coltell, O., Martínez-González, M. T., Salas-Salvadó, J., et al. (2012). Associations of the FTO rs9939609 and the MC4R rs17782313 polymorphisms with type 2 diabetes are modulated by diet, being higher when adherence to the Mediterranean diet pattern is low. *Cardiovascular Diabetology, 11*, 137. https://doi.org/10.1186/1475-2840-11-137.

Özdemir, V., & Kolker, E. (2016). Precision nutrition 4.0: A big data and ethics foresight analysis-convergence of agrigenomics, nutrigenomics, nutriproteomics, and nutrimetabolomics. *OMICS: A Journal of Integrative Biology, 20*, 69–75. https://doi.org/10.1089/omi.2015.0193.

O'Donovan, C. B., Walsh, M. C., Gibney, M. J., Gibney, E. R., & Brennan, L. (2016). Can metabotyping help deliver the promise of personalised nutrition? *Proceedings of the Nutrition Society, 75*, 106–114. https://doi.org/10.1017/S0029665115002347.

O'Donovan, C. B., Walsh, M. C., Nugent, A. P., Mcnulty, B., Walton, J., Flynn, A., et al. (2015). Use of metabotyping for the delivery of personalised nutrition. *Molecular Nutrition and Food Research, 59*, 377–385. https://doi.org/10.1002/mnfr.201400591.

O'Donovan, C. B., Walsh, M. C., Woolhead, C., Forster, H., Celis-Morales, C., Fallaize, R., et al. (2017). Metabotyping for the development of tailored dietary advice solutions in a European population: The Food4Me study. *British Journal of Nutrition, 118*, 561–569. https://doi.org/10.1017/S0007114517002069.

O'Gorman, A., Morris, C., Ryan, M., O'Grada, C. M., Roche, H. M., Gibney, E. R., et al. (2014). Habitual dietary intake impacts on the lipidomic profile. *Journal of Chromatography B: Analalytical Technology Biomedicine and Life Science, 966*, 140–146. https://doi.org/10.1016/j.jchromb.2014.01.032.

O'Sullivan, A., Gibney, M. J., & Brennan, L. (2011). Dietary intake patterns are reflected in metabolomic profiles: Potential role in dietary assessment studies. *American Journal of Clinical Nutrition, 93*, 314–321. https://doi.org/10.3945/ajcn.110.000950.

O'Sullivan, A., Gibney, M. J., Connor, A. O., Mion, B., Kaluskar, S., Cashman, K. D., et al. (2011). Biochemical and metabolomic phenotyping in the identification of a vitamin D responsive metabotype for markers of the metabolic syndrome. *Molecular Nutrition and Food Research, 55*, 679–690. https://doi.org/10.1002/mnfr.201000458.

O'Sullivan, A., Henrick, B., Dixon, B., Barile, D., Zivkovic, A., Smilowitz, J., et al. (2018). 21st century toolkit for optimizing population health through precision nutrition. *Critical Reviews in Food Science and Nutrition, 58*, 3004–3015. https://doi.org/10.1080/10408398.2017.1348335.

Palmnäs, M., Brunius, C., Shi, L., Rostgaard-Hansen, A., Torres, N. E., González-Domínguez, R., et al. (2020). Perspective: Metabotyping-A potential personalized nutrition strategy for precision prevention of cardiometabolic disease. *Advances in Nutrition, 11*(3), 524–532. https://doi.org/10.1093/advances/nmz121. 31782487.

Paquette, M., Medina Larqué, A. S., Weisnagel, S. J., Desjardins, Y., Marois, J., Pilon, G., et al. (2017). Strawberry and cranberry polyphenols improve insulin sensitivity in insulin-resistant, non-diabetic adults: A parallel, double-blind, controlled and randomised clinical trial. *British Journal of Nutrition, 117*, 519–531. https://doi.org/10.1017/S0007114517000393. Cambridge University Press.

Pavlidis, C., Lanara, Z., Balasopoulou, A., Nebel, J. C., Katsila, T., & Patrinos, G. P. (2015). Meta-analysis of genes in commercially available nutrigenomic tests denotes lack of association with dietary intake and nutrient-related pathologies. *OMICS: A Journal of Integrative Biology, 19*, 512–520. https://doi.org/10.1089/omi.2015.0109.

Pavlidis, C., Nebel, J. C., Katsila, T., & Patrinos, G. P. (2016). Nutrigenomics 2.0: The need for ongoing and independent evaluation and synthesis of commercial nutrigenomics tests' scientific knowledge base for responsible innovation. *OMICS: A Journal of Integrative Biology, 20*, 65–68. https://doi.org/10.1089/omi.2015.0170.

Pavlidis, C., Patrinos, G. P., & Katsila, T. (2015). Nutrigenomics: A controversy. *Applied and Translational Genomics, 4*, 50–53. https://doi.org/10.1016/j.atg.2015.02.003.

Peeters, A., Beckers, S., Verrijken, A., Roevens, P., Peeters, P., Van Gaal, L., et al. (2008). Variants in the FTO gene are associated with common obesity in the Belgian population. *Molecular Genetics and Metabolism, 93*, 481–484. https://doi.org/10.1016/j.ymgme.2007.10.011.

Peré-Trepat, E., Ross, A. B., Martin, F. P., Rezzi, S., Kochhar, S., Hasselbalch, A. L., et al. (2010). Chemometric strategies to assess metabonomic imprinting of food habits in epidemiological studies. *Chemometrics and Intelligent Laboratory Systems, 104*, 95–100. https://doi.org/10.1016/j.chemolab.2010.06.001.

Peregrin, T. (2001). The new frontier of nutrition science: Nutrigenomics. *Journal of the American Dietetic Association, 101*, 1306. https://doi.org/10.1016/S0002-8223(01)00309-1.

Perfilyev, A., Dahlman, I., Gillberg, L., Rosqvist, F., Iggman, D., Volkov, P., et al. (2017). Impact of polyunsaturated and saturated fat overfeeding on the DNA-methylation pattern in human adipose tissue: A randomized controlled trial. *American Journal of Clinical Nutrition, 105*, 991–1000. https://doi.org/10.3945/ajcn.116.143164.

Pessione, E., & Cirrincione, S. (2016). Bioactive molecules released in food by lactic acid bacteria: Encrypted peptides and biogenic amines. *Frontiers in Microbiology, 7*, 876. https://doi.org/10.3389/fmicb.2016.00876.

Picó, C., Serra, F., Rodríguez, A. M., Keijer, J., & Palou, A. (2019). Biomarkers of nutrition and health: New tools for new approaches. *Nutrients, 11*, 1092. https://doi.org/10.3390/nu11051092.

Poste, G. (2011). Bring on the biomarkers. *Nature, 469*, 156–157. https://doi.org/10.1038/469156a.

Pranavchand, R., & Reddy, B. M. (2016). Genomics era and complex disorders: Implications of GWAS with special reference to coronary artery disease, type 2 diabetes mellitus, and cancers. *Journal of Postgraduate Medicine, 62*, 188–198. https://doi.org/10.4103/0022-3859.186390.

Praticò, G., Gao, Q., Scalbert, A., Vergères, G., Kolehmainen, M., Manach, C., et al. (2018). Guidelines for Biomarker of Food Intake Reviews (BFIRev): How to conduct an extensive literature search for biomarker of food intake discovery. *Genes and Nutrition, 13*, 3. https://doi.org/10.1186/s12263-018-0592-8.

Putignani, L., & Dallapiccola, B. (2016). Foodomics as part of the host-microbiota-exposome interplay. *Journal of Proteomics, 147*, 3–20. https://doi.org/10.1016/j.jprot.2016.04.033.

Raubenheimer, D., & Simpson, S. J. (2016). Nutritional ecology and human health. *Annual Review of Nutrition, 36*, 603–626. https://doi.org/10.1146/annurev-nutr-071715-051118.

Reamon-Buettner, S. M., Mutschler, V., & Borlak, J. (2008). The next innovation cycle in toxicogenomics: Environmental epigenetics. *Mutation Research: Reviews in Mutation Research, 659*, 158–165. https://doi.org/10.1016/j.mrrev.2008.01.003.

Riccardi, G., Aggett, P., Brighenti, F., Delzenne, N., Frayn, K., Nieuwenhuizen, A., et al. (2004). PASSCLAIM - body weight regulation, insulin sensitivity and diabetes risk. *European Journal of Nutrition, 43*, II17–46. https://doi.org/10.1007/s00394-004-1202-7.

Richard, C., Couture, P., Desroches, S., Nehmé, B., Bourassa, S., Droit, A., et al. (2014). Effect of an isoenergetic traditional mediterranean diet on the high-density lipoprotein proteome in men with the metabolic syndrome. *Journal of Nutrigenetics and Nutrigenomics, 7*, 48–60. https://doi.org/10.1159/000363137.

Riedl, A., Gieger, C., Hauner, H., Daniel, H., & Linseisen, J. (2017). Metabotyping and its application in targeted nutrition: An overview. *British Journal of Nutrition, 117*, 1631–1644. https://doi.org/10.1017/S0007114517001611.

Rohrmann, S., Overvad, K., Bueno-de-Mesquita, H. B., Jakobsen, M. U., Egeberg, R., Tjønneland, A., et al. (2013). Meat consumption and mortality - results from the European prospective investigation into cancer and nutrition. *BMC Medicine, 11*, 63. https://doi.org/10.1186/1741-7015-11-63.

Roncero-Ramos, I., Rangel-Zuñiga, O. A., Lopez-Moreno, J., Alcala-Diaz, J. F., Perez-Martinez, P., Jimenez-Lucena, R., et al. (2018). Mediterranean diet, glucose homeostasis, and inflammasome genetic variants: The CORDIOPREV study. *Molecular Nutrition and Food Research, 62*. https://doi.org/10.1002/mnfr.201700960.

de Roos, B., & Brennan, L. (2017). Personalised interventions—a precision approach for the next generation of dietary intervention studies. *Nutrients, 9*. https://doi.org/10.3390/nu9080847.

de Roos, B., Duivenvoorden, I., Rucklidge, G., Reid, M., Ross, K., Lamers, R., et al. (2005). Response of apolipoprotein E*3-Leiden transgenic mice to dietary fatty acids: Combining liver proteomics with physiological data. *The FASEB Journal, 19*, 1–26. https://doi.org/10.1096/fj.04-2974fje.

Roper, J. A. (1960). Genetic determination of nutritional requirements. *Proceedings of the Nutrition Society, 19*, 39–45. https://doi.org/10.1079/pns19600012.

Rowland, I., Gibson, G., Heinken, A., Scott, K., Swann, J., Thiele, I., et al. (2018). Gut microbiota functions: Metabolism of nutrients and other food components. *European Journal of Nutrition, 57*, 1–24. https://doi.org/10.1007/s00394-017-1445-8.

Rubio-Aliaga, I., Kochhar, S., & Silva-Zolezzi, I. (2012). Biomarkers of nutrient bioactivity and efficacy. *Journal of Clinical Gastroenterology, 46*, 545–554. https://doi.org/10.1097/MCG.0b013e3182548df2.

Sabarneh, A., Ereqat, S., Cauchi, S., AbuShamma, O., Abdelhafez, M., Ibrahim, M., et al. (2018). Common FTO rs9939609 variant and risk of type 2 diabetes in Palestine. *BMC Medical Genetics, 19*. https://doi.org/10.1186/s12881-018-0668-8.

Santoru, M. L., Piras, C., Murgia, A., Palmas, V., Camboni, T., Liggi, S., et al. (2017). Cross sectional evaluation of the gut-microbiome metabolome axis in an Italian cohort of IBD patients. *Scientific Reports, 7*. https://doi.org/10.1038/s41598-017-10034-5.

Schork, N. J., & Goetz, L. H. (2017). Single-subject studies in translational nutrition research. *Annual Review of Nutrition, 37*, 395–422. https://doi.org/10.1146/annurev-nutr-071816-064717.

Schroll, M. M., & Hummon, A. B. (2018). Employing proteomics to understand the effects of nutritional intervention in cancer treatment. *Analytical and Bioanalytical Chemistry, 410*, 6371–6386. https://doi.org/10.1007/s00216-018-1219-z.

Sébédio, J. L. (2017). Metabolomics, nutrition, and potential biomarkers of food quality, intake, and health status. *Advances in Food and Nutrition Research, 82*, 83–116. https://doi.org/10.1016/bs.afnr.2017.01.001. Academic Press Inc.

Sevilla-Villanueva, B., Gibert, K., Sanchez-Marre, M., Fito, M., & Covas, M. I. (2017). Evaluation of adherence to nutritional intervention through trajectory analysis. *IEEE Journal of Biomedical and Health Informatics, 21*, 628–634. https://doi.org/10.1109/JBHI.2016.2634698.

Sikalidis, A. K. (2019). From food for survival to food for personalized optimal health: A historical perspective of how food and nutrition gave rise to nutrigenomics. *Journal of the American College of Nutrition, 38*, 84–95. https://doi.org/10.1080/07315724.2018.1481797.

Simpson, S. J., Le Couteur, D. G., James, D. E., George, J., Gunton, J. E., Solon-Biet, S. M., et al. (2017). The Geometric Framework for Nutrition as a tool in precision medicine. *Nutrition and Healthy Aging, 4*, 217–226. https://doi.org/10.3233/NHA-170027.

Simpson, S. J., & Raubenheimer, D. (2005). Obesity: The protein leverage hypothesis. *Obesity Reviews, 6*, 133–142. https://doi.org/10.1111/j.1467-789X.2005.00178.x.

Skrypnik, K., Suliburska, J., Skrypnik, D., Pilarski, Ł., Reguła, J., & Bogdański, P. (2017). The genetic basis of obesity complications. *Acta Scientiarum Polonorum, Technologia Alimentaria, 16*, 83–91. https://doi.org/10.17306/J.AFS.2017.0442.

Smith, A. J. P., Deloukas, P., & Munroe, P. B. (2018). Emerging applications of genome-editing technology to examine functionality of GWAS-associated variants for complex traits. *Physiological Genomics, 50*, 510–522. https://doi.org/10.1152/physiolgenomics.00028.2018.

Solon-Biet, S. M., McMahon, A. C., Ballard, J. W. O., Ruohonen, K., Wu, L. E., Cogger, V. C., et al. (2014). The ratio of macronutrients, not caloric intake, dictates cardiometabolic health, aging, and longevity in ad libitum-fed mice. *Cell Metabolism, 19*, 418–430. https://doi.org/10.1016/j.cmet.2014.02.009.

Sotos-Prieto, M., Luben, R., Khaw, K. T., Wareham, N. J., & Forouhi, N. G. (2014). The association between Mediterranean Diet Score and glucokinase regulatory protein gene variation on the markers of cardiometabolic risk: An analysis in the European Prospective Investigation into Cancer (EPIC)-Norfolk study. *British Journal of Nutrition, 112*, 122–131. https://doi.org/10.1017/S0007114514000580.

Spanakis, G., Weiss, G., Boh, B., Lemmens, L., & Roefs, A. (2017). Machine learning techniques in eating behavior e-coaching: Balancing between generalization and personalization. *Personal and Ubiquitous Computing, 21*, 645–659. https://doi.org/10.1007/s00779-017-1022-4.

Stenne, R., Hurlimann, T., & Godard, B. (2012). Are research papers reporting results from nutrigenetics clinical research a potential source of biohype? *Accountability in Research, 19*, 285–307. https://doi.org/10.1080/08989621.2012.718681.

Stenne, R., Hurlimann, T., & Godard, B. (2013). Benefits associated with nutrigenomics research and their reporting in the scientific literature: Researchers' perspectives. *Accountability in Research, 20*, 167–183. https://doi.org/10.1080/08989621.2013.788381.

Stover, P. J. (2004). Nutritional genomics. *Physiological Genomics, 16*, 161–165. https://doi.org/10.1152/physiolgenomics.00204.2003.

Sun, Z., Biela, L. M., Hamilton, K. L., & Reardon, K. F. (2012). Concentration-dependent effects of the soy phytoestrogen genistein on the proteome of cultured cardiomyocytes. *Journal of Proteomics, 75*, 3592–3604. https://doi.org/10.1016/j.jprot.2012.04.001.

Sundsten, T., & Ortsäter, H. (2009). Proteomics in diabetes research. *Molecular and Cellular Endocrinology, 297*, 93–103. https://doi.org/10.1016/j.mce.2008.06.018.

Tebani, A., & Bekri, S. (2019). Paving the way to precision nutrition through metabolomics. *Frontiers in Nutrition, 6*, 41. https://doi.org/10.3389/fnut.2019.00041.

Tiffon, C. (2018). The impact of nutrition and environmental epigenetics on human health and disease. *International Journal of Molecular Sciences, 19*, 3425. https://doi.org/10.3390/ijms19113425.

Tönjes, A., Zeggini, E., Kovacs, P., Böttcher, Y., Schleinitz, D., Dietrich, K., et al. (2010). Association of FTO variants with BMI and fat mass in the self-contained population of Sorbs in Germany. *European Journal of Human Genetics, 18*, 104–110. https://doi.org/10.1038/ejhg.2009.107.

de Toro-Martín, J., Arsenault, B. J., Després, J. P., & Vohl, M. C. (2017). Precision nutrition: A review of personalized nutritional approaches for the prevention and management of metabolic syndrome. *Nutrients, 9*, 913. https://doi.org/10.3390/nu9080913.

Valitutti, F., & Fasano, A. (2019). Breaking down barriers: How understanding celiac disease pathogenesis informed the development of novel treatments. *Digestive Diseases and Sciences, 64*, 1748–1758. https://doi.org/10.1007/s10620-019-05646-y.

Van Velzen, E. J. J., Westerhuis, J. A., Van Duynhoven, J. P. M., Van Dorsten, F. A., Grün, C. H., Jacobs, D. M., et al. (2009). Phenotyping tea consumers by nutrikinetic analysis of polyphenolic end-metabolites. *Journal of Proteome Research, 8*, 3317–3330. https://doi.org/10.1021/pr801071p.

Vasan, R. S. (2006). Biomarkers of cardiovascular disease: Molecular basis and practical considerations. *Circulation, 113*, 2335–2362. https://doi.org/10.1161/CIRCULATIONAHA.104.482570.

Vasan, S. K., Karpe, F., Gu, H. F., Brismar, K., Fall, C. H., Ingelsson, E., et al. (2014). FTO genetic variants and risk of obesity and type 2 diabetes: A meta-analysis of 28,394 Indians. *Obesity, 22*, 964–970. https://doi.org/10.1002/oby.20606.

Villalobos-Comparán, M., Teresa Flores-Dorantes, M., Teresa Villarreal-Molina, M., Rodríguez-Cruz, M., García-Ulloa, A. C., Robles, L., et al. (2008). The FTO gene is associated with adulthood obesity in the Mexican population. *Obesity, 16*, 2296–2301. https://doi.org/10.1038/oby.2008.367.

Wang, T. T. Y., Edwards, A. J., & Clevidence, B. A. (2013). Strong and weak plasma response to dietary carotenoids identified by cluster analysis and linked to beta-carotene 15,15'-monooxygenase 1 single nucleotide polymorphisms. *The Journal of Nutritional Biochemistry, 24*, 1538–1546. https://doi.org/10.1016/j.jnutbio.2013.01.001.

Wang, D. D., & Hu, F. B. (2018). Precision nutrition for prevention and management of type 2 diabetes. *The Lancet Diabetes and Endocrinology, 6*, 416–426. https://doi.org/10.1016/S2213-8587(18)30037-8.

Wang, T. J., Larson, M. G., Vasan, R. S., Cheng, S., Rhee, E. P., McCabe, E., et al. (2011). Metabolite profiles and the risk of developing diabetes. *Nature Medicine, 17*, 448–453. https://doi.org/10.1038/nm.2307.

Westerman, K., Reaver, A., Roy, C., Ploch, M., Sharoni, E., Nogal, B., et al. (2018). Longitudinal analysis of biomarker data from a personalized nutrition platform in healthy subjects. *Scientific Reports, 8*, 1–10. https://doi.org/10.1038/s41598-018-33008-7.

Wilson, C. P., Ward, M., McNulty, H., Strain, J. J., Trouton, T. G., Horigan, G., et al. (2012). Riboflavin offers a targeted strategy for managing hypertension in patients with the MTHFR 677TT genotype: A 4-y follow-up. *American Journal of Clinical Nutrition, 95*, 766–772. https://doi.org/10.3945/ajcn.111.026245.

Wittenbecher, C., Mühlenbruch, K., Kröger, J., Jacobs, S., Kuxhaus, O., Floegel, A., et al. (2015). Amino acids, lipid metabolites, and ferritin as potential mediators linking red meat consumption to type 2 diabetes. *American Journal of Clinical Nutrition, 101*, 1241–1250. https://doi.org/10.3945/ajcn.114.099150.

World Health Organization. (2016). *Major non communicable diseases and their risk factors*. WHO accessed March 17, 2020. https://www.who.int/ncds/introduction/en/

Wu, G. D., Chen, J., Hoffmann, C., Bittinger, K., Chen, Y. Y., Keilbaugh, S. A., et al. (2011). Linking long-term dietary patterns with gut microbial enterotypes. *Science (80), 334*, 105–108. https://doi.org/10.1126/science.1208344.

Young, K. L., Graff, M., Fernandez-Rhodes, L., & North, K. E. (2018). Genetics of obesity in diverse populations. *Current Diabetes Reports, 18*, 145. https://doi.org/10.1007/s11892-018-1107-0.

Younus, L. A., Algenabi, A. H. A., Abdul-Zhara, M. S., & Hussein, M. K. (2017). FTO gene polymorphisms (rs9939609 and rs17817449) as predictors of Type 2 Diabetes Mellitus in obese Iraqi population. *Gene, 627*, 79–84. https://doi.org/10.1016/j.gene.2017.06.005.

Zeevi, D., Korem, T., Zmora, N., Israeli, D., Rothschild, D., Weinberger, A., et al. (2015). Personalized nutrition by prediction of glycemic responses. *Cell, 163*, 1079–1094. https://doi.org/10.1016/j.cell.2015.11.001.

Zhang, N., Ju, Z., & Zuo, T. (2018). Time for food: The impact of diet on gut microbiota and human health. *Nutrition, 51–52*, 80–85. https://doi.org/10.1016/j.nut.2017.12.005.

Zhao, Y., Barrere-Cain, R. E., & Yang, X. (2015). Nutritional systems biology of type 2 diabetes. *Genes and Nutrition, 10*, 31. https://doi.org/10.1007/s12263-015-0481-3.

Zhao, L., Du, M., Gao, J., Zhan, B., & Mao, X. (2019). Label-free quantitative proteomic analysis of milk fat globule membrane proteins of yak and cow and identification of proteins associated with glucose and lipid metabolism. *Food Chemistry, 275*, 59–68. https://doi.org/10.1016/j.foodchem.2018.09.044.

Zinöcker, M. K., & Lindseth, I. A. (2018). The western diet–microbiome-host interaction and its role in metabolic disease. *Nutrients, 10*. https://doi.org/10.3390/nu10030365.

Chapter 3

Innovations in functional foods development

Burcu Guldiken[1], Zehra Gulsunoglu[2], Sena Bakir[2,3], Gizem Catalkaya[2], Esra Capanoglu[2], Michael Nickerson[1]

[1]*Department of Food and Bioproduct Sciences, University of Saskatchewan, Saskatoon, SK, Canada;* [2]*Department of Food Engineering, Faculty of Chemical and Metallurgical Engineering, Istanbul Technical University, Maslak, Istanbul, Turkey;* [3]*Department of Food Engineering, Recep Tayyip Erdogan University, Merkez, Rize, Turkey*

3.1 Introduction

New and improved ingredients and technologies are the key aspects of functional food development. Innovative food designs and technologies provide the adaptability of the food industry to new trends. In general, the expectations from functional food concepts are to fulfill the need for healthy products, to replace an ingredient with more sustainable, emerging sources, and to serve multiple needs at one time. In this concept, the role of carbohydrates, proteins, lipids, bioactive compounds, and minerals will be discussed from the point of view of innovative technologies, new ingredients, and the challenges in functional food development.

3.2 Role of carbohydrates in functional food development

Carbohydrates comprise the most prevalent class of organic compounds and are originated from simple sugars with the empirical formula $C_nH_{2n}O_n$ (Shukla & Tiwari, 2012). They can be classified into three main groups depending on their degree of polymerization: (i) sugars (consisted of mono- and disaccharides) (ii) oligosaccharides (formed by three to ten monosaccharide units) (iii) polysaccharides (comprised from ten or more monosaccharide units). Glucose and fructose are the most abundant dietary monosaccharides that arise from fruits, and berries, while the presence of the other monosaccharide, galactose, in nature is quite rare except in fermented milk products (Asp, 1995). Natural occurring oligosaccharides can be divided into two groups: primary (e.g., α,α-trehalose, selaginose) and secondary (e.g., melibiose, gentiobiose) oligosaccharides (Eggleston & Côté, 2003). The primary

Food Technology Disruptions. https://doi.org/10.1016/B978-0-12-821470-1.00008-2

oligosaccharides include certain oligosaccharides synthesized *in vivo* by the activity of glycosyltransferase from a mono- or oligosaccharide and a glycosyl donor. They occur freely in plants in significant quantities and are of metabolic importance, such as primer function, conservation of energy, translocation, and frost resistance.

On the other hand, secondary oligosaccharides include all oligosaccharides resulting from *in vivo* or *in vitro* hydrolysis of higher oligosaccharides, polysaccharides, glycolipids, and glycoproteins. They function in the tissue as structural components. Even if they occur in plants in free form due to *in vivo* hydrolysis, they are further metabolized into monosaccharides instead of being accumulated (Kandler & Hopf, 1980). The polysaccharides can be separated into starches and nonstarch polysaccharides (NSPs). Starches contain linear (amylose) or branched (amylopectin) glucose units with α-glucosidic bonds (α-glucans), whereas cellulose (a linear β-glucan) can be considered as the significant NSP. Other storage polysaccharides include fructans (e.g., inulin), glucomannans, galactomannans (e.g., guar gum), pectic substances, mucilages, alginates, and exudate gums. Dietary NSP is mainly provided by plant cell walls (Asp, 1995, 1996; Tester, Karkalas, & Qi, 2004). As shown in Table 3.1, NSPs can be (i) plant origin (cellulose and derivatives, hemicelluloses, pectins, exudate gums, mucilage gums, fructans), (ii) seaweed origin (alginates, carrageenans, agar), (iii) microbial origin (xanthan gum, pullulan, and gellan gum) and (iv) crustacean origin (chitin and chitosan).

Carbohydrates -the primary substrates for energy metabolism- affect satiety, blood glucose, and insulin levels, cholesterol and triglyceride metabolism, protein glycosylation, bile acid dehydroxylation. Furthermore, fermentation of indigestible carbohydrates in the colon affects the colonic functions, including gastric emptying, bowel habit/laxation/motor activity, metabolism, and balance the gut microbiota, as well as colonic epithelial cell health (Cummings & Stephen, 2007; Wolever & Wahlqvist, 1997). On the other hand, some dietary carbohydrates are known to contribute to the prevention of obesity, coronary heart disease, type 2 diabetes, and some cancers (Wolever, 2001).

Carbohydrates are essential compounds for living organisms. Apart from their biological significance, they are of great importance to the food industry. As listed by Codex Alimentarius (FAO, 2019) carbohydrates can be used as an emulsifier, a stabilizer, a thickener, a carrier agent, a gelling agent, a glazing agent, a humectant, a bulking agent, a foaming agent, an anticaking agent, a sweetener, and a firming agent in food systems. Other functions not listed by Codex Alimentarius can be ordered as follows: cryoprotectant, drying aid, chelating agent, fat mimetic, flavor carrier, substrate in fermentation process, flavor, and color developers via Maillard reactions (Voragen, 1998).

As a result of the existing food demand and production methods, the growing understanding of the value of effect on the environment has motivated researchers to explore new efficient, economically viable, and

TABLE 3.1 Classification of dietary fibers.

Main source	Dietary fiber	Subclass	Example sources
Plant polysaccharides	Cellulose and derivatives	Microcrystalline cellulose Carboxymethyl cellulose Methylcellulose	Cell walls of higher plants such as fruit and vegetables
	Hemicelluloses	Mannans and galactomannans Xyloglucans Glucomannans Arabinoxylans β-*D*-Glucans Arabinogalactan	Locust bean Tamarind seed Kernels of oat, barley, wheat, and rye
	Pectins	Homogalacturonan Xylogalacturonans Rhamnogalacturonans	Apple pulp, citrus peels, Sugarbeet pulp
	Exudate gums	Gum Arabic Taragacanth gum Gum Karaya Gum Ghatti	Acacia tree *Astragalus gummifer* Labillardiere *Sterculia urens*
	Musilage gums	Okra mucilage Yellow mustard Flax mucilage Psyllium gum.	*Plantago*
	Fructans	Inulin Levan	Dandelions Chicory Lettuce
Seaweed polysaccharides	Alginates		Brown seaweeds
	Carrageenan		Red seaweeds
	Agar		Red seaweeds
	Others	Ulvan Fucoidan Furcellaran Floridean	*Ulva rigida* *Enteromorpha compressa* *Furcellaria lumbricialis*

Continued

TABLE 3.1 Classification of dietary fibers.—cont'd

Main source	Dietary fiber	Subclass	Example sources
Microbial polysaccharides	Xanthan		*Xanthomonas*
	Pullulan		*Aureobasidium*
	Gellan		*Sphingomonas*
	Curdlan		*Agrobacterium*
	Levan		*Leuconostoc*
	Bacterial alginate		*Pseudomonas*
	Bacterial cellulose		*Acetobacter xylinum*
Animal polysaccahrides	Chitin and chitosan		Shrimp Squid Crabs
Resistant starch	RS1: Inaccessible, physically entrapped starch in the cell-matrix of whole or partially milled grains		Whole or partly milled grains
	RS2: Native granular starch		Raw potato, unripe banana
	RS3: Retrograded starch		Cooked and cooled pasta, rice
	RS4: Chemically modified starch		

Adapted from Izydorczyk, M., Cui, S. W., & Wang, Q. (2005). Polysaccharide gums: Structures, functional properties, and applications. *Food Carbohydrates: Chemistry, Physical Properties, and Applications*, 293–299. and Niba, L. L. (2005). *Carbohydrates: Starch. Hui YH Handbook of food science, technology, and engineering.* Taylor & Francis 4.

sustainable processes based on taking advantage of agro-food by-products. Food processing by-products have been extensively investigated and characterized as potential sources of carbohydrate-based ingredients, especially dietary fibers (Gourgue, Champ, Lozano, & Delort-Laval, 1992; Müller-Maatsch et al., 2016; Sharoba, Farrag, & Abd El-Salam, 2013). The American Association of Cereal Chemists (AACC) defined dietary fiber as "the remnants of the edible part of plants and analogous carbohydrates that are resistant to digestion and absorption in the human small intestine with complete or partial fermentation in the human large intestine" (DeVries et al., 2001). The European Food Safety Authority's (EFSA) opinion describes dietary fiber as nondigestible carbohydrates, including NSPs and some other polysaccharides, as well as lignin. Among the others, resistant oligosaccharides also include fructooligosaccharides (FOS) and galactooligosaccharides (GOS). Resistant starch comprises of several raw starch granules, physically enclosed starch, retrograded starch, chemically, and physically modified starches (DeVries et al., 2001).

3.2.1 Application of carbohydrates in functional food products

The advances in fiber ingredients have evolved strongly over the past few years, and the production of fiber-rich products as a functional ingredient has increased. Also, there is a wide variety of high fiber products, such as conventional foods having low-moisture content, such as cereals, snacks, and bread, as well as enriched foods with a high fiber content as can be applied to dairy or meat products and beverages (López-Marcos, Bailina, Viuda-Martos, Pérez-Alvarez, & Fernández-López, 2015). Fibers can modify food consistency, rheological characteristics, texture, and organoleptic properties. For example, in meat and fish products, they are used as a fat replacer, water binder, emulsion stabilizer, cooking yield, and texture enhancer, lipid oxidation reducing agents. Moreover, they are applied in dairy products such as cheese, yogurt, and ice cream to improve body and mouthfeel, as well as decrease syneresis. In beverages, dietary fibers are utilized as bulking agents, viscosity and stability improver, and nutritive additives. They are also used in sauces and fruit products such as jam and marmalades to modify rheological properties of these food products, while in breakfast cereals and extruded foods like pasta, they are used as fortifying agents to provide higher dietary fiber content. On the other hand, incorporation of dietary fibers into confectionery products, such as chocolate, reduces sugar content and calorific value, whereas, in fast food products, they reduce oil holding during the frying process. Last but not least, they have a wide application in baked goods to modify texture and springiness, increase volume, improve shelf life, or softness of the crumb, replace wheat flour, improve nutritional quality (Maphosa & Jideani, 2016).

3.2.1.1 Bakery products

Cakes and muffins are the most widely produced bakery products with high consumer demand. However, there is a growing trend in such products with increased nutritional value since the health-conscious consumers demand high-quality and low-calorie products that have reduced fat and sugar content. Sharlyn melon peels and watermelon rinds were evaluated as fiber-rich food by-products to be used in cake formulations to enhance the fiber content while reducing the oil content. When wheat flour or oil in the cake formulation were substituted with sharlyn melon peel or watermelon rind flours, the resultant cakes were enhanced in terms of the volume, specific volume and lipid oxidation, while staling and free fatty acid formation during storage has been retarded (Al-Sayed & Ahmed, 2013). Sharoba et al. (2013) also investigated the effect of wheat flour replacement with orange waste (OW), green pea peels (GPP), potato peels (PP), and carrot pomace (CP) in cake formulations on mechanical, physical and organoleptic properties of the cakes. According to the results, replacement of wheat flour at 5%, 10%, 15%, and 20% levels of by-products significantly increased the water holding capacity, extension resistance, dietary fiber content, and decreased the caloric value, sensorial properties except in OW and CP samples. As a consequence, the source and content of the dietary fibers significantly affected the overall properties of cakes, and the waste samples have been suggested as good sources of dietary fibers for the fortification of cakes (Sharoba et al., 2013). In another study, peeled raw pumpkin pulp (PRPP) and unpeeled raw pumpkin pulp (URPP), as well as their flours (PPPF and UPPF, respectively), were utilized as resistant starch (RS) sources in crackers (Aziah, Ho, Komathi, & Bhat, 2011). Their findings indicated that raw materials had higher RS in the presence of peel and even higher when they were processed into flour. The crackers incorporated with 15% and 20% of UPPF showed the highest RS content reflecting the potential for the use of pumpkin peels as the RS sources in functional food formulations. In another study, tomato pomace was utilized as a fat substitute in low-fat sponge cake. Tomato pomace was incorporated into the cake blend at a fat replacement ratio of 25%, 50%, and 75%. Dough development, mixing tolerance index, stability, and degree of softening was enhanced as the replacement ratio increased. Furthermore, formulations with tomato pomace showed greater water absorption at all substitution levels (Namir, Suleiman, & Hassanien, 2015).

β-glucans and foods rich in or enriched with these polysaccharides are of great interest due to their biological activity. Foods containing significant amounts of β-glucan have been shown to reduce serum LDL (Low-Density Lipoprotein)-cholesterol levels (Gamel, Abdel-Aal, Ames, Duss, & Tosh, 2014). In a study, wheat flour has been supplemented with soya and rice bran at 10%, 15%, 20%, 25% levels each to develop a functional biscuit. When the physical, sensory, and nutritional characteristics of the final products were evaluated, the supplemented wheat flour: soya flour: rice bran biscuit with the ratio of 70:15:15 was determined as the best rated in general preference of sensory rating (Mishra & Chandra, 2012).

3.2.1.2 Meat products

Several forms of fibers were used in the manufacture of restructured meat products and meat emulsions. Dietary fiber addition affects different sensory features of processed meat products such as texture, color, and juiciness. The general acceptance of meat products enriched with dietary fiber was improved positively. Besides, the health-promoting effects of dietary fibers, such as cholesterol-reducing properties, have been used by meat manufacturers to gain the attention of health-conscious customers (Talukder, 2015). Turhan, Sagir, and Ustun (2005) reported that the introduction of hazelnut pellicle fiber found to be successful in improving beef burger's cooking yield, structural changes, and thickness. However, the use of rice bran fiber in meat batters resulted in higher pH values and decreased emulsion stability (Choi et al., 2007). Incorporation of precooked lemon albedo fibers at 2.5−5% levels into bolognas resulted in higher moisture content than the control treatments. Also, a decrease in the fat content observed in the fiber added products, which was higher in raw albedo compared to precooked ones. An increase in hardness and reduction in fatness and hue perception observed in albedo added bolognas independent from the concentration added. Additionally, raw albedo showed a higher increase in hardness compared to cooked albedo (Fernandez-Gines, Fernandez-Lopez, Sayas-Barbera, Sendra, & Perez-Alvarez, 2004). Functional properties of hemicellulose B (RBHB) and insoluble dietary fiber (RBDF) from defatted rice bran to produce low-fat meatballs with acceptable attributes enriched with high content fiber has been investigated (Hu & Yu, 2015). The results revealed that RBHB and RBDF showed high water-binding and swelling capacities. Total trans fatty acids were lower in the samples with added RBHB, and total unsaturated fatty acid/total saturated fatty acid ratio was greater in these samples compared to the control meatballs. Dietary fiber-rich chicken meat patties were produced by combining wheat and oat bran in chicken meat at 5%, 10%, and 15% levels (Talukder & Sharma, 2010).

3.2.1.3 Soft and alcoholic drinks

Certain cultures have consumed seaweeds for decades. They are used in the human diet principally as sea vegetables. Alginate is the primary polysaccharide form present in brown seaweed. There are several industrial applications using alginate: in the food industry, they are used as a thickener and stabilizing agent in ice cream, for whey separation in dairy products, as an emulsifier in mayonnaise, in medical science as a microencapsulating agent and as microsphere vector for controlled drug delivery (Cajnko, Novak, & Likozar, 2019; Fleurence, 2016). Colloidal stability, along with physicochemical and rheological properties, can be considered as the important factors of the sensory quality of beverages. The use of propylene glycol alginate (PGA) and alginate has been found to be effective in producing fermented whey beverages with a physical viscosity similar to the lactic beverages (Hashemi, Pezeshky, Gharedaghi, Javani, 2014). On the other hand, alginates were also used as functional ingredients in beverages to investigate the restriction in calorie intake. However, the alginate-based preload beverage was not successful in weight loss in obese subjects (Jensen, Kristensen, & Astrup, 2011). Similarly, in a study conducted with 24 healthy adult men, the incorporation of strong sodium alginate gels into the chocolate milk has been found to decrease premeal glycemia, insulinemia, and appetite, but not food intake, in other words, caloric intake, at a meal 2 h later after chocolate milk was consumed (El Khoury, Goff, Berengut, Kubant, & Anderson, 2014).

Among the other polysaccharides, chitosan is a polycationic polymer derived from chitin deacetylation, with the crustacean exoskeletons (for example, shrimp shell wastes) and fungi cell walls as its essential origins. It is a linear polymer constituted of D-glucosamine and *N*-acetyl-D-glucosamine units through ($\beta 1 \rightarrow 4$) linkage. Since their free amino groups present significant physical (e.g., solubility), chemical (e.g., interplay with other functional groups), physicochemical (e.g., mucoadhesive), and biological (e.g., antimicrobial, antioxidant) characteristics, they have broad applications in the alcoholic and nonalcoholic drink industry. For example, they have been found to be very useful in the clarification of fruit juices such as apple, orange, and lemon juices, as well as in beer and wine production. Additionally, they are used as antimicrobial agents in the winery to prevent the growth of spoilage microorganisms. They are also used as coating materials in the encapsulation of bioactive materials to protect them from degradation due to adverse environmental conditions (e.g., encapsulation of anthocyanins to prevent color loss in thermal applications) (Rocha, Coimbra, & Nunes, 2017).

3.2.1.4 Dairy products

Fruit fibers have also been used in the production of dairy products. Yogurt fortified with antioxidant dietary fibers from wine grape pomace. Addition of 1, 2, or 3% grape pomace into the yogurts resulted in delayed oxidation, increased viscosity, dietary fiber content, and decreased pH during 3 weeks of storage (Tseng & Zhao, 2013). According to do Espírito Santo, Perego, Converti, and Oliveira (2012), the addition of fiber-rich passion fruit peel powder increased firmness, cohesiveness, and consistency of skim yogurts. Costa, Lucera, Marinelli, Del Nobile, and Conte (2018) assessed the effect of the addition of dietary fiber-rich by-products such as red and white wine grape pomace, broccoli, and artichoke by-products, and tomato peel on the sensory and physicochemical properties of Primo sale cheese. Although the addition of both grape pomaces increased the nutritional value of the cheese better than the other added by-products, they impaired the physical characteristics. Therefore, artichoke by-product was recommended as the optimum fortifier in terms of nutritional value, weight loss during ripening, and moisture content, and pH of the cheese.

3.2.1.5 Emulsions

Foods that are produced in the form of oil in water emulsions, such as salad dressings, may be ideal candidates for dietary fiber addition. Chatsisvili, Amvrosiadis, and Kiosseoglou (2012) investigated the addition of orange pulp fiber in a dressing-type oil in water emulsion and reported that using 4 g/100 g orange pulp in a dressing-type oil in water emulsion with decreased oil content, enhanced its rheological properties, and improved its stability against creaming.

Similarly, the surface activity properties of dried orange pulp powder, which is known to be rich in pectin and cellulose, were investigated in oil in water emulsions. It has been reported that the preactivated pulps resulted in more delicate emulsions due to the opening of the plant cell wall matrix. Additionally, the cellulose network led to long-term stability by the thickening of the continuous phase during storage (Wallecan, McCrae, Debon, Dong, & Mazoyer, 2015). Moreover, delayed oxidation observed during storage when Italian (0.5% and 1% pomace added) and 1000 Island (1% and 2% pomace added) salad dressings incorporated with wine grape pomace (Tseng & Zhao, 2013).

3.2.1.6 Extruded products

Recovery of processing by-products of the agricultural products provides practical preservation and utilization of the nutritionally valuable wastes to be potentially used as a source of functional ingredients in value-added extruded products. For example, apple pomace and whey, food processing by-products, were used to produce a light-weighted, nutritional, storage stable puffed extrudates by supercritical fluid extrusion. As a result, a novel value-added puffed healthy snack with a high content of dietary fiber (14%) and nutritional antioxidants with protected natural color was successfully developed (Paraman, Sharif, Supriyadi, & Rizvi, 2015).

Selani et al. (2014) evaluated the replacement of corn flour with pineapple pomace at 10.5% and 21% levels to develop a functional extruded food product, which can be claimed as a "good source of fiber." According to their observations, a high percentage (21%) of pineapple pomace adversely affected the physicochemical properties of the extrudate, where 10.5% addition of pineapple pomace did not influence the properties of the final product in terms of hardness, yellowness, water absorption, and bulk density. Recently, carrot pomace and cauliflower trimmings were utilized to develop an extruded ready-to-eat soybean-rice product with high fiber content. 164°C die temperature, 313 rpm screw speed and a blend of rice: defatted soy flour: carrot pomace flour: cauliflower trimmings powder in the ratio of 85:7.5:3.25:3.25 are the optimum conditions to produce a desirable (with desirability rate of 76%) functional snack (Alam, Pathania, & Sharma, 2016).

3.2.1.7 Edible films and food-grade coatings

Polysaccharide-based edible films can be used to prolong the shelf-life of fruits, vegetables, fish, meats, and baked goods by avoiding dehydration, rancidity of oxidation, browning of the surface and diffusion of oil, as well as through their ability to influence the internal atmosphere of fruit and vegetable packaging. Furthermore, it has been shown that microbial safety and shelf life can be improved by the inclusion of antimicrobials, antioxidants, or bacteriocins in edible films and coatings. For example, chitosan has been used to develop natamycin carrying edible film to enhance the storability of Saloio cheese (Fajardo et al., 2010). After coating with chitosan-alginate edible films, figs showed increased firmness, better color retention, and less visual fungal growth compared to uncoated counterparts (Reyes-Avalos et al., 2016). In another study, edible films prepared with pectic extracts from Mexican Lime bagasse and pomace demonstrated higher inhibition against Gram-negative bacteria and also foodborne pathogenic bacteria when combined with Mexican lime essential oil (Aldana, Andrade-Ochoa, Aguilar, Contreras-Esquivel, & Nevárez-Moorillón, 2015). Moreover, incorporation of hydrocolloids such as xanthan gum, carrageenan, and propylene glycol alginate into a food-grade coating protected dry-cured hams against ham mite infestation (Zhao et al., 2016).

3.3 Proteins

Consumers of all ages become aware of the importance of protein in their diet. Prepared meal plans for people with specific requirements for kids, athletes, and the elderly should consider not only the protein amount but also the amino acid profile. Proteins in our food formulations come from animal or plant-based sources. Animal proteins are known as complete proteins as they include all the essential amino acids; however, plant proteins are typically limited in one or two of these essential amino acids (Beran et al., 2018). Blending plant proteins may help to overcome this deficiency and maximizes the favorable functional properties. The design of new functional foods will meet the requirements of both functional properties and nutritional profile. However, sensory properties have a crucial role in the acceptance of foods by the customers. The utilization of different protein sources in food product development and their sensory evaluation are presented in Table 3.2. Using food processing by-products as a source of protein or other emerging proteins has a high potential to be the solution of sustainable sources for the food industry. Processing type, morphological properties, particle size, denaturation degree, extraction method, physicochemical properties of the proteins determines their functional properties (Beran et al., 2018). This part will provide detailed information about selected protein sources in functional food development.

TABLE 3.2 Protein fortified products and their acceptability.

Protein	Applied food	Application goal	Sensory evaluation	References
Whey protein powder	Muffin	Elderly nutrition	Artificial taste, large air bubbles	Wendin et al. (2017)
Whey protein (microparticulate)	Yogurt	Fat replacement	Creamy flavor	Torres et al. (2011)
Rice protein	Fermented milk	Protein fortification	Bitter taste, high odor	Akin Ozcan (2017)
Rice bran protein	Biscuit	Protein fortification	Higher fracture strength overall accepteable	Yadav, Yadav, & Chaudhary, 2011
Rice protein flour	Beef patties	Elderly nutrition	Harder texture	Baugreet et al. (2016)
Soy flour	Muffin	Elderly nutrition	Low sweetness, no air bubbles	Wendin et al. (2017)
Soy protein isolate	Fermented milk	Protein fortification	Highly acceptable	Akin and Ozcan (2017)
Soy protein hydrolysates (microparticulated)	Ice cream	Fat replacement	Overall acceptable	Liu et al. (2018)
Pea protein isolate	Fermented milk	Protein fortification	Highly acceptable	Akin and Ozcan (2017)
Pea protein isolate	Beef patties	Elderly nutrition	Harder texture	Baugreet et al. (2016)
Lentil flour	Beef patties	Elderly nutrition	Soft texture	Baugreet et al. (2016)
Hemp protein concentrate (microparticulated)	Baguette	Protein fortification for gluten free foods	Weak aftertaste, overall acceptable	Beran et al. (2018)
Hemp flour	Rice cracker	Protein fortification for gluten free foods	Increasing darkness with higher concentration	Radočaj et al. (2014)

Continued

TABLE 3.2 Protein fortified products and their acceptability.—cont'd

Protein	Applied food	Application goal	Sensory evaluation	References
Canola protein concentrate (microparticulated)	Baguette	Protein fortification for gluten free foods	Weak aftertaste, overall acceptable	Beran et al. (2018)
Almond flour	Muffin	Elderly nutrition	No muffin like structure	Wendin et al. (2017)
Microalgae biomass	Pasta	Protein fortification	Better acceptability than control	Fradique et al. (2010)
Gelatin from insects	Ice cream	Replacement of commercial gelatin	Less acceptable taste and texture	Mariod and Fadul (2015)

3.3.1 Proteins from oil processing by-products

There is an increasing demand for plant-derived proteins due to economic and environmental factors and changing consumer purchase behavior. An increase in vegetarians, vegans, and flexitarians cause an expansion in the plant-based food market and require new products and technologies. Improving the economic value of proteins from plant sources is essential from an economical, but also an innovation point of view. In this context, the oil-producing plant is free sources of proteins suitable for human consumption. Soybean (*Glycine max* L.), canola/rapeseed (*Brassica napus*), cottonseed (*Gossypium hirsutum* L.), sunflower seed (*Helianthus annuus* L.) and peanut (*Arachis hypogaea* L.) are currently the world's primary oil crops, and their protein meal supply is relatively high (about 200 million tonnes in 2015) (Pojić, Mišan, & Tiwari, 2018). Oil processing by-products (e.g., filter cakes/defatted meals) of soybean, canola/rapeseed, sunflower, safflower (*Carthamus tinctorius* L), peanut, corn (*Zea mays* L.), cottonseed, sesame, flax (*Linum usitatissimum* L.), and hemp (*Cannabis sativa*) can be used to produce proteins (Arntfield, 2018).

Defatted sunflower meal contains protein ranging from 40% to 66% after the oil extraction process (González-Pérez & Vereijken, 2007). Chlorogenic acid amount is higher in sunflower meal than other oil crops. Green pigmentation related to sunflower protein association with oxidized chlorogenic acid in alkaline conditions may affect the acceptability of the sunflower protein-fortified products (Wildermuth, Young, & Were, 2016). For some products, green color can be desirable; however, applications of the sunflower protein are restricted with the resultant color. Color improvement in sunflower fortified bread was achieved with the addition of succinic acid precipitation/purification step to protein extraction, which can provide increased food applications with a label-friendly protein source instead of allergenic and GMO plant proteins (Shchekoldina & Aider, 2014).

Moreover, defatted and texturized soybean, flaxseed, and sunflower meals produced by extrusion were used to fortify noodles with their abundant proteins (Bhise, Kaur, & Aggarwal, 2015). The authors reported that protein fortification with texturized sunflower, flaxseed, and soybean flour could be acceptable up to 10%, 10%, and 20%, respectively. The observed dark color in protein fortified noodles increased with increasing texturized protein content. Limited protein hydrolysates of rapeseed ranging from 3.1% to 7.7% demonstrated improved functional properties compared to the initial protein isolate. These enhanced functional properties of rapeseed protein hydrolysates increase their usage in food products, including bread, ice creams, cakes, desserts, meat products, and salad dressings (Vioque, Sánchez-Vioque, Clemente, Pedroche, & Millán, 2000).

Nowadays, changes in regulations in many countries expanded the potential usage of hemp seed proteins and hydrolysates in functional food development. Dabija, Codină, Gâtlan, Sănduleac, and Rusu (2018) reported that the sensory characteristics of yogurts with hemp seed proteins had high overall acceptability in addition to rheological and physicochemical evaluation. Moreover, further improvements are necessary to increase oilseed protein utilization in functional food development. The impact of defatting and extraction technology on the functional and physicochemical properties were investigated in hemp, flax, and canola seeds (Teh, Bekhit, Carne, & Birch, 2014). Alkali extraction of proteins from these oil seeds yielded higher amounts than acid extraction, as well as better water holding capacity, oil holding capacity, and emulsifying stability. Functional properties of hemp seed protein isolate (84%, protein content) and hemp seed protein meal (44%, protein content) were evaluated by Malomo, He, and Aluko (2014). The results showed that hemp seed protein isolates provided better foaming capacity and formed smaller oil droplets in emulsions at higher levels of fortification. Also, hemp flour fortification of brown rice crackers with 20% hemp flour with 4 g of green tea leaves was highly accepted in sensory evaluation and provided a healthy snack profile; however, up to 30% of hemp flour can be added to fulfill the nutritional profile with low sensory properties (Radočaj, Dimić, & Tsao, 2014). Besides, bread fortified with hemp flour at different ratios were found high in protein content (13–19% dw) in regards to wheat bread (11% dw), and the overall effect can be seen in Fig. 3.1 (Mikulec et al., 2019). On the other hand, owing to a balanced amino acid

FIGURE 3.1 Transversal cut of bread. (A) wheat bread, (B) bread with 15% addition of hempseed flour, (C) bread with 30% addition of hempseed flour, (D) bread with 50% addition of hempseed flour. *Reprinted by permission from Mikulec, A., Kowalski, S., Sabat, R., Skoczylas, Ł., Tabaszewska, M., & Wywrocka-Gurgul, A. (2019). Hemp flour as a valuable component for enriching physicochemical and antioxidant properties of wheat bread.* Lebensmittel-Wissenschaft and Technologie, 102, *164–172.*

profile and possible functional properties, including emulsifying, gelling, and foaming, canola protein isolate has been proposed as an alternative to other proteins for human food use (Beran et al., 2018).

3.3.2 Proteins from marine by-products and algae

Alternative protein sources and processing technologies are necessary to meet the customers' needs and global protein demand (Bleakley & Hayes, 2017). The utilization of seafood wastes is an excellent opportunity to produce protein-fortified foods. For instance, shrimp processing provides shrimp head and carapace as a by-product, which includes nutritious and functional compounds such as chitin, protein, minerals, and natural pigments (Guo, Sun, Zhang, & Mao, 2019). Protein extraction can be performed from shrimp heads via enzymatic extraction with commercial enzymes (45%) (Mizani, Aminlari, & Khodabandeh, 2005), autolysis and fermentation (66%) (Guo et al., 2019), and autolysis by gradual temperature (44–87%) (Cao et al., 2009). The use of endogenous enzymes for protein extraction provides a reasonable way to obtain protein from shrimp waste. In a study, protein hydrolysates from cephalothorax of shrimp utilized to fortify biscuits up to 5% protein with high acceptability (Sinthusamran, Benjakul, Kijroongrojana, & Prodpran, 2019). Besides, the authors reported that flavoring amino acids from shrimp head and Maillard reaction products during baking contributed to the biscuit flavor. However, allergic reactions due to shellfish consumption are known, and shrimp contains a few proteins (tropomyosin, parvalbumin) that bind Immunoglobulin E in atopic patients (Yu, Lin, Chiang, & Chow, 2003).

Seaweed (macroalgae) and microalgae are regarded as rich protein sources, and even some species include protein levels comparable to those conventional sources of protein, such as soybean, egg, milk, and meat (Bleakley & Hayes, 2017). The two most common genera of microalgae are *Chlorella* and

Spirulina, which are eukaryotic and prokaryotic microorganisms, respectively (Andrade, Andrade, Dias, Nascimento, & Mendes, 2018). Microalgae proteins from *Chlorella pyrenoidosa* has a higher essential amino acid index (1.35) than soy protein (0.657) and close to that of casein (1.66) (Waghmare, Salve, LeBlanc, & Arya, 2016). *Spirulina* sp. contains phycobiliproteins that have health-promoting effects, such as hepatoprotective, anticancer, anti-inflammatory, immunomodulatory, and antioxidant properties (Andrade et al., 2018). Microalgae can provide high amounts of essential amino acids, including lysine, valine, leucine, isoleucine, and tryptophan, depending on daily consumption (Andrade et al., 2018; Mišurcová et al., 2014). Therefore, algal proteins can be used during functional food development to improve the amino acid profile. Microalgae biomass or its protein fraction can be applied to bakery products to provide both nutritional value enhancement and natural coloring. Microalgae biomasses from *Chlorella vulgaris* with 38% protein content and *Spirulina maxima* with 45% protein content were applied to enrich spaghetti flour. The pasta produced in green color was sensorily accepted by the panelists (Fradique et al., 2010). Also, Fig. 3.2 shows the coloring effect of *Chlorella vulgaris* in cookies according to algae concentration (Gouveia, Batista, Miranda, Empis, & Raymundo, 2007).

Besides, macroalgae (seaweed) have high potential as sustainable protein sources for animal feed and human diet. The geographical location, species, and the harvest season alter their protein content (Garcia-Vaquero & Hayes, 2016). Red macroalgae include high protein content up to 47% w/w dry basis, green macroalgae contain reasonable levels as 9–26% w/w dry basis, whereas brown macroalgae include significantly lower protein contents as 3–15% (w/w) dry basis (Harnedy & FitzGerald, 2011). The highest levels of protein among red macroalgae found in *Palmaria palmata,* and *Porphyra tenera* as 35% and 47% of the dry matter, respectively, and these amounts were comparable to those present in high-protein vegetables such as soybeans (35% of the dry mass) (Burtin, 2003).

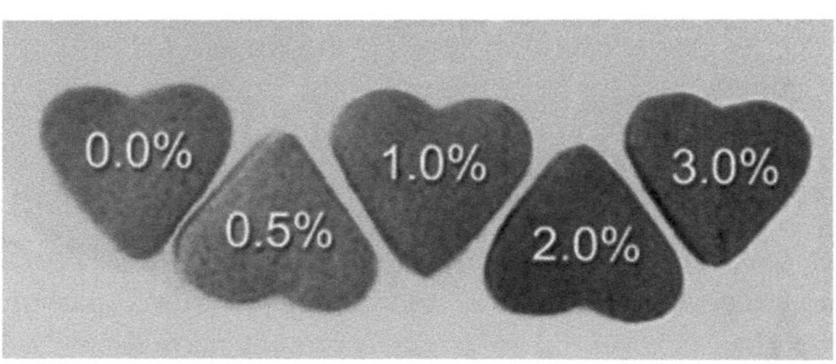

FIGURE 3.2 Cookies fortified with Chlorella vulgaris biomass (0.0–3.0% w/w). *Reprinted by permission from Gouveia, L., Batista, A. P., Miranda, A., Empis, J., & Raymundo, A. (2007). Chlorella vulgaris biomass used as colouring source in traditional butter cookies.* Innovative Food Science and Emerging Technologies, 8*(3), 433–436.*

3.3.3 Insect proteins

Insects are sustainable protein sources that can be applied to many food products. Based on the application, species-specific health factors, such as potential microbial, allergenic, and toxicological risks, must be taken into consideration when choosing the insect species and their developmental stages (Schlüter et al., 2017). The most frequently consumed species are beetles (Coleoptera), ants, bees and wasps (Hymenoptera), caterpillars (Lepidoptera), grasshoppers and locusts (Orthoptera), true bugs, aphids and leafhoppers (Hemiptera), termites (Isoptera) and flies (Diptera) (Yi et al., 2013).

There are a variety of insect-derived food powders available, but they often contain roasted, ground whole insects, which are typically milled directly into flours (Loveday, 2019). Besides, enzymatic hydrolysis has a positive effect on the bioactive content and allergenicity of insect proteins. Allergenic risks of cricket protein reduced with hydrolyzation up to 60–85%, which prevented the Immunoglobulin E (IgE) reactivity to tropomyosin (Hall, Johnson, & Liceaga, 2018). Furthermore, the effect of extraction solvents on protein yield, characterization, and functionality of cricket proteins was investigated by (Ndiritu, Kinyuru, Kenji, & Gichuhi, 2017) who reported that higher yield and less colored protein obtained using hexane as an extraction solvent; however, better functionality, including emulsifying and foaming properties determined with aqueous extraction. On the other hand, it was found that the protein fractions of five insect species (*Tenebrio molitor, Alphitobius diaperinus, Acheta domesticus, Zophobas morio,* and *Blaptica dubia*) have similar essential amino acid levels with soy protein isolate contrarily less than casein. They could form gels at a concentration of 30%; however, their foaming capacity is low at pH 3, 5, 7, and 10 (Yi et al., 2013). On the other hand, gelatin is widely used in food products worldwide. An insect alternative to gelatin may prevent diseases related to the sources and provides a sustainable option. Mariod and Fadul (2015) extracted gelatin from the melon bug (*Coridius viduatus*) and sorghum bug (*Agonoscelis versicoloratus*) using hot water extraction. The extracted gelatin was applied to ice cream, which had a different taste and texture than ice cream produced with commercial gelatin. The organoleptic properties of insect gelatin found to be lower than the commercial sample.

3.3.4 Microparticulated proteins

Lately, more focus has been directed to the development of protein particles with specific properties. Control of morphology, surface, particle size, and intrinsic properties is essential to produce protein particles with the characteristics required for emerging applications (Beran et al., 2018). Microparticulated proteins are widely used as a fat replacer in dairy foods, including yogurt (Torres, Janhøj, Mikkelsen, & Ipsen, 2011), ice cream (Liu, Wang, Liu, Wu, & Zhang, 2018), dressings, sauces, desserts (Ipsen, 2017), and cheese (Sturaro, De Marchi, Zorzi, & Cassandro, 2015). Microparticulated proteins can be produced by simultaneous heating and shearing, extrusion cooking at acidic pHs, or microfluidization (Ipsen, 2017; Torres et al., 2011).

Low-fat yogurt, which was produced with high native-to-denatured whey ratio microparticulates, had great creaminess, viscosity, and a similar texture to full-fat yogurt (Torres et al., 2011). Furthermore, cheese yield decreased while the fat content was reduced; however, at high concentrations of microparticulate whey protein, cheese yield was found higher in low-fat cheese (Sturaro et al., 2015). On the other hand, an increase in protein levels was determined in curd, but on the contrary, there was no significant difference in ripened cheese. New carbon dioxide assisted spray nebulization drying (CASND) technology has been used to obtain microparticulated protein concentrate, which succeeded to enhance foaming and emulsification properties, from renewable plant sources, including canola and hemp for human nutrition (Beran et al., 2018). Application of microparticulated canola or hemp protein concentrates in gluten-free baguettes provided good customer acceptance; however, the aftertaste of protein concentrates was reported in the study. Utilization of microparticulated soy protein hydrolysates and xanthan gum for the production of 50% fat-substituted ice cream led to similar sensory characteristics such as appearance, texture, and taste with 10% full-fat ice cream (Liu et al., 2018).

3.3.5 Pulse and cereal proteins

As sustainable protein sources, pulses (dry beans, peas, lentils, and chickpeas) and cereals (oats, wheat, rice, and corn) have an essential role in the human diet. Pulse proteins contain considerable amounts of lysine, leucine, aspartic acid, glutamic acid, and arginine, but are deficient in sulfur-containing amino acids, as well as tryptophan.

In contrast, cereal proteins are rich in sulfur-containing amino acids and deficient in lysine and are often consumed as part of a complementary diet with the pulses (Boye, Zare, & Pletch, 2010). Fortification of beef patties with pea protein isolate, rice protein, or lentil flour at two levels (3% or 7%) was studied by Baugreet, Kerry, Botineştean, Allen, and Hamill (2016) to increase protein intake in the elderly population. The results showed that fortification with rice protein at 7% increased protein content significantly, as well as protected the redness of beef patties, and lentil flour fortification provided a softer texture. Therefore, the authors mentioned that the combination of rice protein and lentil flour might provide protein enrichment without any hardening effect. Furthermore, cereals and pulse proteins can be used to develop gluten-free food products. Enzymatic crosslinking provides additional linkage that mimics the gluten network in breadmaking. Use of crosslinking enzymes, transglutaminase, and different oxidases, provides improvements in final bread volume and crumb properties for cereals, especially for buckwheat, rice, and oat (Renzetti & Rosell, 2016).

Moreover, cereal side streams, including milling, starch extraction, or bioethanol production, create a protein-rich fraction, which is mostly used in animal feed (Sozer, Nordlund, Ercili-Cura, & Poutanen, 2017). For instance, rice bran protein was utilized to fortify biscuits up to 15% protein content without diminishing its overall acceptability (Yadav, Yadav, & Chaudhary, 2011). Also, pulse flours can be used to develop protein-rich pasta products. Faba bean, black gram, and green lentil flours (24–28%, protein content) were used to produce gluten-free pasta with a low glycemic index with a smooth surface but lower springiness (Laleg, Cassan, Barron, Prabhasankar, & Micard, 2016).

3.3.6 Proteins in extruded products

Developing new functional foods without using additives to enhance some characteristics, including water hydration capacity, digestibility, and solubility can be performed via extrusion, which is a high-temperature, short-time processing technique where the application of temperature and mechanical shear under pressure plasticizes and processes foodstuffs (Patil & Kaur, 2018). However, protein fortification of extruded products with whey proteins, soy protein, or gluten causes a harder texture and a reduction in the expansion (Day & Swanson, 2013). This result was explained by the fact that under a particular concentration, protein behaves like a filler in the starch matrix, which decreases the extensibility of the mixture and thus limits the expansion (Zhu et al., 2010).

Field pea has a balanced amino acid profile with high lysine content and commonly used as a replacement of soybean or animal proteins due to its high accessibility, nutritional value, and health benefits (Lu, He, Zhang, & Bing, 2019). The off-taste pea protein isolate could be masked with acid degraded waxy potato starch during the production of protein-rich food products (Assié, Caussette, & Chen, 2019). The effect of pea protein concentration in feed composition was investigated regarding extrusion conditions by Beck et al. (2018), who reported that protein content of 25% and the pea fiber content of 16% still could provide high expansion. These studies can show the structural limit for protein fortification in extruded products. In addition, the nutritional quality of plant-based extruded products can be enriched using different plant sources in the feed blend. Enhanced protein quality was determined in the cereal-pulse blend compared to cereal flour after the extrusion process (Guldiken et al., 2020).

3.3.7 Protein fortification in dairy products

Protein fortification in dairy-based beverages is common and highly acceptable among customers due to their suitable sensory properties such as consistency and mouthfeel. However, innovations in the dairy food industry should meet the expectations of customers for nutritional quality, taste, and healthiness. Besides, the limit for protein fortification can be determined according to customers' acceptance (Wendin, Höglund, Andersson, & Rothenberg, 2017). In this concept, fortification of fermented milk drinks with soy protein isolate, pea protein isolate, wheat gluten, and rice protein was studied by Akin and Ozcan (2017). They reported that fermented bowls of milk fortified with soy protein isolate had the highest essential amino acid content, and all plant proteins except rice protein enhanced the overall acceptability of sensory properties of fermented kinds of milk.

Furthermore, yogurt with 10% fat fortified with pumpkin seed flour or wheat gluten to 4% protein content was found to have similar overall acceptance with the control sample (Dabija et al., 2018).

3.3.8 Protein fortification in bakery products

Bakery products can be good alternatives to place more protein in our diet. In a study, the sensory and physical characteristics of almond flour, soy flour, and whey protein fortified muffins were determined from the perspective of older adults (age 85+) to fight against sarcopenia with protein-fortified bakery products (Wendin et al., 2017). Overall, whey protein fortified muffins were perceived as artificial but acceptable, almond flour fortified ones were not accepted as muffins due to their texture, and soy flour fortified muffins were found less sweet but highly acceptable.

Malnutrition is an essential problem in low-income families. Effect of protein malnutrition was shown on Wistar female weanling rats fed on casein diet or tortilla diet as high growth performance or dermatitis and hair loss, respectively (Amaya-Guerra, Alanis-Guzman, & Saldivar, 2004). Enrichment of flours with plant-based proteins might be a solution. For instance, protein digestibility, net protein utilization, biological value results were found higher for tortillas produced from masa flour and maize flour fortified with soybean meal (Amaya-Guerra et al., 2004). Furthermore, tortillas and bread fortified with soybean protein led to weight gain in male Wistar rats, as well as higher net protein utilization and biological values (Acevedo-Pacheco & Serna-Saldivar, 2016).

On the other hand, according to the findings of Amirshaghaghi, Rezaei, and Rezaei (2017), wild almond protein isolates can be used in functional food development under alkaline conditions up to 80°C due to its high oil absorption capacity, foaming, and emulsifying properties.

3.4 Dietary lipids

Lipids are water-insoluble organic compounds that include fats, waxes, monoglycerides, diglycerides, triglycerides, phospholipids, sterols, fat-soluble vitamins (A, D, E, K) and others. They consist of an essential part of the human diet and have critical biological functions in the human body, such as storage and transport of metabolic fuel, structural components of membrane, cell surface components, and protective coatings (Marchello, 2016). They aid in overall well-being and health when ingested in appropriate amounts (Meynier & Genot, 2017). When consumers prefer unhealthy lipids (i.e., trans and saturated lipids), they can increase the risk of various health diseases such as heart attack, inflammation, oxidative stress, and LDL cholesterol (Jala & Ganesh Kumar, 2018; Jones & Jew, 2016).

Lipids take an active part in our food acceptability due to their contribution to desirable characteristics such as taste, flavor, smoothness, tenderness, and crispness. Lipids can also prevent the drying out of food, allowing liquids to rise above the boiling point of water. Other significant characteristics of lipids are to enhance cooking and baking characteristics by improving gas retention, aeration, and heat transfer (Rios, Pessanha, Almeida, Viana, & Lannes, 2014). They help to preserve carbon dioxide in the dough, thus increase the final volume of the bakery products (Ali, Abed, Korma, & Hassan, 2016).

Dietary lipids can be extracted from plants, animals, and marine organisms. Lipids are classified into different categories according to their structure, and significant sources of bioactive lipids are summarized in Table 3.3. Dietary choices have a significant effect on the total amount and quality of lipids that are consumed. Health beneficial lipids, for example, polyunsaturated fatty acids (PUFAs), are known to be an essential component of the human diet, but it cannot be synthesized by humans themselves and need to be taken from external sources (Patel, Lecerf, Schenker, & Dewettinck, 2016). PUFAs have various fatty acid groups, including n-3 PUFAs, i.e., eicosapentaenoic acid (EPA), docosahexaenoic acid (DHA), α-linolenic acid (ALA) and n-6 PUFAs, i.e., linoleic acid (LA), arachidonic acid (ARA). PUFAs help to minimize cholesterol levels and to reduce the risk of hyperlipidemia, atherosclerosis, and heart disease (Jala & Ganesh Kumar, 2018; Katiyar & Arora, 2020).

PUFA-rich oils can be found in different sources including fish, microalgae, silkworm, and emu oil (Jala & Ganesh Kumar, 2018). Fish oil exhibits high level of PUFAs content, especially n-3 PUFAs; which play a significant role in promoting the development of neonatal brain cells and eyes, and also appear to prevent adult cardiac disorders. Additionally, the intake of the appropriate amounts of DHA and EPA is essential for cellular tissue metabolism and regulation, growth, and development of the fetus's brain. However, it should be considered that a healthy diet should contain a ratio of 1:4 between

TABLE 3.3 Major dietary bioactive lipids and their sources.

Fatty acids	Dietary sources	References
Eicosapentaenoic acid (C20:5 n-3)	Fish oil (anchovy, bluefish, herring, mackerel, mullet, sardines, salmon, sturgeon, tuna, and trout), marine oil (mussels, oyster, shrimp, crab), algal oil (*Chlorella, Phaeophyceae, Gonyaulox, Phytophthora*)	Katiyar and Arora (2020), Siriwardhana et al. (2012)
Docosapentaenoic acid (C22:5 n-3)	Human milk, fish oil (salmon, mackerel, pompano, herring, sablefish, whitefish, bluefin tuna, rainbow trout), bearded seal oil, poultry, meat, and dairy products	Byelashov et al. (2015)
Docosahexaenoic acid (C22:6 n-3)	Fish oil (anchovy, bluefish, herring, mackerel, mullet, sardines, salmon, sturgeon, tuna, and trout), marine oil (mussels, oyster, shrimp, crab), algal oil (*Gonyaulox, Crypthecodinium, Schizochytrium, Phaeophyceae*)	Katiyar and Arora (2020), Siriwardhana et al. (2012)
Linoleic acid (C18:2 n-6)	Leafy vegetables, seed, nuts, grains, sunflower, corn, cottonseed, soybean, and sunflower oil	Sharma et al. (2012)
Conjugated linoleic acid (C18:2 n-6)	Animal fats such as beef, lamb, and dairy products	Ghazani and Marangoni (2016)
α-linolenic acid (C18:3 n-3)	Canola oil, flaxseed oil, rapeseed oil, soybean oil, and walnut oil	Sharma et al. (2012)
Stearidonic acid (C18:4 n-3)	Flaxseed oil, hemp seed oil, and perilla seed oil	Hernandez (2016)
Arachidonic acid (C20:4 n-6)	Fungal oil extracted from *Mucor javanicus,* and *Mortierella alpina*	Hernandez (2016)

Continued

TABLE 3.3 Major dietary bioactive lipids and their sources.—cont'd

Fatty acids	Dietary sources	References
Carotenoids		
β-carotene	Buriti (*Mauritia vinifera* Mart.), tucuma (*Acrocomia mokayáyba* Barb. Rodr.), acerola, bocaiuva, camu-camu, carrot, mango, nuts, pumpkin, rosehip fruit	Mezzomo and Ferreira (2016)
Lycopene	Tomato fruit, cherry, guava, watermelon, papaya	Mezzomo and Ferreira (2016)
Lutein	Green and dark green leafy vegetables such as broccoli, spinach, parsley, Brussels sprouts, microalgae (*Chlorella vulgaris, Chlorella sorokiniana, Scenedesmus obliquus*)	Mezzomo and Ferreira (2016)
Zeaxanthin	Green and dark green leafy vegetables such as broccoli, Brussels sprouts, parsley, pequi (*Caryocar villosum*), spinach, and microalgae (*Chlorella saccharophila*)	Mezzomo and Ferreira (2016)
Phytosterols		
Stigmasterol, β-sitosterol, Campestrol	Rice bran, wheat germ, peanut, corn oil, soybean, canola oil, sesame oil	Ghazani & Marangoni (2016), Rakel (2018)
Fat-soluble vitamins		
Vitamin A	Dairy products, cod liver oil, liver, dark green and yellow vegetables, fruits and fish	Fathima et al. (2017)
Vitamin D	Egg yolk, liver, cod liver oil, fish, organ meat, and sunlight	Fathima et al. (2017)
Vitamin E	Green leafy vegetables, whole-wheat cereals, nuts, egg yolk, seed oils, liver, and organ meat	Fathima et al. (2017)
Vitamin K	Green leafy vegetables and soya beans	Fathima et al. (2017)

n-3 and n-6. The unbalanced relationship between n-3 and n-6 leads to some health-related risks such as high LDL cholesterol, obesity, hypertension, and cancer (Katiyar & Arora, 2020).

The other essential PUFAs are LA and ALA and their isomers which are belonging to the n-6 and n-3 family. Conjugated linoleic acid (CLA) and conjugated linolenic acid (CLNA) are the most common isomers with the same chemical composition but the different spatial configuration of atoms. Clinical studies indicate that CLA's primary function is antithrombotic and immune-modulating, and it contributes to lean mass muscle development. The primary source of CLA and CLNA are dairy products. However, the amount of these lipids is meager and insufficient to meet the daily requirement. Fermentation can increase the concentration of these lipids in dairy products by the use of CLA and CLNA producer starters, including *Propionibacterium*, *Lactobacillus*, *Lactococcus*, and *Bifidobacterium* (Gao et al., 2020).

3.4.1 Structured lipids

Lipids can be structured to achieve the desired qualities and unique functions (Ang, 2018). Chemical and enzymatic synthesis and genetic engineering can be used to produce the desired lipids. Chemical and enzymatic transesterification and interesterification are generally preferred over genetic engineering, which is time-consuming and overwhelming. Hydrolysis, interesterification, esterification, or physical fractionation can be used to produce structured lipids; however, enzymes are required to change the distribution of fatty acids at specific positions (Jala & Ganesh Kumar, 2018). The enzymatic synthesis of structured lipids, including both medium-chain fatty acid (MCFA) and long-chain fatty acid (LCFA) in one triacylglycerol molecule (MLCT), can be attractive for researchers. These products can meet the market demand for food products with health benefits, which could target specific diseases and metabolic disorders such as Crohn's disease or short bowel syndrome (Yuksel-Bilsel & Sahin-Yesilcubuk, 2019). Diets, including MLCT, restrict the accumulation of body fat and reduce serum cholesterol levels, blood triglycerides, and obesity. Throughout the food industry, MLCT may be used for cooking oils, vegetable oil spreads, salad dressings, pastry fats, coating fats, or dietary supplements (Jala & Ganesh Kumar, 2018).

The possible application of structured lipids contains MCFA in the sn-1 and sn-3 positions and PUFAs in the sn-2 position. This fact allows the rapid hydrolysis of MCFA and its use for energy purposes, as well as the efficient absorption of PUFA, usually ARA, EPA or DHA. This characteristic is critical for patients with malabsorption syndrome (Meynier & Genot, 2017). Many structured lipids are commercially available. Betapol is an example of structured lipids that exhibits similar triacylglycerol structure with that of human milk. This structure facilitates the absorption of palmitic acid, which is esterified in the sn-2 position and decreases its fecal loss. The other example of structured lipid is Caprenin (caprocaprylobehenin), which contains C8:0, C10:0, and C22:0 fatty acids esterified to glycerol moiety and can be used in reduced-calorie and reduced-fat chocolate bars as it provides only 5 kcal/g. Another example of structured lipid is Benefat that contains at least one short-chain fatty acid (SCFA) (C2:0, C3:0 or C4:0) and at least one LCFA (C18:0) randomly positioned to the glycerol backbone. Benefat has low calorie similar to Caprenin, and it can be used in chocolate-flavored coatings, baking chips, baked and dairy products, dips, dressings, or as a cocoa butter substitute in foods. Olestra is a polyester of acylated sucrose with six to eight fatty acids obtained from vegetable oils. Because of the big size and number of nonpolar fatty acids, it can not be hydrolyzed by gastric or pancreatic enzymes; therefore, it is nondigestible, noncaloric, and nontoxic. It can act as a zero-calorie substitute due to its unique characteristics (Zam, 2015).

In infant cookies, structured lipids prepared from milk fat, rapeseed oil, and fish oil concentrate was used. These cookies demonstrated better oxidative stability compared to the use of original fat during baking (Jala & Ganesh Kumar, 2018). Encapsulation technology can be widely used to produce highly stable structured lipids to avoid unfavorable conditions such as high temperature and humidity, as well as mechanical stress during processing and storage period. Furthermore, the interaction between lipids and other food components can result in the development of undesirable tastes and smells in the foods. Yuksel-Bilsel and Sahin-Yesilcubuk (2019) studied complex coacervation techniques to encapsulate structured lipids, and they successfully developed a functional kefir product fortified with MCFA and LCFA.

It is essential to identify economically feasible approaches that can remove or reduce the fat content of food products without affecting their sensory and nutritional properties. The production of new lipophilic functional foods has been considerable interest due to their health benefits such as antioxidant, anti-inflammatory, wound healing, and anticancer properties (Shin, Kim, & Park, 2015).

3.4.2 Fortification of meat products with functional lipids and fats

Meat is a source of several nutritious components, including proteins, L-carnitine, histidyl dipeptides, creatine, CLA, minerals such as iron, zinc and selenium, taurine, vitamins (B, E), glutathione, lipoic acid, and ubiquinone. Hence, meat may be treated as a functional food without further processing (Olmedilla-Alonso, Jimenez-Colmenero, Sanchez-Muniz, 2013). However, meat products have relatively high salt, animal fat, and curing salt contents (nitrites and nitrates) (Ansorena & Astiasarán, 2013). Products with these features are not always among healthier dietary choices, particularly those vulnerable to cardiovascular diseases or obesity. There have been extensive studies on enhancing the nutritional profile of meat products for a healthier diet. The meat industry may explore various implications, including controlling the composition of fresh and manufactured products through fatty acid profile reformulation, reduced fat and cholesterol content (Siro, Kapolna, Kapolna, & Lugasi, 2008).

Processed meat products can be fortified to contain high proportions of n-3 PUFAs or CLA, better n-6/n-3 ratio, lower saturated fatty acid proportions, and lower cholesterol content by the replacement of animal fat with plant or marine oils (Selani et al., 2016). Several scientists have researched meat products where animal fat is replaced with vegetable oils, including perilla oil (Asif, 2011), olive oil (Beriain, Gomez, Petri, Insausti, & Sarries, 2011), avocado oil, olive oil and sunflower oil (Rodriguez-Carpena, Morcuende, & Estevez, 2012), sunflower seed oil (Choi et al., 2013), rapeseed and sunflower oil (Asuming-Bediako et al., 2014), chia oil (Cofrades et al., 2014), canola and flaxseed oil (Baek, Utama, Lee, An, & Lee, 2016), and canola oil (Selani et al., 2016). Fish oil has also been used for this purpose (Josquin, Linssen, & Houben, 2012; Salcedo-Sandoval et al., 2015). Replacing animal fat with plant or marine oils that contain lower saturated fatty acids and higher PUFAs has a beneficial effect on the fatty acid profiles of meat products (Bolger, Brunton, & Monahan, 2018). The use of vegetable oils has shown the potential to create stable meat emulsions, but the small size of the fat globules results in products that have a much firmer texture compared to animal fat usage (Barbut, Wood, & Marangoni, 2016). The study conducted by Lopez-Lopez et al. (2009) reported oxidation problems related to the stability of low-fat frankfurters (12% fat) enriched with n-3 PUFAs from seaweed. Due to the textural changes and oxidation problems, this type of replacement and fortification is rarely used by the industry.

Encapsulation or preemulsifying could be excellent approaches for delivering bioactive lipids into functional meat products. Encapsulation of oils has extra benefits, such as the prevention of lipid oxidation and disguising off-flavors (Ansorena & Astiasarán, 2013). Alejandre, Poyato, Ansorena, and Astiasaran (2016) reported that dry fermented sausages with emulsified linseed oil, which is rich in ALA, did not cause any oxidation problems. Bolger et al. (2018) showed that chicken sausages fortified with encapsulated flaxseed oil had the most significant impact on physical characteristics compared to direct or preemulsified oil addition. Utama, Jeong, Kim, Barido and Lee (2019) reported that perilla-canola oil emulsion not only promotes health benefits but also maintains the technological and sensory properties of chicken sausage. Moschakis, Panagiotopoulou, and Katsanidis (2016) demonstrated that the use of organogel emulsions of sunflower oil in frankfurter sausages showed no significant differences in the physical, chemical, and sensory properties of the product.

3.4.3 Fortification of eggs with functional lipids and fats

Eggs are nutritious foods due to their proteins, vitamins, and lipids of high quality. Based on the diet of chicken, the fatty acid composition of yolk lipids may be changed. Chickens are getting standard feed to tend to lay eggs relatively high in n-6 PUFAs, but the amount of n-3 PUFAs in eggs can be effectively increased by feeding of poultry either directly using fish oil or indirectly by increasing the precursor levels. Also, the n-3 PUFAs from eggs are more bioavailable and may have the potential to reduce heart disease risk (Khan et al., 2017).

Enrichment methods could contribute to dietary intake of n-3 PUFAs. An essential advantage of this strategy is its wide acceptability as a human food and food component. For increasing the n-3 PUFAs in eggs, hens can be fed with ALA-rich seeds or oil from canola (Gul, Yoruk, Aksu, Kaya, & Kaynar, 2012), soybean (Elkin, Ying, & Harvatine, 2015), walnuts (Burns-Whitmore, Haddad, Sabaté, & Rajaram, 2014) and flaxseed (Ehr, Persia, & Bobeck, 2017), or EPA/DHA-rich fish oil (Feng et al., 2020) and microalgae (Świątkiewicz et al., 2020). However, fish oil can give an unpleasant odor (slight fishy) and palatability to eggs, and it affects consumer acceptability. For avoiding the fish flavor, the supplementation ratio should not exceed 1.5% of fish oil and fish meal, which restricted the levels of n-3 PUFAs (Fraeye et al., 2012). Besides, due to the decline of fish stocks and pollutants such as heavy metals and polychlorinated biphenyls possibly found in fish, microalgae can be used as an alternative source for the supplementation.

Moreover, the eggs enriched with n-3 PUFAs from microalgae are still acceptable for human consumption from a sensory perspective (Lemahieu, Bruneel, Muylaert, Buyse, & Foubert, 2017). A comparative study conducted by Feng et al. (2020) reported that microalgae oil would be more promising for egg PUFAs enrichment owing to better sensory quality over the fish oil. Moreover, a flaxseed diet was found to be similar to the regular country eggs in terms of odor, taste, and appearance (Khan et al., 2017).

n-3 PUFAs contain more double bonds than ALA, and n-3 PUFAs enriched eggs are highly perishable compared to conventional eggs, and their quality starts to decrease right away after production at the farm. n-3 PUFAs not only increase the unsaturated fatty acids in egg yolk but also increase the sensitivity against lipid oxidation (Liang, Zu, & Wang, 2020). Therefore, an increase in the oxidation sensitive fatty acids in egg yolk can bring about a greater extent of lipid oxidation that could affect the sensorial quality, damage the biological tissues, and may cause the formation of several diseases such as cancer and atherosclerosis (Fraeye et al., 2012). To enhance lipid stability, researchers have investigated the antioxidant supplementation of feed; however, no significant effect was found (Hayat, Cherian, Pasha, Khattak, & Jabbar, 2010). Storage at refrigeration temperature could be a possible way to decrease the lipid oxidation of eggs. Liang et al. (2020) reported that n-3 PUFAs enriched eggs are stable at 4°C for 24 days.

3.4.4 Fortification of dairy products with functional lipids and fats

Milk and dairy products are great candidates for fortification to improve the nutrients' quality and quantity. Due to their high consumption rate and refrigerated storage, they can be very effective for public health intervention. While there are more than 400 minor fatty acids in milk, the main cis-PUFA is ALA (almost 10 mg/g fat) with much lower levels of EPA (almost 0.9 mg/g fat) and DHA (almost 1 mg/g fat) (Ganesan, Brothersen, & McMahon, 2014). CLA is a natural constituent in milk fat, and many factors such as animal's breed, age, and diet influence the amount of CLA in milk and dairy products. The diet of animals can be manipulated easily to produce high CLA containing milk. Furthermore, milk and dairy products can be fortified with CLA by direct addition (Meraz-Torres & Hernandez-Sanchez, 2012), for example, Abd El-Salam et al. (2011) developed functional probiotic labneh fortified with CLA.

Commercial dairy products such as yogurt and cheese are currently fortified with n-3 PUFAs obtained from vegetable oil, marine microalgae, or fish oil. Fortified dairy products attract a large variety of customers and have the potential to increase market sales and help increase the consumption of n-3 PUFAs. The main ways to add these fatty acids into dairy products are by directly adding n-3 PUFA rich oils and bio-supplying through the meat. Bio-supplying of PUFAs through the cow feed was investigated by Zwyrzykowska and Kupczynski (2014), Fauteux, Gervais, Rico, Lebeuf, and Chouinard (2016), and Santos et al. (2017) to produce fortified dairy products. Researchers have reported that flaxseed is a good source of energy and protein for lactating dairy cows, and it can also influence the fatty acid profile of milk (Oeffner et al., 2013; Petit, 2010).

The most cost-effective and most straightforward choice to produce fortified dairy products is the direct application of the nutrient to the milk before the manufacturing process. Direct addition of n-3 PUFAs rich oil obtained from blackcurrant, *Camelina sativa*, *Echium plantagineum*, flaxseed and raspberry in dairy products were investigated by different researchers for different foods such as oil in yogurt (Dal Bello, Torri, Piochi, & Zeppa, 2015), flaxseed oil in cheeses (Bermúdez-Aguirre & Barbosa-Cánovas, 2011), flaxseed oil enriched with vitamin D_3 in cheese (Stratulat et al., 2015), rapeseed and fish oil in yogurt and cheese (Dawczynski, Martin, Wagner, & Jahreis, 2010), linseed oil in dairy beverages (Giroux, Houde, & Britten, 2010). Milk fortified with vitamin D_3 can be used in cheese making. However, the final product does not contain as much vitamin as the milk alone due to the loss of vitamin D_3 in the whey separation step (Stratulat et al., 2015). Studies have shown that the addition of microalgal species in dairy products will promote the growth of beneficial probiotic bacteria in fermented milk, cheese, and yogurt. In Europe, infant milk supplemented with purified PUFAs from microalgae are found in the market (Katiyar & Arora, 2020).

The differentiation of the added oil on the milk surface makes it difficult for the direct addition of nutrients into milk. Additionally, milk proteins and lipids are prone to oxidation reactions, so modifying the composition of fatty acid to increase unsaturated fats increases oxidation risks (Duncan & Webster, 2010). Studies have reported that microencapsulation of oils protected oxidation and masks undesirable flavor in dairy products (Alfaro et al., 2015; Bermúdez-Aguirre & Barbosa-Cánovas, 2011; Estrada, Boeneke, Bechtel, & Sathivel, 2011; Patrick et al., 2013). The novel nonthermal technologies have also shown essential results in the production of new functional foods. A novel technology, high hydrostatic pressure (HHP), could be an alternative method for the fortification of dairy products. Calligaris et al. (2013) and Bermúdez-Aguirre and Barbosa-Cánovas (2012) reported HHP as an exciting way to incorporate oils rich in n-3 PUFAs into cheeses effectively.

3.4.5 Fortification of bakery products with functional lipids and fats

Bakery products such as bread, biscuits, cakes, which contain wheat flour as the main ingredient, are consumed in large amounts in our daily diet and have an essential role in human nutrition. They have an ideal matrix for the production of functional bakery products. It is necessary to realize that achieving functional food quality does not necessarily entail delivering the active principle at the required level for physiological efficacy, but also providing a product that meets the requirements of the customers in terms of appearance, taste, and texture (Siro et al., 2008).

Bakery products are associated with the presence of saturated fatty acids and trans fatty acids. Moreover, excessive amounts of n-6 PUFAs and a very high n-6/n-3 ratio may cause several diseases such as the increased risk of cardiovascular diseases and cholesterol. However, the increase in the levels of n-3 PUFAs (a low n-6/n-3 ratio) has been shown to exert suppressive effects. The choice of fat for the preparation of bakery products is determined both by the nutritional requirements of the consumers and obtaining the desired rheological properties. Despite the adverse effects on rheological and sensory properties of bakery products, researches are focused on replacing saturated fats with vegetable oils or blends to develop healthier foods (Osuna, Romero, Avallone, Judis, & Bertola, 2018). The use of natural vegetable oils could be an alternative to replace animal fat. Replacing 50% of butter in cookies with emulsified olive oil demonstrated comparable fracture properties and acceptance by consumers, resulting in a product with healthier properties (Giarnetti, Paradiso, Caponio, Summo, & Pasqualone, 2015). Osuna et al. (2018) showed a decrease in saturated fatty acid profile and increased in PUFAs of bread by replacing bovine fat with a mixture of canola and olive oil, resulting in a very good sensory acceptance.

Encapsulation of n-3 PUFA significantly decreased lipid oxidation during the baking process. Increasing the amount of encapsulated oils in dough significantly reduced the formation of acrylamide and hydroxymethylfurfural in bread. Encapsulation can be used for reducing the off-flavor, improving sensorial quality, and reducing the term oxidation of fatty acids during baking (Gökmen et al., 2011). Umesha, Manohar, Indiramma, Akshitha, and Naidu (2015) reported that the supplementation of biscuits with microencapsulated garden cress seed oil rich in ALA enhanced the nutritional, sensory, shelf-life, and storage stability of biscuits. In the study of Jala and Ganesh Kumar (2018), biscuits were supplemented with microalgal biomass to produce PUFA luxurious products. Gökmen et al. (2011) reported that functional bread enriched with encapsulated n-3 PUFAs from flaxseed oil increased the final product quality and safety by decreasing lipid oxidation and the generation of harmful compounds during the bread-making process. Takeungwongtrakul and Benjakul (2017) fortified biscuits with microencapsulated shrimp oil to improve the nutritional and sensory characteristics. They reported that this fortification could be acceptable; however, it must be stored in the dark to extend the oxidative stability.

3.4.6 Fortification of margarine, spreads, baking fats, and shortenings

Margarine is one of the emulsion-based products with a range of classifications depending on its full applications, mainly for spreading, baking, and cooking processes. People have started to pay more attention to healthy margarine because of the high-fat content of typical margarine (Li et al., 2018). Margarine is used as an alternative for butter. Products are available in many countries that are blends of butter and vegetable oil. Usually, soybean oil called spreadable butter. Most of the margarine is made from hydrogenated oils or partial hydrogenation of oil (Ajmal, Nadeem, Batool, & Khan, 2018). After partial hydrogenation, margarine and shortenings tend to produce crystals. This tendency can be inhibited or prevented by stabilizing the crystals in the β' form when incorporated with some cottonseed oil, hydrogenated palm oil or palm olein, modified lard, tallow, or hydrogenated fish oil (Gunstone, 2002). Li et al. (2018) suggested developing a low trans margarine fat analog to beef tallow for healthier formulations by interesterification using soybean oil and fully hydrogenated palm oil.

Margarine manufacturers started using new technologies to produce margarine with much less saturated fatty acids, more PUFAs, and significantly reduced trans fatty acids. Margarine can be used as a carrier for water and lipid-soluble ingredients to increase the nutritional quality of the final products. In many European countries, margarine is fortified with vitamins A and D for contributing to meet vitamin requirements (Patel et al., 2016). Enrichment of phytosterols in mayonnaise, salad dressing, and spreads was associated with greater efficacy to reduce the LDL-cholesterol. However, phytosterols and their derivatives are prone to oxidation, particularly when being subjected to heat treatments and longer shelf-life (Panpipat, Chaijan, & Guo, 2018). Cholesterol-lowering may become increasingly important due to the market launch of a functional variety of margarine containing phytosterol esters, which are supposed to lower the cholesterol level, including camelina oil as a source of n-3 fatty acids (Siro et al., 2008).

3.5 Bioactive compounds

Phenolic compounds, including phenolic acids and flavonoids are considered as suitable functional compounds, particularly, food colorants and antioxidants (Espin, Soler-Rivas, Wichers, & Garcia-Viguera, 2000; Prior, 2003; Stintzing, Stintzing, Carle, Frei, & Wrolstad, 2002). Also, they demonstrate several biological activities such as antiallergic, antibacterial, and anticarcinogenic effects, etc. (Yamamoto & Gaynor, 2001), and this strengthens the idea of using phenolic compounds as food additives in developing functional foods.

Nowadays, nutraceuticals and particularly antioxidants taken from natural sources have attracted much attention, as they help to protect cells from oxidative damage and prevent chronic diseases (Aliakbarian, Dehghani, & Perego, 2009). Fruits and vegetables produce secondary metabolites with antioxidant properties as a defense mechanism (Aliakbarian, Casazza, Montoya, & Convert, 2010; Ben Hamissa et al., 2012; Latoui et al., 2012). Among these secondary metabolites, phenolic compounds have drawn attention in recent years because of their antioxidant, anticarcinogenic, antimutagen, anti-inflammatory, and anticlotting properties (Fresco, Borges, Marques, & Diniz, 2010; Loke et al., 2010; Ostertag, O'Kennedy, Kroon, Duthie, & de Roos, 2010). Fruits are considered as a primary dietary source for phenolics (Record, Dreosti, & McInerney, 2001); hence fruit juices (Coisson, Travaglia, Piana, Capasso, & Arlorio, 2005), powders (Wallace & Giusti, 2008), and extracts have been suggested to be used in the food industry as functional ingredients.

3.5.1 Techniques to fortify foods with bioactive compounds

3.5.1.1 Microencapsulation

This technology is defined as the packing of solid, liquid, and gas materials in mini capsules that loosen their content at supervised ratios over an extended period (Thies, 2009). This technology finds a new place for itself in the food industry in the last years besides its use in the pharmaceutical sector (Champagne & Fustier, 2007). The addition of bioactive compounds in food materials brings many challenges itself such as keeping the stability of bioactive compounds, preventing undesirable interaction with the food matrix during food processing, and storage (Champagne, Gardner, & Roy, 2005). Encapsulation may enhance the functional properties of food materials by providing a controlled release of bioactive compounds, and encapsulation of these bioactive compounds was suggested to be applied using water-in-oil-in-water poly-emulsions (Choi, Decker, & McClements, 2009; Esfanjani, Jafari, Assadpoor, & Mohammadi, 2015; Jimenez-Alvarado, Beristain, Medina-Torres, Roman-Guerrero, & Vernon-Carter, 2009; Lutz, Aserin, Wicker, & Garti, 2009). Designed encapsulation systems should supply the protection of bioactive compounds from chemical degradation to keep it fully functional (Betoret, Betoret, Vidal, & Fito, 2011). The protective effect of encapsulation on antioxidant activity can be seen in Fig. 3.3. Retention of antioxidant activity in cookies enriched with crude and encapsulated pomegranate peel extract was investigated by Kaderides, Mourtzinos and Goula (2020) during 21 days of storage period. The highest antioxidant activity of cookies was observed in the following order; encapsulated extract, crude extract, and the control sample.

Despite the protection of bioactive compounds from hazardous circumstances during food processing and the storage period, deleterious incidents may occur throughout the gastrointestinal tract when these compounds are consumed (de Vos, Faas, Spasojevic, & Sikkema, 2010). The utilization of

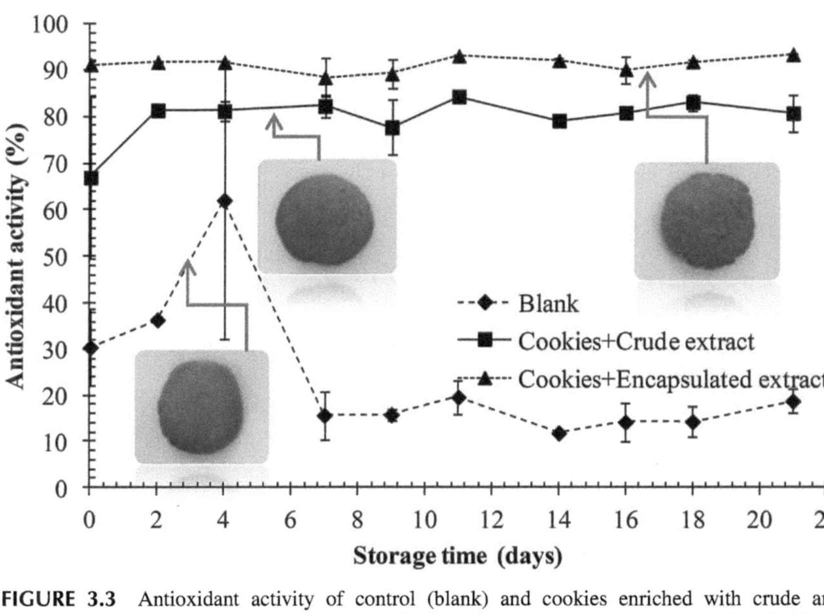

FIGURE 3.3 Antioxidant activity of control (blank) and cookies enriched with crude and encapsulated pomegranate peel extract during storage at room temperature. *Reprinted by permission from Kaderides, K., Mourtzinos, I., & Goula, A. M. (2020). Stability of pomegranate peel polyphenols encapsulated in orange juice industry by-product and their incorporation in cookies.* Food Chemistry, 310, *125849.*

bioactive compounds with a combination of some surfactants and or swelling agents as coating gastro-resistant polymer (cellulose acetate phthalate) to ensure reluctance at low pH of the gastric fluids was suggested previously with good outcomes (Sansone et al., 2011). The microencapsulation of bioactive compounds enhanced the morphology, size, storage stability, and reserved the antioxidant properties (Betoret et al., 2011). Increment in antioxidant characteristics of α-tocopherol, flavonol, and polyphenols in green tea; increased stability against β-carotene oxidation; improved lycopene solubility and its stability against light were observed in nano-emulsions in the food industry (Flanagan & Singh, 2006).

3.5.1.2 Edible films and coating

Extension of food shelf life is possible with the utilization of any coating or wrapping material, which preferably can be eaten together with the nutrients or further removal of it from the food surface (Pavlath & Orts, 2009). Edible films and coating materials are applicable to many food products to control moisture transport, gas transfer, and also oxidation process. Two main processes are required in the formation of edible films; dispersion or solubilization of biopolymers in a film-forming solution, and following this, the evaporation of solvent -wet process- and exhibition of thermoplastic behavior of some proteins and polysaccharides during compression, molding, and extrusion at low moisture degrees –dry process- (Liu, Kerry, & Kerry, 2006). Colorants, antibrowning agents, flavors, spices, and antimicrobial components may carry active ingredients. They can also be used in edible films and coatings and can improve the product's shelf life, decrease the risk of pathogen growth on the food surface and exert bioactivity which may have a positive effect on human health (Betoret et al., 2011).

3.5.1.3 Vacuum impregnation

Vacuum impregnation counted as a beneficial way to introduce desirable solutes into the porous structure of foods and to alter their original composition conveniently as a tool for the development of new items (Betoret et al., 2011). Fruit and vegetable products can be modified with bioactive compounds without changing their unity via vacuum impregnation (Mavroudis, Gekas, & Sjoholm, 1998a, 1998b).

3.5.2 Fortification of dairy products

Milk and dairy products such as yogurt, fermented milk, cheese are the uttermost functional foods in the market (Servili et al., 2011). Although ruminant milk contains considerable amounts of phenolic compounds, the pasteurization process, and bacterial decomposition of milk proteins have been declared to decrease the phenolics in dairy products (O'Connell & Fox, 2001). Phenolic compounds exhibit antioxidant activity, which prevents the oxidation of molecules caused by free radicals, and these compounds are significant in dairy products for the shelf life of products, as well as protection of the human body from oxidative damage (Alenisan, Alqattan, Tolbah, & Shori, 2017). The use of food by-products provided promising results because of their bioactive compounds to improve the health-promoting properties of dairy products, although there are miles of stones on the road to face the difficulties about consumer acceptance of these novel and sustainable dairy products (Iriondo-DeHond, Miguel, & del Castillo, 2018).

Rashidinejad, Birch, Sun-Waterhouse, and Everett (2015) worked on the fortification of low-fat cheese with 125, 250, and 500 mg per kg catechin to monitor the changes in the antioxidant activity and phenolic content. They concluded that the recovery rate of catechin in fortified cheese after *in vitro* digestion ranged from 0.61 to 0.75 depending on the concentration of added catechin. Whey protein isolates were fortified with a stevia fraction, extracted from *Stevia rebaudiana* leaves, to enhance the antioxidant activity, and the results suggested the use of a stevia fraction to increase the antioxidant and antidiabetic features of whey protein isolates (Milani et al., 2017). Whey protein isolates with pectin were fortified with encapsulated phenolic compounds obtained from olive leaf, and the authors recommended the enrichment of food products with olive leaf extracts (Mohammadi, Jafari, Assadpour, & Esfanjani, 2016).

In another study, skimmed milk was fortified with natural phenolic compounds obtained from olive and grape pomaces to produce a novel functional product (Aliakbarian et al., 2015). Another fruit, persimmon, was also studied by Hernandez-Carrion, Varela, Hernando, Fiszman and Quiles (2015) since it is a useful source of antioxidant compounds and produced a functional milkshake. They prepared six different persimmon milkshakes from the whole and skimmed milk and evaluated the sensory characteristics of the products, and they concluded that the persimmon milkshakes had a high nutritional quality and high acceptability by consumers. Functional milk beverages fortified with olive vegetation water's phenolic compounds were prepared by Servili et al. (2011).

Four grape varieties, Cabernet Sauvignon, Chardonnay, Syrah, and Merlot, were extracted with acidified ethanol and were added into yogurt as functional ingredients (Karaaslan, Ozden, Vardin, & Turkoglu, 2011). Researchers pointed out that grape callus cultures could be used as a food supplement to detract chronic diseases. Similar to this study, the addition of diverse levels of grape pomace to yogurt products was investigated (Mohamed, Zayan, & Shahein, 2014) and physicochemical properties, including total phenolic contents, antioxidant activities and sensory characteristics of yogurt products were evaluated during 21 days of storage at 4°C. Mohamed et al. (2014)

emphasized that the concentration of dried grape pomace in fortified yogurt affects the overall acceptance of novel products compared with control yogurt, which does not contain any grape pomace at all. On the other hand, bilberry residues were extracted, and different amounts of extracts were added to drinking yogurt and condensed milk to produce a fortified product. The results highlighted the utilization of agro-industrial wastes, which are good sources of phenolic antioxidants, in dairy products (Fidaleo, Lavecchia, Maffei, & Zuorro, 2015).

3.5.3 Fortification of fruit beverages

Functional fruit drinks include enhanced levels of assumed and or proven health-related bioactive components and receive great interest due to their potential at lowering the risk of specific chronic and degenerative diseases (Popkin, 2011; Torronen et al., 2012).

In a study, the spent grains of brewers' extract was used in the fortification of fruit beverages due to its high hydroxycinnamic acid content (McCarthy et al., 2013), and as a result, the antioxidant activity of cranberry juice was increased. Torronen et al. (2012) worked on the fortification of black currant juice with crowberry powder and examined the existence of polyphenol-derived metabolites in plasma and urine to approve their bioavailability. Consequently, fortification with crowberry powder doubled the polyphenol content and developed postprandial glycemic control in healthy subjects (Torronen et al., 2012).

During the production of plum juice and pulp, the skins of plum, which are rich in polyphenols, are sent to waste (de Beer, Steyn, Joubert, & Muller, 2012). Considering this, de Beer et al. (2012) suggested extracting plum skin polyphenols to provide concentrated or dried functional ingredients, not only for plum nectars but also for other functional fruit juice-based drinks and enriched ice teas. The quality of color-labile fruit juices can be improved with the utilization of rose-petal by-products, and it is suggested to use copigments of distilled rose petals at industrial-scale (Mollov, Mihalev, Shikov, Yoncheva, & Karagyozov, 2007).

Kranz, Braun, Schulze, and Kunz (2010) worked with olive leaf extract because of its high amounts of oleuropein and hydroxytyrosol, which make olive leaf extracts an excellent functional ingredient with high antioxidant activity. Results indicated that the addition of olive leaf extract in fruit smoothies at low concentrations was accepted by 11-panelists involved in the study, and olive leaf extracts were suggested to be used as food ingredients due to their polyphenol content to produce value-added foods (Kranz et al., 2010).

3.5.4 Fortification of bakery products

Bread and bakery products have an essential place in the human diet, and what is generally considered to be a good energy supply (Rozylo, 2013).

Although, many bioactive components are present in the grain, especially in the bran and aleurone layer (Mateo Anson, Havenaar, Bast, & Haenen, 2010; Mateo Anson et al., 2011), during the processing of grain, these bioactive components, including antioxidants are significantly reduced (Gani, Wani, Masoodi, & Hameed, 2012).

Interaction of phenolics with other components may influence the antioxidant efficacy of flavonoids, as well as the protein digestibility in enriched bread (Swieca, Gawlik-Dziki, Dziki, Baraniak, & Czyz, 2013). The bioaccessibility of food components depends heavily on the associations and interactions among them, and it should be noted that several factors decide the future bioactivity of enriched foods and there is no precise method for developing functional goods with a consistent nutritional and nutraceutical consistency (Swieca, Seczyk, Gawlik-Dziki, & Dziki, 2014). Consumers have recently become more aware of the need to consume high quality and healthy foods that contain additives that provide positive health benefits beyond basic nutritional requirements, and enrichment of bread with natural antioxidants (Pop et al., 2016) and phytochemicals (Mildner-Szkudlarz & Bajerska, 2016) found a place themselves in this area more than before.

Gawlik-Dziki et al. (2015) worked on the enrichment of bread with quinoa leaves that promise good antioxidant potential without compromising the sensory quality of bread. In another study performed by Dziki et al. (2015), green coffee beans from Ethiopia, Kenya, Colombia, and Brazil were evaluated as a functional ingredient, and phenolic content and sensory properties of bread fortified with green coffee beans were investigated. The phenolics in bread enriched with coffee was highly bioavailable during *in vitro* simulated digestion. The antiradical activity of coffee enriched bread was found to be higher than the control sample. Moreover, a partial replacement of flour in bread with as much as 3% ground green coffee beans powder was acceptable based on the sensory characteristics.

Onion skin addition in bread was also investigated for its antioxidant properties and sensory value. The stimulated gastrointestinal system was used for the determination of bioaccessibility and bioavailability *in vitro*. Results indicated that the antioxidant activity of bread enriched with onion skins was notably higher than the control. Besides, the replacement of wheat flour with up to 3% of onion skin in bread was considered suitable by consumers in the sensory analysis (Gawlik-Dziki et al., 2013). Phenolic-protein interactions on protein digestibility and antioxidant activity of bread enriched with onion skins were also investigated, and the result indicated that the levels of onion skin substitute were correlated with phenolic content and antiradical capacity, however, the quantities calculated were significantly lower than expected (Swieca et al., 2013). The association of phenolics with bread proteins influences the activity of antioxidants and the digestibility of proteins; therefore, they have several effects on the value of food and health (Swieca et al., 2013).

Antioxidant activity and organoleptic characteristics of sourdough mixed rye bread enriched with grape by-products were evaluated, and the results indicated that increase in the level of grape by-product addition in sourdough caused a significant increase in hardness and gumminess of the bread (Mildner-Szkudlarz, Zawirska-Wojtasiak, Szwengiel, & Pacynski, 2011).

The addition of caffeic and ferulic acids in the flour was investigated by Han and Koh (2011). According to this study, the antioxidant activity of added phenolic acids in flour was close to that of bread, and there was less than 26% loss of free phenolic acids throughout the bread production process. Phenolic acids exhibit antioxidant properties also after the cooking process and hold their benefits for the consumers. On the other hand, grape pomace, a by-product of grape juice processing, could be used in baked products for the aim of preparing value-added goods (Hayta, Özuğur, Etgü, & Seker, 2014).

Wholegrain foods, bread, cookie, and muffin were enriched with lutein, and these products had the ability of scavenging ABTS, DPPH, and peroxyl radicals (Abdel-Aal & Rabalski, 2013). Despite these, some other bakery products were also investigated in phenolic enrichment studies. Spirulina, a multicellular and filamentous microalga, was used to prepare healthy and high nutritious biscuit blends in a study and based on that, dried *Spirulina* algae were considered as a fortification agent and increased the bioactive compounds and antioxidative activity (Barakat, El-Kewaisny, & Salama, 2016). Fortification of biscuits with citrus peel was also the subject of another study, and enriched biscuits with up to 10% orange peel powders were suggested for calorie-reduced diet for obese, diabetic, and overweight persons (Youssef, Youssef, & Mousa, 2014). Biscuits were also enriched with blueberry, defatted grape seed powder, and defatted poppy seeds, and the sensory analysis results indicated that fortified biscuits were not suitable for consumption after 5 months (Aksoylu, Cagindi, & Kose, 2015).

Cereals are accepted as the best enrichment tool in developing countries since 95% of the consumers expend cereals as their dietary staple. Almost 600 million tons of wheat and maize flour is milled commercially annually and spent in nearly every nation of the world as various meal items (Saeed, Anjum, & Akbar Anjum, 2011). As mentioned above, because of their ability to control degenerative diseases related to oxidative damages, polyphenolic compounds are connected with the positive health effect of whole-grain products (Gawlik-Dziki, Swieca, & Dziki, 2012; Lim, Park, Ghafoor, Hwang, & Park, 2011). The use of non-traditional cereals and pseudocereals has started to become widespread as a functional food component since they have some nutritional and physiological attributes that are lacking or deficient in traditional cereal grains (Rayas-Duarte, Majewska, & Doetkott, 1998; Steadman, Burgoon, Lewis, Edwardson, & Obendorf, 2001).

Biney and Beta (2014) investigated the potential use of standard buckwheat flour and bran for improving the phenolic and antioxidant features of durum spaghetti. The addition of buckwheat increased the digestibility of spaghetti along with the improvement of the cooking quality. Even though it is possible to recuperate the phenolic content and antioxidant activity of spaghetti with buckwheat fortification, researchers pointed out the need for further studies to understand the phenolic and antioxidant losses suffered during processing and cooking. Adding bananas to pasta caused an increase in the total phenolic content, and this improvement was even better when bananas were pretreated by ascorbic acid, which is also an antioxidant (Radoi et al., 2015).

In another study, bean pasta was fortified with onion skin, and observations indicated that the interactions of phenolics with the food matrix play an essential role in the nutritional and nutraceutical quality of the produced food (Seczyk, Swieca, & Gawlik-Dziki, 2015).

3.5.5 Fortification of other products

In the literature, there are some other examples that are rich in bioactive compounds and can be used for fortification of other foods that were not categorized here. For instance, Contini, Baccelloni, Frangipane, Merendino, and Massantini (2012) worked on the analysis of both postbrewed and pre-brewed espresso coffees enriched with hazelnut skin extracts. The result indicated that phenolic enriched espresso coffees raised the *in vivo* and *in vitro* antiradical scavenging activity proportional to the hazelnut skin phenolic extract added. Evidence of the potential synergistic effect of espresso coffee and hazelnut skin phenolic extract has also been observed *in vivo*.

Meat and meat products have considerable potential to provide necessary nutrients in the diet. The nutritional composition of meats can be modified by adding bioactive food ingredients directly, or by incorporating bioactive compounds into animal diets (Decker & Park, 2010).

The beneficial properties of coffee are continuously being discovered. In addition, aromatic ingredients such as cinnamon can modify the antioxidant properties of coffee extract, and antioxidant interactions can be established using methods of different difficulty levels (Durak, Gawlik-Dziki, & Pecio, 2014).

White and black mulberry leaves were subjected to several studies because of their high polyphenolic content, and they were suggested as a fortification tool in the food industry such as muesli bars, and yogurt (Sanchez-Salcedo, Mena, Garcia-Viguera, Hernandez, & Martinez, 2015). Similarly, because of its organic acid, vitamins, and phenolic contents, vinegar was also suggested to be used to provide functional properties to foods (Xia, Zhang, Duan, Zhang, & Wang, 2020).

Vu, Scarlett, and Vuong (2018) recommended the use of banana peel as a potentially valuable and cheap source of phenolic compounds in functional foods. Elderberry, on the other hand, was used as a natural flavor compound in alcoholic, non-alcoholic beverages, fruit brandies, and various spirits, tea, yogurt, or ice cream and provided additional benefits because of its phenolic compounds (Fazio, Plastina, Meijerink, Witkamp, & Gabriele, 2013). The antioxidant activity of coffee by-products, including pulp, husk, silver skin, and spent coffee, was reported to vary from 1.5 to 2 mmol trolox per 100 g that is similar to most of the fresh fruits and vegetables. So, isolated bioactive and functional compounds of coffee by-products could be a good source for developing value-added products (Murthy & Naidu, 2012).

3.6 Minerals

Minerals are inorganic chemical elements (or, more specifically, their dissociated ions) necessary for biological or biochemical processes, including the accumulation of electrolytes for the human diet. They have essential roles in our body to perform functions that are necessary—from building strong bones to transmitting nerve impulses—for a stable and long life. The presence of a set of minerals does not only make various hormones but can also control a normal heartbeat (Gharibzahedi & Jafari, 2017).

There are 17 crucial minerals, but 11 of them are needed and/or abundant in foods and drinking water. The remaining are present in limiting amounts in many foods, so a monotonous diet can easily result in deficiency (Gomez-Galera et al., 2010). Calcium (Ca), sodium (Na), magnesium (Mg), potassium (K), phosphorus (P), chloride (Cl), and sulfur (S), and trace minerals, including iodine (I), selenium (Se), zinc (Zn), iron (Fe), copper (Cu), cobalt (Co), manganese (Mn), fluoride (F), chromium (Cr), molybdenum (Mo) are grouped as significant minerals (Gharibzahedi & Jafari, 2017). Many macroelements and microelements are found in the structures of teeth (Ca, F, and P) and bones (Ca, Mg, P, B, Mn, and F). On the other hand, most microelements (Cu, Fe, Se, Zn, Mn, and Mg) play a vital role in many enzymes as a structural component (Gharibzahedi & Jafari, 2017).

Malnutrition of micronutrients impacts more than half of the world's population, especially in developing countries. Fe and I deficiencies are global problems, and manifestations of these deficiencies in micronutrients are often anemia and goiter, and indeed, these are the most significant risk factors for disease and death, affecting two billion people worldwide (Mayer, Pfeiffer, & Beyer, 2008).

Food fortification is one of the most economically effective long term mineral management approaches (Horton, 2006). Biofortification is an excellent breeding method to increase the amount and consistency of nutrients in a specific crop, including rice, maize, wheat, common beans, and some other legumes and cereals and for broadly speaking, in the growing phase, crops are primarily aimed at implementing mineral biofortification (Dwivedi et al., 2012). Dwivedi et al. (2012) mentioned three main biofortification techniques; agricultural (soil modifications or foliar applications), traditional breeding (radiation and chemical mutagenesis), and transgenesis (genetic engineering and recombinant DNA). Although most mineral biofortification activities carried out via conventional breeding built on the use of transgenic technology, breeding pipelines use standard types, and non-genetically modified organism practices are used to shape the first delivery items (Pfeiffer & McClafferty, 2007).

Nanoencapsulation, on the other hand, is accepted as a novel technology to cover bioactive substances in a matrix of less than 1000 nm. This process can offer new mineral delivery systems, as well as other functional ingredients with improved physicochemical strength, bioaccessibility, water-solubility, and bioavailability (Esfanjani & Jafari, 2016; Katouzian & Jafari, 2016).

3.6.1 Food products fortified with minerals

Consumers benefit from equivalent goods that offer adequate sources of micronutrients to satisfy their nutrient needs, and the dairy products are inexpensive and consumed in small amounts, making them a convenient tool for the dietary fortification of minerals. Milk is generally enriched with Ca, Fe, Zn, and Se. Although these minerals are essential for the human body, excessive addition of these minerals may also decrease the quality of dairy products. It should be kept in mind that the fortification of dairy products requires the optimum mixture of micronutrients to maintain the nutritional value and sensory experience of the consumption (Ocak & Rajendram, 2013). The requirements for labeling and safe dose management during the manufacture of fortified goods in complex matrices involve a reliable, accurate, and precise dosage analysis. On-line or real-time spectroscopic approaches may provide valuable insight into in-process food production to enable the rate of production optimization and food product quality assurance (Hassel & Rodriguez-Saona, 2012). Yogurt, a dairy product, is creamy with protein and Ca; however, it is deficient in iron (Fe) and some other minerals (Kunz, Rodriguez-Palmero, Koletzko, & Jensen, 1999). The deficiency of Fe in dairy products can be solved with the enrichment of products by Fe. However, lipid oxidation, taste, shelf life, sensory quality, and microbial activity of yogurt must be assessed before the fortification of samples with Fe (Jackson & Lee, 1991; Zhang & Mahoney, 1989, 1990). Ocak and Köse (2010) investigated the effect of the content of Cu, Fe, and Zn in yogurt production. The fortification level of these elements affected the firmness and cohesiveness. Besides, Cu and Zn increased the required incubation time for fermentation due to their inhibitory effect on fermenting capacity. Also, fortification of Fe and Zn salts in Domiati cheese with buffalo milk was investigated by El-Din, Hassan, El-Behairy, and Mohamed (2012) who reported that higher salt levels (more than 40 mg/kg milk) negatively affected the sensory profile in terms of aroma, body, and texture of cheese.

Fortified bread with Zn, Fe, and Ca assessed to monitor their bioavailability by using 64 rats for a diet period of 28 days. Zn, Ca, and Fe retention measured in plasma, femur, and liver rat tissues. The results indicated that body weight and feed intake of the experimental animals raised significantly during the experiment period. Absorption levels of Ca, Fe, and Zn were higher in fortified bread rather than unfortified ones. Besides, the interaction of Ca, Fe, and Zn with each other were observed in enriched foods. Hence, enrichment of wheat flour with Zn, Fe, and Ca be suggested to help the target group, and it is a possible means of overcoming deficiencies of a few micronutrients cost-effectively (Ahmed et al., 2012).

Das, Salam, Kumar, and Bhutta (2013) reviewed the fortification of foods with diverse minerals and their effect on children and women. They summarized different studies, and briefly, they emphasized some bullet points. Hematological test results, including concentrations of hemoglobin, showed a substantial rise by the fortification of foods with vitamin A, Fe, and multiple micronutrients. Consumption of Ca and vitamin D fortified foods in women resulted in a significant impact on the menopausal period. Fe fortification of foods led to an increase in serum ferritin and hemoglobin levels in reproductive-aged and pregnant women.

Se, which is an essential trace element, is critical in a very narrow concentration ratio when deficiencies or toxicity exist outside of this range from a biological perspective (Sager, 2006). Worldwide, dietary deficiency of Se is more severe than its excessive amounts, and dietary supplements to improve daily intake have been advised (Domokos-Szabolcsy, Barnóczki, Prokisch, Sztrik, & Fári, 2011). Increasing the biofortified Se level in plants had both advantages and disadvantages. The correct exogenous amount of Se biofortification varies between plants since different species can accumulate and withstand different rates of Se. Se concentration that can assist or trigger growth to the initial stage of stress is crucial to have the desired impact on phytochemicals' improvement. Also, Se enriched molecules serve as protective agents against heat, UV, heavy metals, cold, and salt. Despite these advantages, an overdose of Se fortification is toxic to plants at certain levels (Chomchan, Siripongvutikorn, & Puttarak, 2017). Production of Se fortified bread and their nutritional activities were reviewed by Serna-Saldivar and Lazo-Vélez (2015). Onion cultivars in the greenhouse were investigated for the comparisons on the distinction of their Se tolerance (Domokos-Szabolcsy et al., 2011).

On the other hand, fortification of gluten-free pasta with fresh and dried banana was investigated, and promising results were obtained by the use of these fortified gluten-free pasta products for people who suffer from iron deficiency anemia (Radoi et al., 2015). Moreover, wheat flour has gained considerable attention as an effective medium for the fortification of micronutrients. For instance, Fe, Zn, and vitamin A malnutrition have been a priority of governments and world organizations for years. Significant efforts are being made to resolve this deficiency. Akhtar, Anjum, and Anjum (2011) worked on wheat flour fortification as it is a potential medium for mineral enrichment. Due to the environmental conditions, and the lack of modern storage centers, enriched flours would also require high storage stability and customer acceptability.

References

Abd El-Salam, M. H., Hippen, A. R., El-Shafie, K., Assem, F. M., Abbas, H., Abd El-Aziz, M., et al. (2011). Preparation and properties of probiotic concentrated yoghurt (labneh) fortified with conjugated linoleic acid. *International Journal of Food Science and Technology, 46*(10), 2103–2110.

Abdel-Aal, E. S. M., & Rabalski, I. (2013). Antioxidant Properties of high-lutein grain-based functional foods in comparison with ferulic acid and lutein. *American Journal of Biomedical Science, 5*(2), 109–125.

Acevedo-Pacheco, L., & Serna-Saldivar, S. O. (2016). In vivo protein quality of selected cereal-based staple foods enriched with soybean proteins. *Food and Nutrition Research, 60*(1), 31382.

Ahmed, A., Anjum, F. M., Randhawa, M. A., Farooq, U., Akhtar, S., & Sultan, M. T. (2012). Effect of multiple fortification on the bioavailability of minerals in wheat meal bread. *Journal of Food Science and Technology-Mysore, 49*(6), 737–744.

Ajmal, M., Nadeem, M., Batool, M., & Khan, I. T. (2018). Review probable ingredients for trans free margarine with omega-3 fatty acids. *Biological Sciences-PJSIR, 61*(3), 182–186.

Akhtar, S., Anjum, F. M., & Anjum, M. A. (2011). Micronutrient fortification of wheat flour: Recent development and strategies. *Food Research International, 44*(3), 652–659.

Akin, Z., & Ozcan, T. (2017). Functional properties of fermented milk produced with plant proteins. *Lebensmittel-Wissenschaft and Technologie, 86*, 25–30.

Aksoylu, Z., Cagindi, O., & Kose, E. (2015). Effects of blueberry, grape seed powder and poppy seed incorporation on physicochemical and sensory properties of biscuit. *Journal of Food Quality, 38*(3), 164–174.

Al-Sayed, H. M., & Ahmed, A. R. (2013). Utilization of watermelon rinds and sharlyn melon peels as a natural source of dietary fiber and antioxidants in cake. *Annals of Agricultural Sciences, 58*(1), 83–95.

Alam, M. S., Pathania, S., & Sharma, A. (2016). Optimization of the extrusion process for development of high fibre soybean-rice ready-to-eat snacks using carrot pomace and cauliflower trimmings. *Lebensmittel-Wissenschaft and Technologie, 74*, 135–144.

Aldana, D. S., Andrade-Ochoa, S., Aguilar, C. N., Contreras-Esquivel, J. C., & Nevárez-Moorillón, G. V. (2015). Antibacterial activity of pectic-based edible films incorporated with Mexican lime essential oil. *Food Control, 50*, 907–912.

Alejandre, M., Poyato, C., Ansorena, D., & Astiasaran, I. (2016). Linseed oil gelled emulsion: A successful fat replacer in dry fermented sausages. *Meat Science, 121*, 107–113.

Alenisan, M. A., Alqattan, H. H., Tolbah, L. S., & Shori, A. B. (2017). Antioxidant properties of dairy products fortified with natural additives: A review. *Journal of the Association of Arab Universities for Basic and Applied Sciences, 24*(1), 101–106.

Alfaro, L., Hayes, D., Boeneke, C., Xu, Z., Bankston, D., Bechtel, P. J., et al. (2015). Physical properties of a frozen yogurt fortified with a nano-emulsion containing purple rice bran oil. *Lebensmittel-Wissenschaft and Technologie- Food Science and Technology, 62*(2), 1184–1191. https://doi.org/10.1016/j.lwt.2015.01.055

Ali, A. H., Abed, S. M., Korma, S. A., & Hassan, H. M. (2016). Lipids: Functions, applications in food industry and oxidation. *American Journal of Food Science and Nutrition Research, 3*(6), 162.

Aliakbarian, B., Casale, M., Paini, M., Casazza, A. A., Lanteri, S., & Perego, P. (2015). Production of a novel fermented milk fortified with natural antioxidants and its analysis by NIR spectroscopy. *Lebensmittel-Wissenschaft and Technologie-Food Science and Technology, 62*(1), 376–383.

Aliakbarian, B., Casazza, A. A., Montoya, E. J. O., & Convert, A. (2010). Valorisation of olive oil solid wastes: Valuable compounds recovery using high pressure-high temperature. *Journal of Biotechnology, 150*. S332–S332.

Aliakbarian, B., Dehghani, F., & Perego, P. (2009). The effect of citric acid on the phenolic contents of olive oil. *Food Chemistry, 116*(3), 617–623.

Amaya-Guerra, C. A., Alanis-Guzman, M. G., & Saldivar, S. O. S. (2004). Effects of soybean fortification on protein quality of tortilla-based diets produced from regular and quality protein maize. *Plant Foods for Human Nutrition, 59*(2), 45–50.

Amirshaghaghi, Z., Rezaei, K., & Rezaei, M. H. (2017). Characterization and functional properties of protein isolates from wild almond. *Journal of Food Measurement and Characterization, 11*(4), 1725–1733.

Andrade, L. M., Andrade, C. J., Dias, M., Nascimento, C. A. O., & Mendes, M. A. (2018). Chlorella and spirulina microalgae as sources of functional foods. *Nutraceuticals, and Food Supplements*, 45–58.

Ang, X. Q., Chen, S.-Y., & Hong. (2018). *Structured lipid functionality and application.*

Ansorena, D., & Astiasarán, I. (2013). Enrichment of meat products with omega-3 fatty acids by methods other than modification of animal diet. In *Food enrichment with omega-3 fatty acids* (pp. 299–318).

Arntfield, S. D. (2018). Proteins from oil-producing plants. In *Proteins in food processing* (pp. 187–221). Elsevier.

Asif, M. (2011). Health effects of omega-3, 6, 9 fatty acids: Perilla frutescens is a good example of plant oils. *Oriental Pharmacy and Experimental Medicine, 11*(1), 51–59.

Asp, N.-G. (1995). Classification and methodology of food carbohydrates as related to nutritional effects. *American Journal of Clinical Nutrition, 61*(4), 930S–937S.

Asp, N.-G. (1996). Dietary carbohydrates: Classification by chemistry and physiology. *Food Chemistry, 57*(1), 9–14.

Assié, M. E. M.-L., Caussette, M. P. D., & Chen, Z. (2019). Protein-fortified food product. In *Google patents.*

Asuming-Bediako, N., Jaspal, M. H., Hallett, K., Bayntun, J., Baker, A., & Sheard, P. R. (2014). Effects of replacing pork backfat with emulsified vegetable oil on fatty acid composition and quality of UK-style sausages. *Meat Science, 96*(1), 187–194.

Aziah, A. N., Ho, L., Komathi, C., & Bhat, R. (2011). Evaluation of resistant starch in crackers incorporated with unpeeled and peeled pumpkin flour. *American Journal of Food Technology, 6*(12), 1054–1060.

Baek, K. H., Utama, D. T., Lee, S. G., An, B. K., & Lee, S. K. (2016). Effects of replacing pork back fat with canola and flaxseed oils on physicochemical properties of emulsion sausages from spent layer meat. *Asian-Australasian Journal of Animal Sciences, 29*(6), 865–871.

Barakat, E. H., El-Kewaisny, N. M., & Salama, A. A. (2016). Chemical and nutritional evaluation of fortified biscuits with dried Spirulina Algae. *Journal of Food and Dairy Sciences, 7*(3), 1667–1177.

Barbut, S., Wood, J., & Marangoni, A. (2016). Potential use of organogels to replace animal fat in comminuted meat products. *Meat Science, 122*, 155–162.

Baugreet, S., Kerry, J. P., Botineştean, C., Allen, P., & Hamill, R. M. (2016). Development of novel fortified beef patties with added functional protein ingredients for the elderly. *Meat Science, 122*, 40–47.

Beck, S. M., Knoerzer, K., Foerster, M., Mayo, S., Philipp, C., & Arcot, J. (2018). Low moisture extrusion of pea protein and pea fibre fortified rice starch blends. *Journal of Food Engineering, 231*, 61–71.

de Beer, D., Steyn, N., Joubert, E., & Muller, N. (2012). Enhancing the polyphenol content of a red-fleshed Japanese plum (*Prunus salicina* Lindl.) nectar by incorporating a polyphenol-rich extract from the skins. *Journal of the Science of Food and Agriculture, 92*(13), 2741–2750.

Beran, M., Drahorad, J., Vltavsky, O., Urban, M., Laknerova, I., Fronek, M., et al. (2018). Pilot-scale production and application of microparticulated plant proteins. *Journal of Nutrition and Food Sciences, 8*(1), 655–662.

Beriain, M. J., Gomez, I., Petri, E., Insausti, K., & Sarries, M. V. (2011). The effects of olive oil emulsified alginate on the physico-chemical, sensory, microbial, and fatty acid profiles of low-salt, inulin-enriched sausages. *Meat Science, 88*(1), 189–197.

Bermúdez-Aguirre, D., & Barbosa-Cánovas, G. V. (2011). Quality of selected cheeses fortified with vegetable and animal sources of omega-3. *Lebensmittel-Wissenschaft and Technologie-Food Science and Technology, 44*(7), 1577–1584. https://doi.org/10.1016/j.lwt.2011.01.023

Bermúdez-Aguirre, D., & Barbosa-Cánovas, G. V. (2012). Fortification of queso fresco, cheddar and mozzarella cheese using selected sources of omega-3 and some nonthermal approaches. *Food Chemistry, 133*(3), 787–797. https://doi.org/10.1016/j.foodchem.2012.01.093

Betoret, E., Betoret, N., Vidal, D., & Fito, P. (2011). Functional foods development: Trends and technologies. *Trends in Food Science and Technology, 22*(9), 498–508.

Bhise, S., Kaur, A., & Aggarwal, P. (2015). Development of protein enriched noodles using texturized defatted meal from sunflower, flaxseed and soybean. *Journal of Food Science and Technology, 52*(9), 5882–5889.

Biney, K., & Beta, T. (2014). Phenolic profile and carbohydrate digestibility of durum spaghetti enriched with buckwheat flour and bran. *Lebensmittel-Wissenschaft and Technologie-Food Science and Technology, 57*(2), 569–579.

Bleakley, S., & Hayes, M. (2017). Algal proteins: Extraction, application, and challenges concerning production. *Foods, 6*(5), 33.

Bolger, Z., Brunton, N. P., & Monahan, F. J. (2018). Impact of inclusion of flaxseed oil (pre-emulsified or encapsulated) on the physical characteristics of chicken sausages. *Journal of Food Engineering, 230*, 39–48. https://doi.org/10.1016/j.jfoodeng.2018.02.026

Boye, J., Zare, F., & Pletch, A. (2010). Pulse proteins: Processing, characterization, functional properties and applications in food and feed. *Food Research International, 43*(2), 414–431.

Burns-Whitmore, B., Haddad, E., Sabaté, J., & Rajaram, S. (2014). Effects of supplementing n-3 fatty acid enriched eggs and walnuts on cardiovascular disease risk markers in healthy free-living lacto-ovo-vegetarians: A randomized, crossover, free-living intervention study. *Nutrition Journal, 13*(1), 29.

Burtin, P. (2003). Nutritional value of seaweeds. *Electronic Journal of Environmental, Agricultural and Food Chemistry, 2*(4), 498–503.

Byelashov, O. A., Sinclair, A. J., & Kaur, G. (2015). Dietary sources, current intakes, and nutritional role of omega-3 docosapentaenoic acid. *Lipid Technology, 27*(4), 79–82.

Cajnko, M. M., Novak, U., & Likozar, B. (2019). Cascade valorization process of brown alga seaweed Laminaria hyperborea by isolation of polyphenols and alginate. *Journal of Applied Phycology, 31*(6), 3915–3924.

Calligaris, S., Gulotta, A., Ignat, A., Bermúdez-Aguirre, D., Barbosa-Cánovas, G. V., & Nicoli, M. C. (2013). Milk pre-treatment by high pressure homogenization in the manufacturing of "queso fresco" fortified with omega-3 fatty acids. *Lebensmittel-Wissenschaft and Technologie- Food Science and Technology, 50*(2), 629–633. https://doi.org/10.1016/j.lwt.2012.07.035

Cao, W., Zhang, C., Hong, P., Ji, H., Hao, J., & Zhang, J. (2009). Autolysis of shrimp head by gradual temperature and nutritional quality of the resulting hydrolysate. *Lebensmittel-Wissenschaft and Technologie-Food Science and Technology, 42*(1), 244–249.

Champagne, C. P., & Fustier, P. (2007). Microencapsulation for the improved delivery of bioactive compounds into foods. *Current Opinion in Biotechnology, 18*(2), 184–190.

Champagne, C. P., Gardner, N. J., & Roy, D. (2005). Challenges in the addition of probiotic cultures to foods. *Critical Reviews in Food Science and Nutrition, 45*(1), 61–84.

Chatsisvili, N. T., Amvrosiadis, I., & Kiosseoglou, V. (2012). Physicochemical properties of a dressing-type o/w emulsion as influenced by orange pulp fiber incorporation. *Lebensmittel-Wissenschaft and Technologie-Food Science and Technology, 46*(1), 335–340.

Choi, S. J., Decker, E. A., & McClements, D. J. (2009). Impact of iron encapsulation within the interior aqueous phase of water-in-oil-in-water emulsions on lipid oxidation. *Food Chemistry, 116*(1), 271–276.

Choi, Y.-S., Jeong, J.-Y., Choi, J.-H., Han, D.-J., Kim, H.-Y., Lee, M.-A., et al. (2007). Quality characteristics of meat batters containing dietary fiber extracted from rice bran. *Food Science of Animal Resources, 27*(2), 228–234.

Choi, Y. S., Park, K. S., Kim, H. W., Hwang, K. E., Song, D. H., Choi, M. S., et al. (2013). Quality characteristics of reduced-fat frankfurters with pork fat replaced by sunflower seed oils and dietary fiber extracted from makgeolli lees. *Meat Science, 93*(3), 652–658.

Chomchan, R., Siripongvutikorn, S., & Puttarak, P. (2017). Selenium bio-fortification: An alternative to improve phytochemicals and bioactivities of plant foods. *Functional Foods in Health and Disease, 7*(4), 263–279.

Cofrades, S., Santos-López, J. A., Freire, M., Benedí, J., Sánchez-Muniz, F. J., & Jiménez-Colmenero, F. (2014). Oxidative stability of meat systems made with W1/O/W2 emulsions prepared with hydroxytyrosol and chia oil as lipid phase. *Lebensmittel-Wissenschaft and Technologie- Food Science and Technology, 59*(2), 941–947. https://doi.org/10.1016/j.lwt.2014.06.051

Coisson, J. D., Travaglia, F., Piana, G., Capasso, M., & Arlorio, M. (2005). Euterpe oleracea juice as a functional pigment for yogurt. *Food Research International, 38*(8–9), 893–897.

Contini, M., Baccelloni, S., Frangipane, M. T., Merendino, N., & Massantini, R. (2012). Increasing espresso coffee brew antioxidant capacity using phenolic extract recovered from hazelnut skin waste. *Journal of Functional Foods, 4*(1), 137–146.

Costa, C., Lucera, A., Marinelli, V., Del Nobile, M. A., & Conte, A. (2018). Influence of different by-products addition on sensory and physicochemical aspects of Primosale cheese. *Journal of Food Science and Technology, 55*(10), 4174–4183.

Cummings, J., & Stephen, A. (2007). Carbohydrate terminology and classification. *European Journal of Clinical Nutrition, 61*(1), S5–S18.

Dabija, A., Codină, G. G., Gâtlan, A.-M., Sănduleac, E. T., & Rusu, L. (2018). Effects of some vegetable proteins addition on yogurt quality. *Scientific Study and Research. Chemistry and Chemical Engineering, Biotechnology, Food Industry, 19*(2), 181–192.

Dal Bello, B., Torri, L., Piochi, M., & Zeppa, G. (2015). Healthy yogurt fortified with n-3 fatty acids from vegetable sources. *Journal of Dairy Science, 98*(12), 8375–8385.

Das, J. K., Salam, R. A., Kumar, R., & Bhutta, Z. A. (2013). Micronutrient fortification of food and its impact on woman and child health: A systematic review. *Systematic Reviews, 2*, 67.

Dawczynski, C., Martin, L., Wagner, A., & Jahreis, G. (2010). n-3 LC-PUFA-enriched dairy products are able to reduce cardiovascular risk factors: A double-blind, cross-over study. *Clinical Nutrition, 29*(5), 592–599.

Day, L., & Swanson, B. G. (2013). Functionality of protein-fortified extrudates. *Comprehensive Reviews in Food Science and Food Safety, 12*(5), 546–564.

Decker, E. A., & Park, Y. (2010). Healthier meat products as functional foods. *Meat Science, 86*(1), 49–55.

DeVries, J., Camire, M., Cho, S., Craig, S., Gordon, D., Jones, J., et al. (2001). The definition of dietary fiber. *Cereal Foods World, 46*(3), 112–129.

Domokos-Szabolcsy, É., Barnóczki, A., Prokisch, J., Sztrik, A., & Fári, M. G. (2011). Variation in selenium tolerance among two onion cultivars in closed fortification system. *International Journal of Horticultural Science, 17*(1–2), 75–77.

Duncan, S. E., & Webster, J. B. (2010). Oxidation and protection of milk and dairy products. In *Oxidation in foods and beverages and antioxidant applications* (pp. 121–155).

Durak, A., Gawlik-Dziki, U., & Pecio, L. (2014). Coffee with cinnamon - impact of phytochemicals interactions on antioxidant and anti-inflammatory in vitro activity. *Food Chemistry, 162*, 81–88.

Dwivedi, S. L., Sahrawat, K., Rai, K. N., Blair, M. W., Andersson, M. S., & Pfeiffer, W. (2012). Nutritionally enhanced staple food crops. In J. Janick (Ed.), *Plant breeding reviews* (Vol. 36). Hoboken, NJ, USA: John Wiley & Sons, Inc.,.

Dziki, D., Gawlik-Dziki, U., Pecio, Ł., Różyło, R., Świeca, M., Krzykowski, A., et al. (2015). Ground green coffee beans as a functional food supplement–Preliminary study. *Lebensmittel-Wissenschaft and Technologie-Food Science and Technology, 63*(1), 691–699.

Eggleston, G., & Côté, G. L. (2003). Oligosaccharides in food and agriculture. In *ACS publications.*

Ehr, I., Persia, M., & Bobeck, E. (2017). Comparative omega-3 fatty acid enrichment of egg yolks from first-cycle laying hens fed flaxseed oil or ground flaxseed. *Poultry Science, 96*(6), 1791–1799.

El Khoury, D., Goff, H., Berengut, S., Kubant, R., & Anderson, G. (2014). Effect of sodium alginate addition to chocolate milk on glycemia, insulin, appetite and food intake in healthy adult men. *European Journal of Clinical Nutrition, 68*(5), 613–618.

El-Din, A. G., Hassan, A. S. H., El-Behairy, S. A., & Mohamed, E. A. (2012). Impact of zinc and iron salts fortification of buffalo's milk on the dairy product. *World Journal of Dairy and Food Sciences, 7*(1), 21–27.

Elkin, R. G., Ying, Y., & Harvatine, K. J. (2015). Feeding laying hens stearidonic acid-enriched soybean oil, as compared to flaxseed oil, more efficiently enriches eggs with very long-chain n-3 polyunsaturated fatty acids. *Journal of Agricultural and Food Chemistry, 63*(10), 2789–2797.

Esfanjani, A. F., & Jafari, S. M. (2016). Biopolymer nano-particles and natural nano-carriers for nano-encapsulation of phenolic compounds. *Colloids and Surfaces B: Biointerfaces, 146*, 532–543.

Esfanjani, A. F., Jafari, S. M., Assadpoor, E., & Mohammadi, A. (2015). Nano-encapsulation of saffron extract through double-layered multiple emulsions of pectin and whey protein concentrate. *Journal of Food Engineering, 165*, 149–155.

Espin, J. C., Soler-Rivas, C., Wichers, H. J., & Garcia-Viguera, C. (2000). Anthocyanin-based natural colorants: A new source of antiradical activity for foodstuff. *Journal of Agricultural and Food Chemistry, 48*(5), 1588–1592.

do Espírito Santo, A., Perego, P., Converti, A., & Oliveira, M. N. (2012). Influence of milk type and addition of passion fruit peel powder on fermentation kinetics, texture profile and bacterial viability in probiotic yoghurts. *Lebensmittel-Wissenschaft and Technologie, 47*(2), 393–399.

Estrada, J. D., Boeneke, C., Bechtel, P., & Sathivel, S. (2011). Developing a strawberry yogurt fortified with marine fish oil. *Journal of Dairy Science, 94*(12), 5760–5769.

Fajardo, P., Martins, J., Fuciños, C., Pastrana, L., Teixeira, J., & Vicente, A. (2010). Evaluation of a chitosan-based edible film as carrier of natamycin to improve the storability of Saloio cheese. *Journal of Food Engineering, 101*(4), 349–356.

FAO. (2019). Codex stan. In *Codex Alimentarius general standard for food additives* (pp. 192–1995).

Fathima, S. J., Nallamuthu, I., & Khanum, F. (2017). Vitamins and minerals fortification using nanotechnology: Bioavailability and recommended daily allowances. In *Nutrient delivery* (pp. 457–496).

Fauteux, M.-C., Gervais, R., Rico, D., Lebeuf, Y., & Chouinard, P. (2016). Production, composition, and oxidative stability of milk highly enriched in polyunsaturated fatty acids from dairy cows fed alfalfa protein concentrate or supplemental vitamin E. *Journal of Dairy Science, 99*(6), 4411–4426.

Fazio, A., Plastina, P., Meijerink, J., Witkamp, R. F., & Gabriele, B. (2013). Comparative analyses of seeds of wild fruits of Rubus and Sambucus species from Southern Italy: Fatty acid composition of the oil, total phenolic content, antioxidant and anti-inflammatory properties of the methanolic extracts. *Food Chemistry, 140*(4), 817–824.

Feng, J., Long, S., Zhang, H. J., Wu, S. G., Qi, G. H., & Wang, J. (2020). Comparative effects of dietary microalgae oil and fish oil on fatty acid composition and sensory quality of table eggs. *Poultry Science, 99*(3), 1734–1743.

Fernandez-Gines, J., Fernandez-Lopez, J., Sayas-Barbera, E., Sendra, E., & Perez-Alvarez, J. (2004). Lemon albedo as a new source of dietary fiber: Application to bologna sausages. *Meat Science, 67*(1), 7–13.

Fidaleo, M., Lavecchia, R., Maffei, G., & Zuorro, A. (2015). Phenolic extracts from bilberry (Vaccinium myrtillus L.) residues as new functional food ingredients. *International Journal of Applied Engineering Research, 10*(6), 36222–36225.

Flanagan, J., & Singh, H. (2006). Microemulsions: A potential delivery system for bioactives in food. *Critical Reviews in Food Science and Nutrition, 46*(3), 221–237.

Fleurence, J. (2016). Seaweeds as food. In *Seaweed in health and disease prevention* (pp. 149–167). Elsevier.

Fradique, M., Batista, A. P., Nunes, M. C., Gouveia, L., Bandarra, N. M., & Raymundo, A. (2010). Incorporation of Chlorella vulgaris and Spirulina maxima biomass in pasta products. Part 1: Preparation and evaluation. *Journal of the Science of Food and Agriculture, 90*(10), 1656–1664.

Fraeye, I., Bruneel, C., Lemahieu, C., Buyse, J., Muylaert, K., & Foubert, I. (2012). Dietary enrichment of eggs with omega-3 fatty acids: A review. *Food Research International, 48*(2), 961–969. https://doi.org/10.1016/j.foodres.2012.03.014

Fresco, P., Borges, F., Marques, M. P. M., & Diniz, C. (2010). Extraction of phenolic compounds from Vitex agnus-castus. *Current Pharmaceutical Design, 21*, 114–134.

Gamel, T. H., Abdel-Aal, E.-S. M., Ames, N. P., Duss, R., & Tosh, S. M. (2014). Enzymatic extraction of beta-glucan from oat bran cereals and oat crackers and optimization of viscosity measurement. *Journal of Cereal Science, 59*(1), 33–40.

Ganesan, B., Brothersen, C., & McMahon, D. J. (2014). Fortification of foods with omega-3 polyunsaturated fatty acids. *Critical Reviews in Food Science and Nutrition, 54*(1), 98–114.

Gani, A., Wani, S. M., Masoodi, F. A., & Hameed, G. (2012). Whole- grain cereal bioactive compounds and their health benefits: A review. *Food Processing and Technology, 3*, 1–10.

Gao, H., Yang, B., Stanton, C., Ross, R. P., Zhang, H., Liu, Z., et al. (2020). Characteristics of bifidobacterial conjugated fatty acid and hydroxy fatty acid production and its potential application in fermented milk. *Lebensmittel-Wissenschaft and Technologie, 120*. https://doi.org/10.1016/j.lwt.2019.108940

Garcia-Vaquero, M., & Hayes, M. (2016). Red and green macroalgae for fish and animal feed and human functional food development. *Food Reviews International, 32*(1), 15–45.

Gawlik-Dziki, U., Dziki, D., Świeca, M., Seczyk, Ł., Różyło, R., & Szymanowska, U. (2015). Bread enriched with Chenopodium quinoa leaves powder—The procedures for assessing the fortification efficiency. *Lebensmittel-Wissenschaft and Technologie-Food Science and Technology, 62*(2), 1226–1234.

Gawlik-Dziki, U., Swieca, M., & Dziki, D. (2012). Comparison of phenolic acids profile and antioxidant potential of six varieties of spelt (*Triticum spelta* L.). *Journal of Agricultural and Food Chemistry, 60*(18), 4603–4612.

Gawlik-Dziki, U., Świeca, M., Dziki, D., Baraniak, B., Tomiło, J., & Czyż, J. (2013). Quality and antioxidant properties of breads enriched with dry onion (*Allium cepa* L.) skin. *Food Chemistry, 138*(2–3), 1621–1628.

Gharibzahedi, S. M. T., & Jafari, S. M. (2017). The importance of minerals in human nutrition: Bioavailability, food fortification, processing effects and nanoencapsulation. *Trends in Food Science and Technology, 62*, 119–132.

Ghazani, S. M., & Marangoni, A. G. (2016). Healthy fats and oils. In *Reference module in food science*.

Giarnetti, M., Paradiso, V. M., Caponio, F., Summo, C., & Pasqualone, A. (2015). Fat replacement in shortbread cookies using an emulsion filled gel based on inulin and extra virgin olive oil. *Lebensmittel-Wissenschaft and Technologie- Food Science and Technology, 63*(1), 339–345. https://doi.org/10.1016/j.lwt.2015.03.063

Giroux, H. J., Houde, J., & Britten, M. (2010). Use of heated milk protein–sugar blends as antioxidant in dairy beverages enriched with linseed oil. *Lebensmittel-Wissenschaft and Technologie-Food Science and Technology, 43*(9), 1373–1378.

Gökmen, V., Mogol, B. A., Lumaga, R. B., Fogliano, V., Kaplun, Z., & Shimoni, E. (2011). Development of functional bread containing nanoencapsulated omega-3 fatty acids. *Journal of Food Engineering, 105*(4), 585–591. https://doi.org/10.1016/j.jfoodeng.2011.03.021

Gomez-Galera, S., Rojas, E., Sudhakar, D., Zhu, C. F., Pelacho, A. M., Capell, T., et al. (2010). Critical evaluation of strategies for mineral fortification of staple food crops. *Transgenic Research, 19*(2), 165–180.

González-Pérez, S., & Vereijken, J. M. (2007). Sunflower proteins: Overview of their physicochemical, structural and functional properties. *Journal of the Science of Food and Agriculture, 87*(12), 2173–2191.

Gourgue, C. M., Champ, M. M., Lozano, Y., & Delort-Laval, J. (1992). Dietary fiber from mango byproducts: Characterization and hypoglycemic effects determined by in vitro methods. *Journal of Agricultural and Food Chemistry, 40*(10), 1864–1868.

Gouveia, L., Batista, A. P., Miranda, A., Empis, J., & Raymundo, A. (2007). Chlorella vulgaris biomass used as colouring source in traditional butter cookies. *Innovative Food Science and Emerging Technologies, 8*(3), 433–436.

Gul, M., Yoruk, M. A., Aksu, T., Kaya, A., & Kaynar, O. (2012). The effect of different levels of canola oil on performance, egg shell quality and fatty acid composition of laying hens. *International Journal of Poultry Science, 11*(12), 769–776. https://doi.org/10.3923/ijps.2012.769.776

Guldiken, B., Yovchev, A., Nosworthy, M. G., Stone, A. K., House, J. D., Hood-Niefer, S., et al. (2020). Effect of extrusion conditions on the physical properties of desi chickpea-barley extrudates and quality attributes of their resulting flours. *Journal of Texture Studies, 51*(2), 300–307.

Gunstone, F. D. (2002). Food applications of lipids. In *Food lipids* (pp. 746–767). CRC Press.

Guo, N., Sun, J., Zhang, Z., & Mao, X. (2019). Recovery of chitin and protein from shrimp head waste by endogenous enzyme autolysis and fermentation. *Journal of Ocean University of China, 18*(3), 719–726.

Hall, F., Johnson, P. E., & Liceaga, A. (2018). Effect of enzymatic hydrolysis on bioactive properties and allergenicity of cricket (*Gryllodes sigillatus*) protein. *Food Chemistry, 262*, 39–47.

Ben Hamissa, A. M., Seffen, M., Aliakbarian, B., Casazza, A. A., Perego, P., et al. (2012). Phenolics extraction from *Agave americana* (L.) leaves using high-temperature, high-pressure reactor. *Food and Bioproducts Processing, 90*(C1), 17–21.

Han, H. M., & Koh, B. K. (2011). Antioxidant activity of hard wheat flour, dough and bread prepared using various processes with the addition of different phenolic acids. *Journal of the Science of Food and Agriculture, 91*(4), 604–608.

Harnedy, P. A., & FitzGerald, R. J. (2011). Bioactive proteins, peptides, and amino acids from microalgae 1. *Journal of Phycology, 47*(2), 218–232.

Hashemi, F. S., Pezeshky, A., Gharedaghi, K., & Javani, R. (2014). Effect of hydrocolloids on sensory properties of the fermented whey beverage. In *Paper presented at the 1st International Conference on Natural Food Hydrocolloids, Mashhad, Iran.*

Hassel, S., & Rodriguez-Saona, L. E. (2012). Application of a handheld infrared sensor for monitoring the mineral fortification in whole grain cornmeal. *Food Analytical Methods, 5*(3), 571–578.

Hayat, Z., Cherian, G., Pasha, T. N., Khattak, F. M., & Jabbar, M. A. (2010). Sensory evaluation and consumer acceptance of eggs from hens fed flax seed and 2 different antioxidants. *Poultry Science, 89*(10), 2293–2298.

Hayta, M., Özuğur, G., Etgü, H., & Seker, İ. T. (2014). Effect of grape (*Vitis vinifera* L.) pomace on the quality, total phenolic content and anti-radical activity of bread. *Journal of Food Processing and Preservation, 38*(3), 980–986.

Hernandez, E. M. (2016). Specialty oils. In *Functional dietary lipids* (pp. 69–101).

Hernandez-Carrion, M., Varela, P., Hernando, I., Fiszman, S. M., & Quiles, A. (2015). Persimmon milkshakes with enhanced functionality: Understanding consumers' perception of the concept and sensory experience of a functional food. *Lebensmittel-Wissenschaft and Technologie-Food Science and Technology, 62*(1), 384–392.

Horton, S. (2006). The economics of food fortification. *Journal of Nutrition, 136*(4), 1068–1071.

Hu, G., & Yu, W. (2015). Effect of hemicellulose from rice bran on low fat meatballs chemical and functional properties. *Food Chemistry, 186*, 239–243.

Ipsen, R. (2017). Microparticulated whey proteins for improving dairy product texture. *International Dairy Journal, 67*, 73–79.

Iriondo-DeHond, M., Miguel, E., & del Castillo, M. D. (2018). Food byproducts as sustainable ingredients for innovative and healthy dairy foods. *Nutrients, 10*(10).

Izydorczyk, M., Cui, S. W., & Wang, Q. (2005). Polysaccharide gums: Structures, functional properties, and applications. *Food Carbohydrates: Chemistry, Physical Properties, and Applications*, 293–299.

Jackson, L. S., & Lee, K. (1991). Microencapsulated iron for food fortification. *Journal of Food Science, 56*(4), 1047–1050.

Jala, R. C. R., & Ganesh Kumar, C. (2018). Designer and functional food lipids in dietary regimes: Current trends and future prospects. In *Alternative and replacement foods* (pp. 283–316).

Jensen, M. G., Kristensen, M., & Astrup, A. (2011). Can alginate-based preloads increase weight loss beyond calorie restriction? A pilot study in obese individuals. *Appetite, 57*(3), 601–604.

Jimenez-Alvarado, R., Beristain, C. I., Medina-Torres, L., Roman-Guerrero, A., & Vernon-Carter, E. J. (2009). Ferrous bisglycinate content and release in W-1/O/W-2 multiple emulsions stabilized by protein-polysaccharide complexes. *Food Hydrocolloids, 23*(8), 2425–2433.

Jones, P. J. H., & Jew, S. (2016). Health claims and nutrition marketing. In *Functional dietary lipids* (pp. 287–302).

Josquin, N. M., Linssen, J. P., & Houben, J. H. (2012). Quality characteristics of Dutch-style fermented sausages manufactured with partial replacement of pork back-fat with pure, pre-emulsified or encapsulated fish oil. *Meat Science, 90*(1), 81–86.

Kaderides, K., Mourtzinos, I., & Goula, A. M. (2020). Stability of pomegranate peel polyphenols encapsulated in orange juice industry by-product and their incorporation in cookies. *Food Chemistry, 310*, 125849.

Kandler, O., & Hopf, H. (1980). Occurrence, metabolism, and function of oligosaccharides. In *Carbohydrates: Structure and function* (pp. 221–270). Elsevier.

Karaaslan, M., Ozden, M., Vardin, H., & Turkoglu, H. (2011). Phenolic fortification of yogurt using grape and callus extracts. *Lebensmittel-Wissenschaft and Technologie-Food Science and Technology, 44*(4), 1065–1072.

Katiyar, R., & Arora, A. (2020). Health promoting functional lipids from microalgae pool: A review. *Algal Research, 46*. https://doi.org/10.1016/j.algal.2020.101800

Katouzian, I., & Jafari, S. M. (2016). Nano-encapsulation as a promising approach for targeted delivery and controlled release of vitamins. *Trends in Food Science and Technology, 53*, 34–48.

Khan, S. A., Khan, A., Khan, S. A., Beg, M. A., Ali, A., & Damanhouri, G. (2017). Comparative study of fatty-acid composition of table eggs from the Jeddah food market and effect of value addition in omega-3 bio-fortified eggs. *Saudi Journal of Biological Sciences, 24*(4), 929–935.

Kranz, P., Braun, N., Schulze, N., & Kunz, B. (2010). Sensory quality of functional beverages: Bitterness perception and bitter masking of olive leaf extract fortified fruit smoothies. *Journal of Food Science, 75*(6), S308–S311.

Kunz, C., Rodriguez-Palmero, M., Koletzko, B., & Jensen, R. (1999). Nutritional and biochemical properties of human milk, part I: General aspects, proteins, and carbohydrates. *Clinics in Perinatology, 26*(2), 307–333.

Laleg, K., Cassan, D., Barron, C., Prabhasankar, P., & Micard, V. (2016). Structural, culinary, nutritional and anti-nutritional properties of high protein, gluten free, 100% legume pasta. *PLoS One, 11*(9).

Latoui, M., Aliakbarian, B., Casazza, A. A., Seffen, M., Converti, A., & Perego, P. (2012). Extraction of phenolic compounds from Vitex agnus-castus L. *Food and Bioproducts Processing, 90*(C4), 748–754.

Lemahieu, C., Bruneel, C., Muylaert, K., Buyse, J., & Foubert, I. (2017). Microalgal feed supplementation to enrich eggs with omega-3 fatty acids. In *Egg innovations and strategies for improvements* (pp. 383–391).

Liang, K., Zu, H., & Wang, X. (2020). Effect of storage on n-3 PUFA-enriched eggs. *CyTA Journal of Food, 18*(1), 102–107. https://doi.org/10.1080/19476337.2020.1713896

Lim, H. S., Park, S. H., Ghafoor, K., Hwang, S. Y., & Park, J. (2011). Quality and antioxidant property of bread containing turmeric (*Curcuma longa* L.) cultivated in South Korea. *Food Chemistry, 112*, 1577–1582.

Liu, L., Kerry, J. F., & Kerry, J. P. (2006). Effect of food ingredients and selected lipids on the physical properties of extruded edible films/casings. *International Journal of Food Science and Technology, 41*(3), 295–302.

Liu, R., Wang, L., Liu, Y., Wu, T., & Zhang, M. (2018). Fabricating soy protein hydrolysate/xanthan gum as fat replacer in ice cream by combined enzymatic and heat-shearing treatment. *Food Hydrocolloids, 81*, 39–47.

Li, Y., Zhao, J., Xie, X., Zhang, Z., Zhang, N., & Wang, Y. (2018). A low trans margarine fat analog to beef tallow for healthier formulations: Optimization of enzymatic interesterification using soybean oil and fully hydrogenated palm oil. *Food Chemistry, 255*, 405–413.

Loke, W. M., Proudfoot, J. M., Hodgson, J. M., McKinley, A. J., Hime, N., Magat, M., et al. (2010). Specific dietary polyphenols attenuate atherosclerosis in apolipoprotein E-knockout mice by alleviating inflammation and endothelial dysfunction. *Arteriosclerosis, Thrombosis, and Vascular Biology, 30*(4), 749–U243.

Lopez-Lopez, I., Bastida, S., Ruiz-Capillas, C., Bravo, L., Larrea, M. T., Sanchez-Muniz, F., et al. (2009). Composition and antioxidant capacity of low-salt meat emulsion model systems containing edible seaweeds. *Meat Science, 83*(3), 492–498.

López-Marcos, M. C., Bailina, C., Viuda-Martos, M., Pérez-Alvarez, J. A., & Fernández-López, J. (2015). Properties of dietary fibers from agroindustrial coproducts as source for fiber-enriched foods. *Food and Bioprocess Technology, 8*(12), 2400–2408.

Loveday, S. M. (2019). Food proteins: Technological, nutritional, and sustainability attributes of traditional and emerging proteins. *Annual Review of Food Science and Technology, 10*, 311–339.

Lu, Z. X., He, J. F., Zhang, Y. C., & Bing, D. J. (2019). Composition, physicochemical properties of pea protein and its application in functional foods. *Critical Reviews in Food Science and Nutrition*, 1–13.

Lutz, R., Aserin, A., Wicker, L., & Garti, N. (2009). Release of electrolytes from W/O/W double emulsions stabilized by a soluble complex of modified pectin and whey protein isolate. *Colloids and Surfaces B: Biointerfaces, 74*(1), 178–185.

Malomo, S. A., He, R., & Aluko, R. E. (2014). Structural and functional properties of hemp seed protein products. *Journal of Food Science, 79*(8), C1512–C1521.

Maphosa, Y., & Jideani, V. A. (2016). Dietary fiber extraction for human nutrition—a review. *Food Reviews International, 32*(1), 98–115.

Marchello, J. A. (2016). Animal fat composition and modification. In *Functional dietary lipids* (pp. 47–67).

Mariod, A. A., & Fadul, H. (2015). Extraction and characterization of gelatin from two edible Sudanese insects and its applications in ice cream making. *Food Science and Technology International, 21*(5), 380–391.

Mateo Anson, N., Aura, A. M., Selinheimo, E., Mattila, I., Poutanen, K., van den, B. R., et al. (2011). Bioprocessing of wheat bran in whole wheat bread increases the bioavailability of phenolic acids in men and exerts anti-inflammatory effects ex vivo. *Journal of Nutritional Biochemistry, 141*, 137–143.

Mateo Anson, N., Havenaar, R., Bast, A., & Haenen, G. R. M. M. (2010). Antioxidant and anti-inflammatory capacity of bioaccessible compounds from wheat fractions after gastrointestinal digestion. *Journal of Cereal Science, 51*, 110–114.

Mavroudis, N. E., Gekas, V., & Sjoholm, I. (1998a). Osmotic dehydration of apples - effects of agitation and raw material characteristics. *Journal of Food Engineering, 35*(2), 191–209.

Mavroudis, N. E., Gekas, V., & Sjoholm, I. (1998b). Osmotic dehydration of apples. Shrinkage phenomena and the significance of initial structure on mass transfer rates. *Journal of Food Engineering, 38*(1), 101–123.

Mayer, J. E., Pfeiffer, W. H., & Beyer, P. (2008). Biofortified crops to alleviate micronutrient malnutrition. *Current Opinion in Plant Biology, 11*(2), 166–170.

McCarthy, A. L., O'Callaghan, Y. C., Neugart, S., Piggott, C. O., Connolly, A., Jansen, M. A. K., et al. (2013). The hydroxycinnamic acid content of barley and brewers' spent grain (BSG) and the potential to incorporate phenolic extracts of BSG as antioxidants into fruit beverages. *Food Chemistry, 141*(3), 2567–2574.

Meraz-Torres, L., & Hernandez-Sanchez, H. (2012). Conjugated linoleic acid in dairy products: A review. *American Journal of Food Technology, 7*, 176–179.

Meynier, A., & Genot, C. (2017). Molecular and structural organization of lipids in foods: Their fate during digestion and impact in nutrition. *Oilseeds and Fats Crops and Lipids, 24*(2). https://doi.org/10.1051/ocl/2017006

Mezzomo, N., & Ferreira, S. R. S. (2016). Carotenoids functionality, sources, and processing by supercritical technology: A review. *Journal of Chemistry*, 1–16. https://doi.org/10.1155/2016/3164312, 2016.

Mikulec, A., Kowalski, S., Sabat, R., Skoczylas, Ł., Tabaszewska, M., & Wywrocka-Gurgul, A. (2019). Hemp flour as a valuable component for enriching physicochemical and antioxidant properties of wheat bread. *Lebensmittel-Wissenschaft and Technologie, 102*, 164–172.

Milani, P. G., Formigoni, M., Lima, Y. C., Piovan, S., Peixoto, G. M. L., Camparsi, D. M., et al. (2017). Fortification of the whey protein isolate antioxidant and antidiabetic activity with fraction rich in phenolic compounds obtained from *Stevia rebaudiana* (Bert.). Bertoni leaves. *Journal of Food Science and Technology-Mysore, 54*(7), 2020–2029.

Mildner-Szkudlarz, S., & Bajerska, J. (2016). Phytochemicals as functional bread compounds: Physiological effects. In C. M. Rosell, J. Bajerska, & A. F. E. Sheikha (Eds.), *Bread and its fortification nutrition and health benefits*.

Mildner-Szkudlarz, S., Zawirska-Wojtasiak, R., Szwengiel, A., & Pacynski, M. (2011). Use of grape by-product as a source of dietary fibre and phenolic compounds in sourdough mixed rye bread. *International Journal of Food Science and Technology, 46*(7), 1485–1493.

Mishra, N., & Chandra, R. (2012). Development of functional biscuit from soy flour & rice bran. *International Journal of Agricultural and Food Science, 2*(1), 14–20.

Mišurcová, L., Buňka, F., Ambrožová, J. V., Machů, L., Samek, D., & Kráčmar, S. (2014). Amino acid composition of algal products and its contribution to RDI. *Food Chemistry, 151*, 120–125.

Mizani, M., Aminlari, M., & Khodabandeh, M. (2005). An effective method for producing a nutritive protein extract powder from shrimp-head waste. *Food Science and Technology International, 11*(1), 49–54.

Mohamed, A. G., Zayan, A. F., & Shahein, N. M. (2014). Physiochemical and sensory evaluation of yoghurt fortified with dietary fiber and phenolic compounds. *Life Science Journal, 11*(9), 816–822.

Mohammadi, A., Jafari, S. M., Assadpour, E., & Esfanjani, A. F. (2016). Nano-encapsulation of olive leaf phenolic compounds through WPC-pectin complexes and evaluating their release rate. *International Journal of Biological Macromolecules, 82*, 816–822.

Mollov, P., Mihalev, K., Shikov, V., Yoncheva, N., & Karagyozov, V. (2007). Colour stability improvement of strawberry beverage by fortification with polyphenolic copigments naturally occurring in rose petals. *Innovative Food Science and Emerging Technologies, 8*(3), 318–321.

Moschakis, T., Panagiotopoulou, E., & Katsanidis, E. (2016). Sunflower oil organogels and organogel-in-water emulsions (part I): Microstructure and mechanical properties. *Lebensmittel-Wissenschaft and Technologie, 73*, 153–161.

Müller-Maatsch, J., Bencivenni, M., Caligiani, A., Tedeschi, T., Bruggeman, G., Bosch, M., et al. (2016). Pectin content and composition from different food waste streams. *Food Chemistry, 201*, 37–45.

Murthy, P. S., & Naidu, M. M. (2012). Recovery of phenolic antioxidants and functional compounds from coffee industry by-products. *Food and Bioprocess Technology, 5*(3), 897–903.

Namir, M., Suleiman, A. R., & Hassanien, M. F. R. (2015). Characterization and functionality of alcohol insoluble solids from tomato pomace as fat substitute in low fat cake. *Journal of Food Measurement and Characterization, 9*(4), 557–563.

Ndiritu, A. K., Kinyuru, J. N., Kenji, G. M., & Gichuhi, P. N. (2017). Extraction technique influences the physico-chemical characteristics and functional properties of edible crickets (*Acheta domesticus*) protein concentrate. *Journal of Food Measurement and Characterization, 11*(4), 2013–2021.

Niba, L. L. (2005). *Carbohydrates: Starch. Hui YH Handbook of food science, technology, and engineering*. Taylor & Francis.

O'Connell, J. E., & Fox, P. F. (2001). Significance and applications of phenolic compounds in the production and quality of milk and dairy products: A review. *International Dairy Journal, 11*(3), 103–120.

Ocak, E., & Köse. (2010). The effects of fortifying milk with Cu, Fe and Zn minerals on the production and texture of yoghurt. *Journal of Food Agriculture and Environment, 8*(2), 122–125.

Ocak, E., & Rajendram, R. (2013). Fortification of milk with mineral elements. In *Handbook of food fortification and health*. New York, NY.: Humana Press.

Oeffner, S., Qu, Y., Just, J., Quezada, N., Ramsing, E., Keller, M., et al. (2013). Effect of flaxseed supplementation rate and processing on the production, fatty acid profile, and texture of milk, butter, and cheese. *Journal of Dairy Science, 96*(2), 1177–1188.

Olmedilla-Alonso, B., Jimenez-Colmenero, F., & Sanchez-Muniz, F. J. (2013). Development and assessment of healthy properties of meat and meat products designed as functional foods. *Meat Science, 95*(4), 919–930.

Ostertag, L. M., O'Kennedy, N., Kroon, P. A., Duthie, G. G., & de Roos, B. (2010). Impact of dietary polyphenols on human platelet function - a critical review of controlled dietary intervention studies. *Molecular Nutrition and Food Research, 54*(1), 60–81.

Osuna, M. B., Romero, A. M., Avallone, C. M., Judis, M. A., & Bertola, N. C. (2018). Animal fat replacement by vegetable oils in formulations of breads with flour mixes. *Journal of Food Science and Technology, 55*(3), 858–867.

Panpipat, W., Chaijan, M., & Guo, Z. (2018). Oxidative stability of margarine enriched with different structures of β -sitosteryl esters during storage. *Food Bioscience, 22*, 78–84. https://doi.org/10.1016/j.fbio.2018.01.009

Paraman, I., Sharif, M. K., Supriyadi, S., & Rizvi, S. S. (2015). Agro-food industry byproducts into value-added extruded foods. *Food and Bioproducts Processing, 96*, 78–85.

Patel, A. R., Lecerf, J.-M., Schenker, S., & Dewettinck, K. (2016). The contribution of modern margarine and fat spreads to dietary fat intake. *Comprehensive Reviews in Food Science and Food Safety, 15*(3), 633–645. https://doi.org/10.1111/1541-4337.12198

Patil, S. S., & Kaur, C. (2018). Current trends in extrusion: Development of functional foods and novel ingredients. *Food Science and Technology Research, 24*(1), 23–34.

Patrick, K. E., Lv, Y., Muhamyankaka, V., Denis, O., Ntsama, I., & Zhang, X. (2013). Development of EPA-DHA microcapsules supplemented probiotic fermented milk. *Academic Food Journal/Akademik GIDA, 11*, 6–15.

Pavlath, A. E., & Orts, W. (2009). Edible films and coatings: Why, what, and how? In I. M. E. Embuscado, & K. C. Huber (Eds.), *Edible films and coatings for food applications* (pp. 1–23). New York: Springer.

Petit, H. (2010). Feed intake, milk production and milk composition of dairy cows fed flaxseed. *Canadian Journal of Animal Science, 90*(2), 115–127.

Pfeiffer, W. H., & McClafferty, B. (2007). HarvestPlus: Breeding crops for better nutrition. *Crop Science, 47*, S88–S105.

Pojić, M., Mišan, A., & Tiwari, B. (2018). Eco-innovative technologies for extraction of proteins for human consumption from renewable protein sources of plant origin. *Trends in Food Science and Technology, 75*, 93–104.

Popkin, B. M. (2011). Is the obesity epidemic a national security issue around the globe? *Current Opinion in Endocrinology Diabetes and Obesity, 18*(5), 328–331.

Pop, A., Petrut, G., Muste, S., Paucean, A., Muresan, C., Salanta, L., et al. (2016). Addition of plant materials rich in phenolic compounds in wheat bread in terms of functional food aspects. *Hop and Medicinal Plants, 24*(1–2), 37–44.

Prior, R. L. (2003). Fruits and vegetables in the prevention of cellular oxidative damage. *American Journal of Clinical Nutrition, 78*(3), 570s–578s.

Radočaj, O., Dimić, E., & Tsao, R. (2014). Effects of hemp (Cannabis sativa L.) seed oil press-cake and decaffeinated green tea leaves (Camellia sinensis) on functional characteristics of gluten-free crackers. *Journal of Food Science, 79*(3), C318–C325.

Radoi, P. B., Alexa, E., Radulov, I., Morvay, A., Mihai, C. S. S., Trasca, T. I., et al. (2015). Total phenolic, cinnamic acids and selected microelements in gluten free pasta fortified with banana. *Revista de Chimie, 66*(8), 1162–1165.

Rakel, D. (2018). Benign prostatic hyperplasia. In *Integrative medicine* (pp. 601–607). e601.

Rashidinejad, A., Birch, E. J., Sun-Waterhouse, D., & Everett, D. W. (2015). Total phenolic content and antioxidant properties of hard low-fat cheese fortified with catechin as affected by in vitro gastrointestinal digestion. *Lebensmittel-Wissenschaft and Technologie-Food Science and Technology, 62*(1), 393–399.

Rayas-Duarte, P., Majewska, K., & Doetkott, C. (1998). *Effect of extrusion process parameters on the quality of buckwheat flour mixes Cereal Chemistry* (Vol. 75, pp. 338–345) (3).

Record, I. R., Dreosti, I. E., & McInerney, J. K. (2001). Changes in plasma antioxidant status following consumption of diets high or low in fruit and vegetables or following dietary supplementation with an antioxidant mixture. *British Journal of Nutrition, 85*(4), 459–464.

Renzetti, S., & Rosell, C. M. (2016). Role of enzymes in improving the functionality of proteins in non-wheat dough systems. *Journal of Cereal Science, 67*, 35–45.

Reyes-Avalos, M., Femenia, A., Minjares-Fuentes, R., Contreras-Esquivel, J., Aguilar-González, C., Esparza-Rivera, J., et al. (2016). Improvement of the quality and the shelf life of figs (Ficus carica) using an alginate–chitosan edible film. *Food and Bioprocess Technology, 9*(12), 2114–2124.

Rios, R. V., Pessanha, M. D. F., Almeida, P. F.d., Viana, C. L., & Lannes, S. C.d. S. (2014). Application of fats in some food products. *Food Science and Technology, 34*(1), 3–15.

Rocha, M. A. M., Coimbra, M. A., & Nunes, C. (2017). Applications of chitosan and their derivatives in beverages: A critical review. *Current Opinion in Food Science, 15*, 61–69.

Rodriguez-Carpena, J. G., Morcuende, D., & Estevez, M. (2012). Avocado, sunflower and olive oils as replacers of pork back-fat in burger patties: Effect on lipid composition, oxidative stability and quality traits. *Meat Science, 90*(1), 106–115.

Rozylo, R. (2013). Effect of process modifications in two cycles of dough mixing on physical properties of wheat bread baked from weak flour. *Food and Bioprocess Technology, 7*(3), 774–783. https://doi.org/10.1007/s11947-013-1100-1

Saeed, A., Anjum, F. M., & Akbar Anjum, M. (2011). Micronutrient fortification of wheat flour: Recent development and strategies. *Food Research International, 44*, 652–659.

Sager, M. (2006). Selenium in agriculture, food and nutrition. *Pure and Applied Chemistry, 78*, 111–133.

Salcedo-Sandoval, L., Cofrades, S., Ruiz-Capillas, C., Matalanis, A., McClements, D. J., Decker, E. A., et al. (2015). Oxidative stability of n-3 fatty acids encapsulated in filled hydrogel particles and of pork meat systems containing them. *Food Chemistry, 184*, 207–213.

Sanchez-Salcedo, E. M., Mena, P., Garcia-Viguera, C., Hernandez, F., & Martinez, J. J. (2015). (Poly)phenolic compounds and antioxidant activity of white (*Morus alba*) and black (*Morus nigra*) mulberry leaves: Their potential for new products rich in phytochemicals. *Journal of Functional Foods, 18*, 1039–1046.

Sansone, F., Picerno, P., Mencherini, T., Villecco, F., D'Ursi, A. M., Aquino, R. P., et al. (2011). Flavonoid microparticles by spray-drying Influence of enhancers of the dissolution rate on properties and stability. *Journal of Food Engineering, 103*(2), 188–196.

Santos, N. W., Yoshimura, E. H., Mareze-Costa, C. E., Machado, E., Agustinho, B. C., Pereira, L. M., et al. (2017). Supplementation of cow milk naturally enriched in polyunsaturated fatty acids and polyphenols to growing rats. *PLoS One, 12*(3).

Schlüter, O., Rumpold, B., Holzhauser, T., Roth, A., Vogel, R. F., Quasigroch, W., et al. (2017). Safety aspects of the production of foods and food ingredients from insects. *Molecular Nutrition and Food Research, 61*(6), 1600520.

Seczyk, L., Swieca, M., & Gawlik-Dziki, U. (2015). Nutritional and health-promoting properties of bean paste fortified with onion skin in the light of phenolic-food matrix interactions. *Food and Function, 6*(11), 3560–3566.

Selani, M. M., Brazaca, S. G. C., dos Santos Dias, C. T., Ratnayake, W. S., Flores, R. A., & Bianchini, A. (2014). Characterisation and potential application of pineapple pomace in an extruded product for fibre enhancement. *Food Chemistry, 163*, 23–30.

Selani, M. M., Shirado, G. A., Margiotta, G. B., Rasera, M. L., Marabesi, A. C., Piedade, S. M., et al. (2016). Pineapple by-product and canola oil as partial fat replacers in low-fat beef burger: Effects on oxidative stability, cholesterol content and fatty acid profile. *Meat Science, 115*, 9–15.

Serna-Saldivar, S. O., & Lazo-Vélez, M. A. (2015). Production of selenium-enriched breads and their nutritional and nutraceutical properties. In J. B. Cristina, M. Rosell, F. Aly, & El Sheikha (Eds.), *Bread and its fortification: Nutrition and health benefits*.

Servili, M., Rizzello, C. G., Taticchi, A., Esposto, S., Urbani, S., Mazzacane, F., et al. (2011). Functional milk beverage fortified with phenolic compounds extracted from olive vegetation water, and fermented with functional lactic acid bacteria. *International Journal of Food Microbiology, 147*(1), 45–52.

Sharma, K. D., Karki, S., Thakur, N. S., & Attri, S. (2012). Chemical composition, functional properties and processing of carrot-a review. *Journal of Food Science and Technology-Mysore, 49*(1), 22–32.

Sharoba, A. M., Farrag, M., & Abd El-Salam, A. (2013). Utilization of some fruits and vegetables waste as a source of dietary fiber and its effect on the cake making and its quality attributes. *Journal of Agroalimentary Processes and Technologies, 19*(4), 429–444.

Shchekoldina, T., & Aider, M. (2014). Production of low chlorogenic and caffeic acid containing sunflower meal protein isolate and its use in functional wheat bread making. *Journal of Food Science and Technology, 51*(10), 2331–2343.

Shin, G. H., Kim, J. T., & Park, H. J. (2015). Recent developments in nanoformulations of lipophilic functional foods. *Trends in Food Science and Technology, 46*(1), 144–157. https://doi.org/10.1016/j.tifs.2015.07.005

Shukla, R. K., & Tiwari, A. (2012). Carbohydrate polymers: Applications and recent advances in delivering drugs to the colon. *Carbohydrate Polymers, 88*(2), 399–416.

Sinthusamran, S., Benjakul, S., Kijroongrojana, K., & Prodpran, T. (2019). Chemical, physical, rheological and sensory properties of biscuit fortified with protein hydrolysate from cephalothorax of Pacific white shrimp. *Journal of Food Science and Technology, 56*(3), 1145–1154.

Siriwardhana, N., Kalupahana, N. S., & Moustaid-Moussa, N. (2012). Health benefits of n-3 polyunsaturated fatty acids: Eicosapentaenoic acid and docosahexaenoic acid. In *Advances in food and nutrition research* (Vol. 65, pp. 211–222). Elsevier.

Siro, I., Kapolna, E., Kapolna, B., & Lugasi, A. (2008). Functional food. Product development, marketing and consumer acceptance—a review. *Appetite, 51*(3), 456–467.

Sozer, N., Nordlund, E., Ercili-Cura, D., & Poutanen, K. (2017). Cereal side-streams as alternative protein sources. *Cereal Foods World, 62*(4), 132–137.

Steadman, K. J., Burgoon, M. S., Lewis, B. A., Edwardson, S. E., & Obendorf, R. L. (2001). Buckwheat seed milling fractions: Description, macronutrient composition and dietary fibre. *Journal of Cereal Science, 33*(3), 271–278.

Stintzing, F. C., Stintzing, A. S., Carle, R., Frei, B., & Wrolstad, R. E. (2002). Color and antioxidant properties of cyanidin-based anthocyanin pigments. *Journal of Agricultural and Food Chemistry, 50*(21), 6172–6181.

Stratulat, I., Britten, M., Salmieri, S., Fustier, P., St-Gelais, D., Champagne, C. P., et al. (2015). Enrichment of cheese with vitamin D3 and vegetable omega-3. *Journal of Functional Foods, 13*, 300–307. https://doi.org/10.1016/j.jff.2015.01.004

Sturaro, A., De Marchi, M., Zorzi, E., & Cassandro, M. (2015). Effect of microparticulated whey protein concentration and protein-to-fat ratio on Caciotta cheese yield and composition. *International Dairy Journal, 48*, 46–52.

Swieca, M., Gawlik-Dziki, U., Dziki, D., Baraniak, B., & Czyz, J. (2013). The influence of protein-flavonoid interactions on protein digestibility in vitro and the antioxidant quality of breads enriched with onion skin. *Food Chemistry, 141*, 451–458.

Swieca, M., Seczyk, Q., Gawlik-Dziki, U., & Dziki, D. (2014). Bread enriched with quinoa leaves e the influence of proteinephenolics interactions on the nutritional and antioxidant quality. *Food Chemistry, 162*, 54–62.

Świątkiewicz, S., Arczewska-Włosek, A., Szczurek, W., Calik, J., Bederska-Łojewska, D., Orczewska-Dudek, S., et al. (2020). Algal oil as source of polyunsaturated fatty acids in laying hens nutrition: Effect on egg performance, egg quality indices and fatty acid composition of egg yolk lipids. *Annals of Animal Science, 20*(3), 961–973.

Takeungwongtrakul, S., & Benjakul, S. (2017). Biscuits fortified with micro-encapsulated shrimp oil: Characteristics and storage stability. *Journal of Food Science and Technology, 54*(5), 1126–1136.

Talukder, S. (2015). Effect of dietary fiber on properties and acceptance of meat products: A review. *Critical Reviews in Food Science and Nutrition, 55*(7), 1005–1011.

Talukder, S., & Sharma, D. (2010). Development of dietary fiber rich chicken meat patties using wheat and oat bran. *Journal of Food Science and Technology, 47*(2), 224–229.

Teh, S.-S., Bekhit, A. E.-D., Carne, A., & Birch, J. (2014). Effect of the defatting process, acid and alkali extraction on the physicochemical and functional properties of hemp, flax and canola seed cake protein isolates. *Journal of Food Measurement and Characterization, 8*(2), 92–104.

Tester, R. F., Karkalas, J., & Qi, X. (2004). Starch—composition, fine structure and architecture. *Journal of Cereal Science, 39*(2), 151–165.

Thies, C. (2009). History of microencapsulation. *Abstracts of Papers of the American Chemical Society, 237.*

Torres, I. C., Janhøj, T., Mikkelsen, B.Ø., & Ipsen, R. (2011). Effect of microparticulated whey protein with varying content of denatured protein on the rheological and sensory characteristics of low-fat yoghurt. *International Dairy Journal, 21*(9), 645–655.

Torronen, R., McDougall, G. J., Dobson, G., Stewart, D., Hellstrom, J., Mattila, P., et al. (2012). Fortification of blackcurrant juice with crowberry: Impact on polyphenol composition, urinary phenolic metabolites, and postprandial glycemic response in healthy subjects. *Journal of Functional Foods, 4*(4), 746–756.

Tseng, A., & Zhao, Y. (2013). Wine grape pomace as antioxidant dietary fibre for enhancing nutritional value and improving storability of yogurt and salad dressing. *Food Chemistry, 138*(1), 356–365.

Turhan, S., Sagir, I., & Ustun, N. S. (2005). Utilization of hazelnut pellicle in low-fat beef burgers. *Meat Science, 71*(2), 312–316.

Umesha, S. S., Manohar, R. S., Indiramma, A. R., Akshitha, S., & Naidu, K. A. (2015). Enrichment of biscuits with microencapsulated omega-3 fatty acid (Alpha-linolenic acid) rich Garden cress (Lepidium sativum) seed oil: Physical, sensory and storage quality characteristics of biscuits. *Lebensmittel-Wissenschaft and Technologie- Food Science and Technology, 62*(1), 654–661. https://doi.org/10.1016/j.lwt.2014.02.018

Utama, D. T., Jeong, H. S., Kim, J., Barido, F. H., & Lee, S. K. (2019). Fatty acid composition and quality properties of chicken sausage formulated with pre-emulsified perilla-canola oil as an animal fat replacer. *Poultry Science, 98*(7), 3059–3066.

Vioque, J., Sánchez-Vioque, R., Clemente, A., Pedroche, J., & Millán, F. (2000). Partially hydrolyzed rapeseed protein isolates with improved functional properties. *Journal of the American Oil Chemists' Society, 77*(4), 447–450.

Voragen, A. G. (1998). Technological aspects of functional food-related carbohydrates. *Trends in Food Science and Technology, 9*(8–9), 328–335.

de Vos, P., Faas, M. M., Spasojevic, M., & Sikkema, J. (2010). Encapsulation for preservation of functionality and targeted delivery of bioactive food components. *International Dairy Journal, 20*(4), 292–302.

Vu, H. T., Scarlett, C. J., & Vuong, Q. V. (2018). Phenolic compounds within banana peel and their potential uses: A review. *Journal of Functional Foods, 40*, 238–248.

Waghmare, A. G., Salve, M. K., LeBlanc, J. G., & Arya, S. S. (2016). Concentration and characterization of microalgae proteins from Chlorella pyrenoidosa. *Bioresources and Bioprocessing, 3*(1), 16.

Wallace, T. C., & Giusti, M. M. (2008). Determination of color, pigment, and phenolic stability in yogurt systems colored with nonacylated anthocyanins from Berberis boliviana L. as compared to other natural/synthetic colorants. *Journal of Food Science, 73*(4), C241–C248.

Wallecan, J., McCrae, C., Debon, S., Dong, J., & Mazoyer, J. (2015). Emulsifying and stabilizing properties of functionalized orange pulp fibers. *Food Hydrocolloids, 47*, 115–123.

Wendin, K., Höglund, E., Andersson, M., & Rothenberg, E. (2017). Protein enriched foods and healthy ageing: Effects of protein fortification on muffin characteristics. *Agro Food Industry Hi-Tech, 28*(5), 16–18.

Wildermuth, S. R., Young, E. E., & Were, L. M. (2016). Chlorogenic acid oxidation and its reaction with sunflower proteins to form green-colored complexes. *Comprehensive Reviews in Food Science and Food Safety, 15*(5), 829–843.

Wolever, T. (2001). Carbohydrates and health–the FAO/WHO consultation. *Australian Journal of Nutrition and Dietetics, 58*(2). S3–S3.

Wolever, T., & Wahlqvist, M. (1997). *Carbohydrates in human nutrition-report of a joint FAO/ WHO expert consultation* (Vol. 66).

Xia, T., Zhang, B., Duan, W. H., Zhang, J., & Wang, M. (2020). Nutrients and bioactive components from vinegar: A fermented and functional food. *Journal of Functional Foods, 64*.

Yadav, R. B., Yadav, B. S., & Chaudhary, D. (2011). Extraction, characterization and utilization of rice bran protein concentrate for biscuit making. *British Food Journal, 113*(9), 1173–1182.

Yamamoto, Y., & Gaynor, R. B. (2001). Therapeutic potential of inhibition of the NF-kappa B pathway in the treatment of inflammation and cancer. *Journal of Clinical Investigation, 107*(2), 135–142.

Yi, L., Lakemond, C. M. M., Sagis, L. M. C., Eisner-Schadler, V., van Huis, A., & van Boekel, M. A. J. S. (2013). Extraction and characterisation of protein fractions from five insect species. *Food Chemistry, 141*(4), 3341–3348.

Youssef, M. K. E., Youssef, H. M. K. E., & Mousa, R. M. (2014). Evaluation of antihyperlipidemic activity of citrus peels powders fortified biscuits in albino induced hyperlipidemia. *Food and Public Health, 4*(1), 1–9.

Yuksel-Bilsel, A., & Sahin-Yesilcubuk, N. (2019). Production of probiotic kefir fortified with encapsulated structured lipids and investigation of matrix effects by means of oxidation and in vitro digestion studies. *Food Chemistry, 296*, 17–22.

Yu, C.-J., Lin, Y.-F., Chiang, B.-L., & Chow, L.-P. (2003). Proteomics and immunological analysis of a novel shrimp allergen, Pen m 2. *The Journal of Immunology, 170*(1), 445–453.

Zam, W. (2015). Structured lipids: Methods of production, commercial products and nutraceutical characteristics. *Progress in Nutrition, 17*(3), 198–213.

Zhang, D. J., & Mahoney, A. W. (1989). Bioavailability of iron-milk-protein complexes and fortified cheddar cheese. *Journal of Dairy Science, 72*(11), 2845–2855.

Zhang, D., & Mahoney, A. W. (1990). Effect of iron fortification on quality of cheddar cheese .2. Effects of aging and fluorescent light on pilot scale cheeses. *Journal of Dairy Science, 73*(9), 2252–2258.

Zhao, Y., Abbar, S., Phillips, T., Williams, J., Smith, B., & Schilling, M. (2016). Developing food-grade coatings for dry-cured hams to protect against ham mite infestation. *Meat Science, 113*, 73–79.

Zhu, L.-J., Shukri, R., de Mesa-Stonestreet, N. J., Alavi, S., Dogan, H., & Shi, Y.-C. (2010). Mechanical and microstructural properties of soy protein–high amylose corn starch extrudates in relation to physiochemical changes of starch during extrusion. *Journal of Food Engineering, 100*(2), 232–238.

Zwyrzykowska, A., & Kupczynski, R. (2014). Application of dietary fish oil in dairy cow reproduction. *Turkish Journal of Veterinary and Animal Sciences, 38*(6), 618–624.

Chapter 4

Alternative protein sources

Cátia Saldanha do Carmo[1], Leonor Costa[2], Ana Teresa Serra[2,3], Svein Halvor Knutsen[1], Stefan Sahlstrøm[1], Maria Rosário Bronze[2,3,4]

[1]*Nofima, Norwegian Institute for Food, Fisheries and Aquaculture Research, Aas, Norway;* [2]*iBET, Instituto de Biologia Experimental e Tecnológica, Oeiras, Portugal;* [3]*Instituto de Tecnologia Química e Biológica António Xavier, Universidade Nova de Lisboa, Oeiras, Portugal;* [4]*iMed. Ulisboa, Faculdade de Farmácia, Universidade de Lisboa, Lisboa, Portugal*

4.1 Introduction

During the last years, the protein market has changed due to consumer demand for ethically sourced, sustainable, and healthier protein sources. This fact has resulted in a more diverse range of ingredients, driven by an interest in 'alternative protein' sources, where plant-based ingredients, insects, and more recently cell-cultured protein, compete with more traditional animal proteins-based ingredients. The market of alternative proteins is increasing, and the main key challenges of this industry are consumer demand, sustainability, cost-effective nutritional value, processing capacity, labeling claims and taste and texture of the final products.

This chapter presents an overview of some alternative protein sources, including:

(i) Proteins from land plants, such as pulses and cereals - several pulses are described from lupin to pea, which is recognized as a key opportunity within the protein market. For cereals, only wheat, barley, and oat are presented once they have been processed using emerging and non-chemical sustainable technologies. Additionally, the use of plant by-products in the development of protein-enriched fractions is highlighted as a sustainable, greener and cheaper possibility in the circular economy;

(ii) microalgae – from cyanobacteria to *Chlorella* and *Spirulina* species, the main microalgae protein products are introduced, including protein, concentrates, whole-cell protein, hydrolysates, protein isolates, and bioactive peptides;

(iii) insects – more than 30 species are described, and the challenges faced by the food industry regarding its use, such as safety and consumer acceptability, are discussed.

Food Technology Disruptions. https://doi.org/10.1016/B978-0-12-821470-1.00010-0

Additionally, this chapter describes the recovery of proteins from these alternative sources focusing on dry fractionation and wet extraction/separation/fractionation processes, using conventional and emerging technologies. How these alternative protein products (protein flours, concentrates, and isolates from the presented alternative sources) will assist the disruptive applications in the Food Industry is also discussed, concerning their use in the development of meat analogs (Fig. 4.1).

4.2 Proteins from land plants

4.2.1 Pulses

The Global Pulse Confederation (https://pulses.org/what-are-pulses) defines pulses as «the edible seeds of plants in the legume family » including the 11 types recognized by the United Nations Food and Agriculture Organization (FAO), namely: dry peas and dry beans, lentils, vetches, chickpeas and lupins, as well as broad beans and Bambara beans, cowpeas, pigeon peas and minor pulses (not integrated into one of the previous categories).

Pulses have been gaining interest as an alternative and sustainable protein source (Table 4.1). They are rich in essential amino acids (leucine, lysine, arginine and aspartic acid) and proteins (predominantly albumins, globulins, prolamins, and glutelins) responsible for some functional properties such as water holding capacity, fat binding capacity, foaming, emulsification and gelation (Bessada, Barreira, & Oliveira, 2019; Fasolin et al., 2019; Loveday, 2019; Roy, Boye, & Simpson, 2010; Sá, Moreno, & Carciofi, 2020). All these functional properties are triggering the increasing popularity of pulses like peas, beans, and lupins, which are being appointed as promising sources for new plant-derived proteins (Day, 2013, 2016; Loveday, 2019).

Dry edible seeds are abundant sources of proteins, energy, carbohydrates, fibers, vitamins (B-vitamins), and minerals (iron, calcium, and potassium). Therefore, they are particularly important in areas where resources are scarce and in populations that, due to religious or cultural beliefs, need to replace animal protein (Boye, Aksay et al., 2010; Boye, Zare, & Pletch, 2010; Sá et al., 2020). Conversely, in the western world, the recommended daily intake of

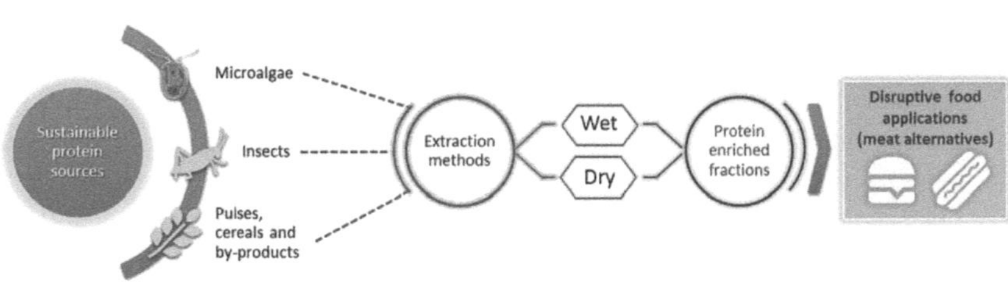

FIGURE 4.1 The value chain for alternative protein sources.

TABLE 4.1 Protein content (% dry matter) of different pulses.

Pulse	Protein content (% dry matter)
Pea (*Pisum sativum* L.)	23.0–31.0
Cowpea (*Vigna unguiculata* L. Walp)	22.0–30.0
Pigeon pea (*Cajanus cajan*)	19.4
Bean (*Phaseolus vulgaris* L.)	16.0–33.0
Faba bean (*Vicia faba* L.)	26.6–31.2
Chickpea (*Cicer arietinum* L.)	15.0–30.0
Lentil (*Lens culinaris* Medik.)	26.4–31.4
Lupin (*Lupinus* spp.)	30.5–55.0
White lupin (*Lupinus albus* L.)	33.0–47.0

Adapted from Sá, A. G. A., Moreno, Y. M. F., & Carciofi, B. A. M. (2020). Plant proteins as high-quality nutritional source for human diet. *Trends in Food Science and Technology, 97,* 170–184. https://doi.org/10.1016/j.tifs.2020.01.011, Bessada, S. M. F., Barreira, J. C. M., & Oliveira, M. B. P. P. (2019). Pulses and food security: Dietary protein, digestibility, bioactive and functional properties. *Trends in Food Science and Technology, 93,* 53–68. https://doi.org/10.1016/j.tifs.2019.08.022, Fasolin, L. H., Pereira, R. N., Pinheiro A. C., Martins, J. T., Andrade, C. C. P., Ramos, O. L., et al. (2019). Emergent food proteins – towards sustainability, health and innovation. *Food Research International, 125,* 108586. https://doi.org/10.1016/j.foodres.2019.108586 and Boye, J. I., Aksay, S., Roufik, S., Ribéreau, S., Mondor, M., Farnworth, E., et al. (2010). Comparison of the functional properties of pea, chickpea and lentil protein concentrates processed using ultrafiltration and isoelectric precipitation techniques. *Food Research International, 43,* 537–546. https://doi.org/10.1016/j.foodres.2009.07.021, Boye, J., Zare, F., & Pletch, A. (2010). Pulse proteins: Processing, characterization, functional properties and applications in food and feed. *Food Research International, 43,* 414–431. https://doi.org/10.1016/j.foodres.2009.09.003.

pulses (which are low in fat) may contribute to prevent metabolic syndrome, and therefore, the predisposition to diabetes, obesity, and coronary heart diseases (Rebello, Greenway, & Finley, 2014). Additionally, it helps to reduce the levels of LDL cholesterol and to reduce the risk of developing cancer, osteoporosis, hypertension, gastrointestinal disorders, and adrenal disease (Boye, Aksay et al., 2010; Boye, Zare et al., 2010). It is also worth mentioning that pulse seeds have comparatively minor components, known as antinutritional factors (ANF), which include enzyme inhibitors (like protease inhibitors), lectins, alkaloids, phytic acid, and some specific phenolic compounds like saponins and tannins. The presence of such components in the human digestive tract decreases nutrient absorption. Moreover, it inhibits the proteolytic activity of trypsin, chymotrypsin, and α-amylase, and subsequently reduces pulses' digestibility and nutrients' bioavailability (Bessada et al., 2019; Rebello et al., 2014). Nonetheless, many of these ANF are generally eliminated or inactivated to a large extent during soaking, cooking, and processing, enabling its safe consumption (Boye, Aksay et al., 2010; Boye, Zare et al., 2010).

Concerning the amino acid profile, despite having substantial amounts of lysine (a limiting amino acid in cereals), pulses have low concentrations of methionine and cysteine (sulfur-containing amino acids), as well as low concentrations of tryptophan (an aromatic amino acid). This profile makes it a suitable complement in cereal-based diets, potentiating a synergistic effect between the nutritional and phytochemical components of both plant groups. This fact highlights the importance of having a balanced diet and the need to gather different protein sources to achieve complementary health benefits (Bessada et al., 2019; Rebello et al., 2014; Sá et al., 2020).

4.2.2 Cereals

Cereals belong to the phylogenetic family *Gramineae* (grasses) and are found on every continent apart from Antarctica, covering around 31%–43% of the Earth's surface (Wilson, 2009). Cereals are a valuable dietary protein source worldwide (Mäkinen, Sozer, Ercili-Cura, & Poutanen, 2017). The major cereals are maize (corn), wheat, rice, barley, and oats constituting essential foods globally. Their consumption has been related to the diminished occurrence of chronic diseases, namely cardiovascular diseases and type-2 diabetes (Wu et al., 2015). Cereal proteins have been classified, according to their solubility fractionation, in albumins, globulins, prolamins, and glutenins (Osborne, 1924). In terms of amino acid profile, cereals lack lysine and contain a high amount of sulfur amino acids and sufficient branched-chain amino acids, namely isoleucine, leucine, and valine (Flambeau, Redl, & Respondek, 2017). On the other hand, pulses, as previously mentioned, have a deficiency in sulfur amino acids. Although pulses and cereals lack some essential amino acids, foods produced by the combination of both sources provide a balanced diet (Petterson, 2011).

Wheat is a valuable grain produced worldwide originating from the Near East. Wheat has been mainly used for food consumption in the form of flour. A small portion is also used in animal feed. The protein content of wheat is around 14% of the whole grain kernel (Brouns, Hemery, Price, & Anson, 2012). A large amount (around eight million tons) of wheat is processed into wheat protein (wheat gluten), and starch corresponding to around 560,000 tones. Different types of proteins can be found in wheat, such as albumins, globulins, and gliadins (alcohol-soluble proteins). The insoluble polymeric proteins existing in wheat are called glutenins. Gliadins and glutenins are the most abundant proteins in wheat (75%–80%) (Flambeau et al., 2017).

Barley (*Hordeum vulgare* L.) is a well adaptable crop and is the fourth most abundant cereal in the world. Barley is mainly used in the brewing and feed industries, being a small part utilized for food production. Barley grains and by-products such as brew spent grain, are inexpensive and abundant sources rich in protein, containing 8%–15% and 20%–30% dry matter protein, respectively (Holtekjølen, Uhlen, & Knutsen, 2008; Wang, Tian, & Chen, 2011). Barley contains bioactive compounds, such as phenolic compounds and vitamins from the complex B and β-glucan, being recognized as having functional and nutraceutical properties. Albumins (water-soluble), globulins (salt-soluble), hordeins (alcohol-soluble), and glutenins (alkali-soluble) are the barley protein types (Sharma & Gujral, 2013). Barley is extensively used as a raw material in the starch production industry, which generates substantial amounts of a by-product rich in protein with value for the food industry (Houde, Khodaei, Benkerroum, & Karboune, 2018).

Oat (*Avena sativa*) is the seventh cereal produced in the world after corn, rice, wheat, barley, sorghum, and millet. Oat has been mainly used as feed for livestock, meaning that approximately 50% of oat production is feed for cattle, and only 10% has been used for food products (Liu, 2014). Oat has higher protein content (15%–20%), soluble fiber (β-glucan), unsaturated fatty acids, and antioxidants being a different cereal, among others (Lásztity, 1998). Oat has a more balanced amino acid profile than other cereals due to its interesting protein composition: 70%–80% of globulins (salt-soluble) (Mäkinen et al., 2017). Approximately 4%–15% of the total proteins in oat correspond to the prolamin fraction (avenins). Most of the cereals from the *Triticeae* family (barley, rye, and wheat) are mainly composed of prolamins (alcohol-soluble). Oat protein can be produced from a protein-rich side stream after β-glucan production (Brückner-Gühmann et al., 2019). The fact that oat lacks gluten allows producing safe ingredients for consumers with celiac disease (Mäkinen et al., 2017). The production of fractions from grains rich in beta-glucan, protein, and starch has received substantial attention (Heusala et al., 2020).

4.2.3 By-products

Over the last decades, by-products derived from cereals, oilseeds, and legumes (including pulses) have been identified as interesting feedstocks for the extraction of edible proteins (Table 4.2), carbohydrates, lipids and other bioactive compounds (dietary fibers, phenolic compounds, alkaloids, and pigments), which have been associated to anti-cholesterol and anti-atherosclerotic properties (Dapčević-Hadnadev, Hadnadev, Pojić, & Balandrán-Quintana, 2018).

In this regard, despite being considered non-utilized and inexpensive materials, brewer's spent grain, distiller's grains and bran (produced during cereal manufacturing), along with potato skins and tomato seeds, have been recently receiving much attention as potential food ingredients (Dhillon, 2016). For instance, during the barley milling process, barley bran is removed from grains bearing a white kernel, used for several food purposes. In this process known as pearling, high amounts of protein (mainly albumin and globulin) are found

TABLE 4.2 Protein content (% dry matter) of different agricultural and industrial by-products.

Source/By-product	Protein content (% dm)	Source/By-product	Protein content (% dm)
Cereals		**Oilseed/Industrial crops**[a]	
Barley		*Canola/Rapeseed*	
Brewers' spent grain	15.4–30.0	Oil meal	34.1–40.6
Straw	2.3–2.4	Oil cake	33.9–35.6
Wheat		*Sesame*	
Bran	13.2–25.0	Oil cake	32.8–35.6
Dried distiller's grains	32.6–38.6	*Sunflower*	
Defatted germ protein	27.6	Oil meal	26.6–42.0
Straw	3.6	Oil cake	34.1
Rice		Stalk	5.2
Bran	11.3–15.0	*Palm Kernel*	
Residue from starch sugar production	62.3	Oil meal	14.9
Legumes		Oil cake	18.6
Soybean		*Soybean*	
Protein concentrate	66.2	Oil meal	44.4–53.8
Protein isolate	92.8	Oil cake	40.1–49.1
Flours and grit	41.0–52.5	*Cotton Seed*	
Cowpea		Oil meal	34.3–44.9
Meal	32.7	Oil cake	21.1–57.3
Vegetables and tubers		*Flaxseed/Linseed*	
Potato		Oil meal	32.6–35.4
Skins	9.5	Oil cake	34.7
Sweet potato		*Mustard seed*	
Filter cake	22.0–32.0	Oil cake	38.5
Centrifuged solids	42.0–57.0	*Peanut*	
Tomato		Oil meal	51.8
Seeds	24.5–39.3	Oil cake	49.5

TABLE 4.2 Protein content (% dry matter) of different agricultural and industrial by-products.—cont'd

Source/By-product	Protein content (% dm)	Source/By-product	Protein content (% dm)
Culled tomatoes	14.0–20.0	*Coconut*	
Tomato pomace	19.0–22.0	Oil cake	25.2
Defatted seed meal	29.9–41.4	*Olive*	
		Oil cake	6.3

[a]*Oil meal - Different defatted raw materials: cold-pressed meal, non-pressed meal, full-pressed meal, pre-pressed meal, and toasted meal. Oil cake - residue obtained after oil extraction.*
Adapted from Dapčević-Hadnadev, T., Hadnadev, M., Pojić, M., & Balandrán-Quintana, R. R. (2018). *The healthy components of cereal by-products and their functional properties.* https://doi.org/10.1016/B978-0-08-102162-0.00005-8, Contreras M del, M., Lama-Muñoz, A., Manuel Gutiérrez-Pérez, J., Espínola, F., Moya, M., & Castro, E., (2019). Protein extraction from agri-food residues for integration in biorefinery: Potential techniques and current status. *Bioresource Technology, 280,* 459–477. https://doi.org/10.1016/j.biortech.2019.02.040, Roth, M., Jekle, M., & Becker, T., (2019). Opportunities for upcycling cereal byproducts with special focus on Distiller's grains. *Trends in Food Science and Technology, 91,* 282–293. https://doi.org/10.1016/j.tifs.2019.07.041, Singh, P., Kumar, R., Sabapathy, S. N., & Bawa, A. S., (2008). Functional and edible uses of soy protein products. *Comprehensive Reviews in Food Science and Food Safety, 7,* 14–28., Dhillon, G. S., (2016). *Protein by products - transformation from environmental burden into value-added products.* Nikki Levy. https://doi.org/10.1016/B978-0-12-802391-4/00018-5, Wu, Y. V., & Bagby, M. O., (1987). Recovery of protein-rich by-products from sweet potato stillage following alcohol distillation. *Journal of Agricultural and Food Chemistry, 35,* 321–325. and Ramachandran, S., Singh, S. K., Larroche, C., Soccol, C. R., & Pandey, A., (2007). Oil cakes and their biotechnological applications – a review. *Bioresource Technology, 98,* 2000–2009. https://doi.org/10.1016/j.biortech.2006.08.002.

in the residue, also called pearling flour (Dapčević-Hadnadev et al., 2018). In a different process, barley endosperm proteins hordein and glutelin (40%–50%) – considered to be defects or contaminants in brewing industries – are usually precipitated out in brewer's spent grain (BSG), generally used as animal feed (Dhillon, 2016; Roth, Jekle, & Becker, 2019). However, as pointed out by Roth et al. (2019), the nutritional benefit in cereals like barley and the high levels of phenolic compounds present in BSG, highlights its application as a valuable source of natural antioxidants, fiber and proteins (Roth et al., 2019). Other cereals by-products identified to present high protein content include wheat by-products such as dried distiller's grains and defatted germ protein, and rice by-products, namely the residues of starch sugar production (Table 4.2).

As already discussed, legumes (which comprehend pulses) provide large amounts of protein and essential amino acids, important to complement cereal proteins. For instance, after oil removal, the defatted flakes obtained from soybean, emerge as a proteinaceous material (ranging from 40% to

90%), used to prepare flours and grits, concentrates and isolates, currently processed into several food products, such as bakery, dairy and meat (Singh, Kumar, Sabapathy, & Bawa, 2008). Besides its protein content (about 9.5% of crude protein (CP)), potato skins also have numerous bioactive compounds like carotenoids, peptides, polyphenols, polyamines, glycoalkaloids, dietary fiber, and suberins, with many potential applications in the food industry.

Another waste material includes the solid by-products largely generated during tomato processing. Tomato waste is obtained from culled tomatoes (14%–20% crude protein), which is discarded from the fresh markets due to damaging/rotting, size, and form. Moreover, tomato pomace containing between 19% and 22% crude protein is obtained from the leftovers of tomato processing (generally including a mixture of seeds, peels, and in some cases, pulp). Even though tomato peel is recognized to be prime raw material for lycopene, tomato seeds noted by Sarkar and Kaul (2014) as a promising protein source (25% CP), high in lysine, glutamic and aspartic acid, explored as a sustainable supplement for future formulations. Interestingly, Sarkar and Kaul (2014) showed that all amino acids were present in tomato seeds, exceeding the values recommended by WHO/FAO/UN (Sarkar & Kaul, 2014; World Health Organization, 2007).

Similarly, oilseeds, such as canola/rapeseed, soybean, sesame, sunflower, among others, are also being highlighted as compelling sources for the extraction of bioactive compounds and proteins. After oil extraction, by either mechanical (pressing) or chemical (using solvents) means, press cakes and oil meals are respectively produced, constituting the main by-products of the oil industry. These are traditionally used to animal feed (especially for ruminants and fish). Still, the high content of proteins (ranging from 15% to 50%), antioxidants, fibers, vitamins, and minerals makes them also suitable for human consumption as edible and palatable food supplements (after processing). However, some of these sources may contain anti-nutrient factors and other contaminants, such as toxins and heavy metals, which must be eliminated before consumption. For instance, in sesame cake, some authors reported the occurrence of trypsin inhibitors, as well as pepsin inhibitors, hemagglutinin, phytates, and tannins (Sunil, Appaiah, Prasanth Kumar, & Gopala Krishna, 2015). Furthermore, according to Dhillon (2016), the high aflatoxin level present in peanut meal limits its application as a food source (Dhillon, 2016). The main oilseeds by-products recognized to present high protein content are oil meal and oil cake from canola/rapeseed, sesame, sunflower, soybean, cottonseed, flaxseed, mustard seed, and peanut oil, as summarized in Table 4.2.

4.2.4 Wet extraction methods for pulses, cereals, and by-products

Wet extraction/fractionation is the conventional way to obtain protein-rich ingredients from pulses, cereals, and by-products. However, this method requires a substantial consumption of enzymes, chemicals, and water (Schutyser, Pelgrom, van der Goot, & Boom, 2015). The major drawback of wet fractioning techniques is related with the enormous variability of isolates' protein content, purity and yield, dependent on processing conditions (i.e., time and temperature, initial matrix state and protein solubility within a specific flour: solvent ratio, as well as the equipment and processes (batch or continuous) chosen, which clearly compromises the scale-up (Boye, Aksay et al., 2010; Boye, Zare et al., 2010).

Wet extraction/separation methods for protein extraction from pulses and cereals have been done by isoelectric precipitation (IEP), membrane filtration or cryo-precipitation (refrigeration), and finally drying. The extraction media used for protein extraction in IEP has been water and alkaline/acid or saline (also known as "micellization") (Mäkinen et al., 2017). It should be noted that, before extraction, samples are usually milled or ground, but, depending on its features/conditions, other pre-treatments like dehulling or dehusking and defatting (with hexane) may also take place (Boye, Aksay et al., 2010; Boye, Zare et al., 2010).

As previously mentioned, pulses are considered promising alternative sources of protein, rich in globulins and albumins. Being the main storage proteins, globulins deserve special attention due to its application in several food products. These peptides are usually extracted in the initial stage of protein isolation. Due to its high solubility in alkaline solutions, they are generally recovered using alkaline extraction followed by IEP (Boye, Aksay et al., 2010; Boye, Zare et al., 2010; Mccurdy & Knipfel, 1990). Through this method, the reported recovery yields ranged between 60% and 65% for bean isolates (protein content > 90% dm), 51% and 62% for lentils concentrates (protein content: 82% dm), and approximately 83% for pea protein concentrates (protein content: 81%−83% dm). In alkaline conditions, the pH of the extracting media ranged from 7.3 for peas and 12.0 for chickpeas, depending, for example, on plant variety and/or growing conditions (Arntfield & Maskus, 2011; Boye, Aksay et al., 2010; Boye, Zare et al., 2010). Despite the exciting recovery rates and purity levels (protein content > 70%), the use of alkaline or acid extraction, followed by IEP, may reduce protein quality, decreasing its digestibility due to the production of lysinoalanine. Furthermore, the loss of essential amino acids is also promoted, as well as the loss of functional properties (Arntfield & Maskus, 2011; Boye, Aksay et al., 2010; Boye, Zare et al., 2010; Singhal, Karaca, Tyler, & Nickerson, 2016). An alternative method to enhance protein recovery is ultrafiltration (UF), often combined with diafiltration (DF), to achieve higher purity levels. This technique is especially applied to obtain protein concentrates/isolates of specific proteins, having the advantage of effectively removing some anti-nutritional components like amylase and protease inhibitors, polyphenols, and lectins. Hence, in order to achieve the maximum separation efficiency, membranes with specific molecular weight cut-offs must be thoroughly selected and the volume concentration ratio, as well as diafiltration conditions, must be previously studied (Arntfield & Maskus, 2011; Boye, Aksay et al., 2010; Boye, Zare et al., 2010; Singhal et al., 2016).

In a comparative study conducted by Boye, Aksay et al. (2010), Boye, Zare et al. (2010), the amount of protein recovered by UF/DF (83.9%, 88.6%, 82.7%, 76.5%, and 68.5% respectively for yellow pea, green lentil, red lentil, and desi and Kabuli chickpea), was higher than the values reported for the concentrates obtained by IEP (81.7%, 79.1%, 78.2%, 73.6%, and 63.9% respectively for the same legume crops) (Boye, Aksay et al., 2010; Boye, Zare et al., 2010). Moreover, the authors observed that concentrates prepared by IEP were composed of globulins, whereas both globulins and albumins composed the concentrates prepared by UF. As for functional properties, though all concentrates exhibited good functional properties and UF extracts presented better gelling properties, suggesting that the manufacturing processes may have an impact on products' functional properties (Boye, Aksay et al., 2010; Boye, Zare et al., 2010).

For cereals, several pre-treatments and extraction processes were applied to recover high protein fractions. The production of protein concentrates from defatted barley flour was recently studied following alkaline and mild enzymatic treatments (Houde et al., 2018). Different approaches for protein extraction were studied: (i) alkaline extraction with NaOH; (ii) alkaline extraction followed by isoelectric precipitation; (iii) Bi-enzymatic extraction for starch removal (alpha-amylase and amyloglucosidase); (iv) tri-enzymatic extraction; starch and glucan removal (alpha-amylase and amyloglucosidase and a glucanase); (v) combined tri-enzymatic extraction (starch and glucan removal) and isoelectric precipitation. The highest protein recovery yield obtained was 78.3% through the last approach containing a protein content of 41.4%. Recovery yield of 51.4% and protein content of 68.9% were achieved by a conventional alkaline extraction methodology followed by isoelectric precipitation. Moreover, the barley protein concentrates presented emulsifying capacity equivalent to that of whey protein isolate.

Regarding wet fractionation of oat, alkaline extraction and acid precipitation were conducted by Wu, Sexson, Cluskey and Inglett (1977) and Ma (1983). Both studies reported obtaining protein fractions with protein contents above 90% dm from defatted oat groats (Wu et al., 1977; Ma, 1983).

More recently, Liu (2014) studied the extraction of protein from ground oat groats and defatted ground groats. First, beta-glucan was extracted with water and precipitated with ethanol or isopropanol. Splitting of protein, starch, and other carbohydrates in the residuals was conducted with alkaline extraction, acid precipitation, and filtration. In this study, 72%–93% dm protein was obtained with a recovery yield between 60% and 72%. It is important to mention that defatting increased the protein content but not protein recovery (Liu, 2014).

Concerning by-products, an alkaline extraction at pH 9.5 for 2 h, followed by isoelectric precipitation (pH 4.2), was performed to durum wheat bran by (Alzuwaid et al., 2020). A wheat bran protein concentrate (WBPC) predominantly constituted by albumins and globulins was obtained. Comparatively, with the initial raw material (wheat bran), WBPC comprised all the amino acids (particularly lysine and threonine). Besides having a protein content of nearly 61% (recovery yield of 20.5%−24.8%), WBPC also showed excellent functional properties, namely high solubility over a wide range of pH and water and fat absorption capacity (Alzuwaid, Sissons, Laddomada, & Fellows, 2020).

Despite the substantial availability of cereal and pulses' by-products (generated during processing), the complexity of these matrixes (e.g., presence of cell walls, extremely tight bonds and interactions between proteins and other biopolymeric constituents) and the fact of being currently explored in small-scale facilities, impose new challenges that drive back the commercial attractiveness of these food processing industries (Pojić, Mišan, & Tiwari, 2018).

To increase the interest in by-products recovery, eco-innovative, and cost-effective wet extraction techniques have been developed and explored over the last years. These include reverse micelles extraction (nanomolecular aggregates of surfactants with inner cores of water, assembled in nonpolar solvents), subcritical water extraction, deep eutectic solvents (DES) and natural deep eutectic solvents (NADES). DES emerged as a new generation of "green" solvents, prepared from the mixture of two or three solid components (e.g., quaternary ammonium salts) capable of establishing hydrogen bonds with each other (for example with amines, alcohols, and acids), thus becoming liquid at the desired temperature. When the components are primary plant metabolites (i.e., organic acids, amino acids, choline derivatives, or sugars), these solvents become NADES (Pojić et al., 2018).

Also, new/emergent cell disruption methods are proposed by Contreras et al. (2019), Fasolin et al. (2019) and Pojić et al. (2018), including microwave-assisted extraction (MWE) and ultrasound-assisted extraction (UAE) − the most widely used in the food industry − as well as pulsed electric energy (PEE, divided in pulsed ohmic heating (POH), pulsed electric fields (PEF) and high-voltage electrical discharges (HVED)), supercritical fluid (SCE) and pressurized liquid (PLE) extraction, and, finally, extrusion (used to transform biomass residues in a single, large scale and continuous step) (Contreras et al., 2019; Fasolin et al., 2019; Ochoa-Rivas, Nava-Valdez, Serna-Saldívar, & Chuck-Hernández, 2017; Pojić et al., 2018). In a comparative study, Ochoa-Rivas et al. (2017) tested the effect of microwaves (8 min at 725W) and ultrasounds (100% amplitude at 24 kHz and 15 min) after the alkaline extraction of defatted peanut flour (produced during oil crushing). When compared with the control, the results showed increases in protein extraction yields of 77% (100% purity) and 136% (86% purity) for

MWE and UAE, respectively. Regarding the techno-functional properties, MWE led to an improvement of foam and emulsifying activity, water absorption, and *in vitro* digestibility. In contrast, UAE had a negative effect on water solubility, emulsifying activity, foam stability, and *in vitro* digestibility. In another study, Zhu, Sun and Zhou (2009) used UAE and reverse micelles to extract proteins from defatted wheat germ. Besides an overall extraction efficiency of 45.6%, the authors also highlight the advantages of this method over alkaline-IEP extraction (Sun, Zhu, & Zhou, 2009). Lower extraction time and better forward extraction efficiency were obtained using UAE and reverse micelles, which contribute to the scale-up and commercial application of these combined techniques (Zhu et al., 2009).

According to with Pojić et al. (2018), besides being a non-thermal treatment that requires low sized equipment, the application of Ultrasound-Assisted Extraction (UAE) allows a more selective and efficient extraction (faster response, energy transfer, and mixing), guaranteeing a quicker start-up and an increased production. However, when sonication is carried for long periods and/or with high energy inputs, it also shows some drawbacks. It can lead to changes in protein structure (and ultimately protein denaturation) and functional properties (e.g., loss of emulsification and foaming capability) (Pojić et al., 2018). As an alternative approach, Gerzhova, Mondor Benali and Aider (2015) studied a novel technology of electro-activation used to extract proteins from canola meal, performed in a three-compartment cell separated by ion exchange membranes, achieving a protein extractability of $34.32 \pm 1.21\%$ (after generating a 0.3A electro-activated solution). In contrast, under the same conditions, the conventional extraction (pH 7−10) yielded only $31.18 \pm 1.89\%$ of proteins. Through the application of this novel extraction method, the authors produced higher purity isolates, containing up to 94% protein with minor denatured spectra (Gerzhova et al., 2015).

Roselló-Soto et al. (2015) reported HVED-assisted technology as a potential sustainable method to enhance the isolation of high-added-value components (like proteins and polyphenols) from olive kernels upon UAE and PEF technologies (Roselló-Soto et al., 2015). These studies demonstrate that the use of electron-based technologies (non-thermal) promotes stabilization and enhance extraction and diffusion of compounds. Nevertheless, despite requiring low energy input, there is no evidence about the feasibility of HVED at the pilot or industrial scales (Contreras et al., 2019; Roselló-Soto et al., 2015).

4.2.5 Dry protein extraction methods for pulses, cereals, and by-products

Dry fractionation is an alternative way to produce protein-enriched fractions, which avoids the addition of chemicals and water, preserving the functionality of the components. Within dry fractionation methodologies, milling and air classification is the combination commonly used to obtain protein and starch-enriched fractions. An alternative to air-classification is sieving with decreasing mesh size (Liu, 2009). Dry fractionation technology also includes other methods apart from milling and air-classification, such as an electrostatic separation that classifies particles by applying electrical forces that act on charged materials. The most common electrostatic charge mechanism is triboelectric separation (Schutyser & van der Goot, 2011).

Pin-milling grinds the raw material, reducing its particle size, and detaching the protein bodies from the starch granules. A subsequent fractionation step (air-classification) splits the smaller particles (protein bodies) from the bigger particles (starch granules) based on size, density, and shape, leading to the production of a protein concentrate (fine-fraction) and a starch concentrate (coarse-fraction) (Boye, Aksay et al., 2010; Boye, Zare et al., 2010). The produced fractions maintain the native structure and functionality (Schutyser & van der Goot, 2011). The main differences between wet- and dry-fractionation methods are highlighted in Fig. 4.2.

Regarding dry fractionation of cereals by milling and air-classification, few studies were reported. Dry fractionation of barley was recently studied by Silventoinen, Sipponen, Holopainen-Mantila, Poutanen and Sozer (2018). In this study, the barley endosperm fraction was mixed with Aerosil 200F as a flow aid. The initial protein content of barley was 8.3%, and the protein content in the enriched fraction resulted in being up to 28.3% with 21.7%

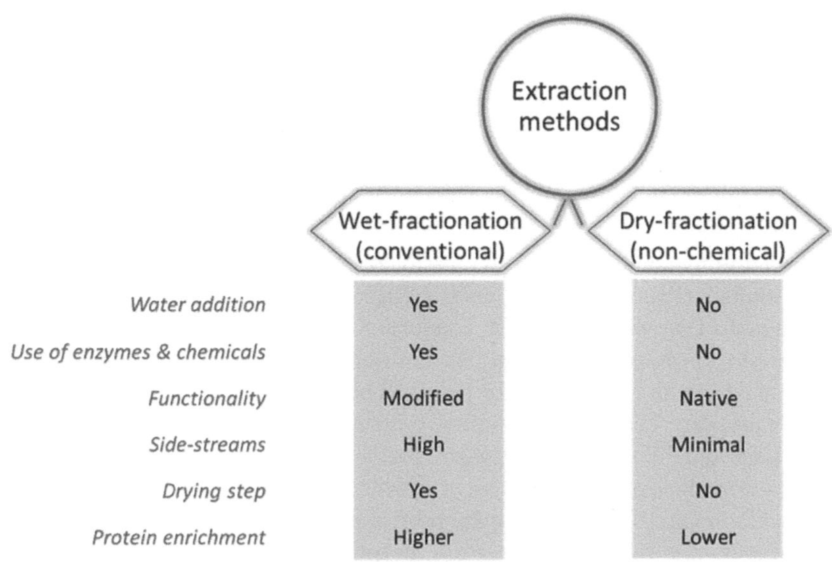

FIGURE 4.2 Comparison between wet- and dry-fractionation methodologies to obtain protein-rich ingredients.

protein separation efficiency (PSE). PSE corresponds to the amount of raw material protein recovered to the fine fraction during air-classification (Pelgrom, Boom, & Schutyser, 2015; Pelgrom, Wang, Boom & Schutyser, 2015). In another experiment, the protein content of 22.3% with a PSE of 59.4% was also obtained (Silventoinen et al., 2018). (Sibakov et al., 2011) also investigated dry fractionation of oat. A protein concentrate from oat having 73% protein content was obtained, but with a very low mass yield of 5% (Sibakov et al., 2011). It is important to mention that the main objective of this study was to extract lipids from oat grain using supercritical carbon dioxide extraction ($ScCO_2$). This method is a non-conventional extraction technique extensively used to extract volatiles and non-polar compounds from natural matrices (do Carmo, Serra, & Duarte, 2017), which avoids the use of harmful organic solvents. It is, therefore, considered an important strategy within green sustainable technologies (Leitner & Poliakoff, 2008). The most commonly used technique to defat oat is hexane extraction, which produces high levels of solvent residue, compromising the safety of defatted oats (Kong, Baeyens, Qin, Zhang, & Tan, 2018; Liu et al., 2019). In this study, the fat was first extracted from oat flakes using $ScCO_2$ technology and then the defatted oat flakes went through the pin mill and air classification leading to a fine fraction (protein-rich) and a coarse-fraction (starch-rich). The coarse-fraction was then pin milled, and air classified a second time leading to a beta-glucan concentrate.

Dry fractionation of pulses such as peas (Pelgrom, Schutyser, & Boom, 2013; Pelgrom, Vissers, Boom, & Schutyser, 2013; Pelgrom, Boom et al., 2015; Pelgrom, Wang et al., 2015; Rempel, Geng, & Zhang, 2019; Sosulski & Youngs, 1979; Vose, Basterrechea, Gorin, Finlayson & Youngs, 1976; Wu & Nichols, 2005), beans (Poel & Aarts, 1990; Sosulski, Elkowicz & Reichert, 1982; Sosulski & Youngs, 1979), lupine (Elkowicz & Sosulski, 1982; Pelgrom, Berghout, van der Goot, Boom, & Schutyser, 2014; Pelgrom, Boom, & Schutyser, 2014; Sosulski & Youngs, 1979), chickpeas (Elkowicz & Sosulski, 1982; Pelgrom, Boom et al., 2015; Pelgrom, Wang et al., 2015; Sosulski & Youngs, 1979; Xing et al., 2020), lentil by-product (Schutyser et al., 2015) through milling combined with air classification or electrostatic separation, has been studied (Pelgrom, Boom et al., 2015; Pelgrom, Wang et al., 2015). Saldanha do Carmo et al. (2020), conducted dry fractionation trials of whole and dehulled peas and fava beans. Protein-rich fractions with a protein content between 44.0% and 46.2% dm for peas and between 60.0% and 60.9% dm for fava beans were obtained. The initial protein content varied between 20.4% and 21.3% dm for peas and between 30.1% and 34.8% dm for fava beans, which means that protein enrichment doubled upon dry fractionation (Saldanha do Carmo et al., 2020). A lupine protein concentrate was also obtained by milling and air classification (Pelgrom, Berghout et al., 2014; Pelgrom, Boom et al., 2014). The protein content varied from 54% to 59% dm, with yields up to 13%. Moreover, flow aids - fused silica particles and potato starch - were tested, and results showed that the yield of the protein fraction doubled. It was suggested that the flow aid was shifted to the protein-rich fraction during air-classification, due to higher ash contents in this fraction. Functionality tests showed that the protein concentrate obtained through dry fractionation, led to the extended half-life of foam compared to the wet fractionated protein powder (intensively heated), due to the preservation of the native functional properties of the dry methodology. (Pelgrom Boom et al., 2015), studied protein enrichment during dry fractionation of starch-rich legumes, including lentils and chickpeas. The initial protein content was 24.9% dm and 21.6% dm for lentils and chickpeas, respectively. The dry fractionation procedure enabled to obtain protein concentrates having 58.5% dm and 45.3% dm for lentils and chickpeas, respectively, with the utilization of Aerosil200 as a flowability aid. Table 4.3 summarizes dry fractionation studies by milling and air-classification of pulses, cereals, and by-products.

4.3 Microalgae

Limited resources of freshwater and land areas, together with the increase in the world population, have driven the search for alternative protein production systems and sources. Microalgae are examples of under-exploited protein sources (Becker, 2007). Some of their species have protein contents (~50% dm) similar to meat (~43% dm), egg (~35% dm), soybean (~37% dm), milk (~26% dm) and bakers' yeast (~39% dm). Moreover, microalgae present promising nutritional properties and low allergenicity (Bleakley & Hayes, 2017; Chiong, Acquah, Lau, Khor, & Danquah, 2016). Microalgae have higher protein yield per unit compared to terrestrial crops, such as pulse legumes, wheat, and soybean (Van Krimpen, Bikker, Van der Meer, Van der Peet-Schwering, & Vereijken, 2013). Several microalgae species contain very high concentrations of protein, but it should be noted in general that microalgae include diverse groups of organisms. Some examples and corresponding protein contents are presented in Table 4.4. The protein content can be up to 58% dm for *Chlorella Vulgaris* and can vary between 42% and 70% dm in certain cyanobacteria (Barkia, Saari, & Manning, 2019). The common/traditional method to cultivate microalgae is in open ponds, mimicking the natural environment of microalgae. These methods benefit from the sunlight. However, the expenses to operate open pond systems should be taken into consideration, namely water consumption (quite significant), energy costs, and land use to analyze the full monetary and environmental costs (Klamczynska & Mooney, 2017). According to the protein content of microalgae and the degree of refining, protein products can be classified as concentrates, isolates, protein hydrolysates, whole-cell protein, and peptides. The whole-cell protein normally contains a protein content of 40%−50% dm, is protein-dense, and holds an intact cellular structure and tissue, is usually consumed directly.

TABLE 4.3 Dry fractionation (by milling and air classification) of pulses, cereals, and by-products to obtain protein concentrates.

Raw material	Initial protein content (% dm)	Protein content of fine fraction (% dm)	Mass yield (%)[a]	Conditions used	Refs.
Pulses					
Whole and dehulled yellow peas (*Pisum sativum* L.)	20.4–21.3	44.0–46.2	29.2–32.1	1 or 2 step pin-milling; milling rotor speed 17,800 rpm; AC wheel speed 12,500–15,000 rpm; AC air flow rate 220 m^3/h	(Saldanha do Carmo et al., 2020)
Green pea flour, split green pea flour, and organic split yellow pea flour (*Pisum sativum* L.)	16–21 (w.b.)	84.7–87.3 (w.b.)	34.9–43.5	Industrial-scale dry fractionation Feeding speed 300 kg/h; milling rotor speed 3,166 rpm; AC wheel speed 20/2700 (Hz/RPM); airflow rate 780 CPM; repetition of milling and air classification of the coarse fraction 7 times	(Rempel et al., 2019)
Yellow peas (*Pisum sativum* L.)	23	55.0	15.5	Feed speed 0.75–1 kg/h; impact milling speed 8,000 rpm; AC wheel speed 5,000–12,000 rpm; AC air flow rate 52 m^3/h	(Pelgrom, Schutyser et al., 2013; Pelgrom, Vissers et al., 2013)
Whole and dehulled faba beans (*Vicia faba* L. var. Kontu)	30.1–34.8	60.0–60.9	24.9–26.5	1 or 2 step pin-milling; milling rotor speed 17,800 rpm; AC wheel speed 12,500–15,000 rpm AC airflow rate of 220 m^3/h	(Saldanha do Carmo et al., 2020)

Beans (*Phaseolus vulgaris* L.)	21.4	40.0–52.6	7.0–7.6	Milling 5000 rpm and pin milling rotor speed 11,000/14,000 rpm; AC wheel speeds 5,400/8,700 rpm; air flow rate 46.3–49.6 m^3/h	(Poel & Aarts, 1990)
Dehulled faba beans (*Vicia faba* L. var. Imposa)	Not referred	64.1%	Not referred	Milling rotor speed 5300 rpm; AC wheel speed 5800 rpm	(Vogelsang-O'Dwyer et al., 2020)
Lupine (*Lupinus angustifolius* L.)	Approx. 38–40	53.7–58.9	6.1–12.8	Feed milling speed 1 kg/h, milling rotor speed 8,000 rpm; AC wheel speed 7,000–13,000 rpm; AC air flow rate 80 m^3/h	(Pelgrom, Berghout et al., 2014; Pelgrom, Boom et al., 2014)
Chickpeas (*Cicer arietinum* L.)	21.6	45.3	n.d.	Impact milling speed 8000 rpm; feed rate 2 rpm (0.5 kg/h); AC wheel 2,200–8,000 rpm; air flow rate 40–52 m^3/h; AC speed 5,000–10,000 rpm; AC feed rate 15 rpm (1.0 kg/h)	(Pelgrom, Boom et al., 2015; Pelgrom, Wang et al., 2015)
Chickpeas (*Cicer arietinum* L.)	Circa 21.0	Circa 38.0	28.4	Impact milling feed rate 2–5 rpm (circa 0.5 kg/h); AC feed rate 2–5 rpm (circa 0.5 kg/h); AC wheel speed 10,000 rpm; air flow rate 52 m^3/h	(Xing et al., 2020)
Lentil (*Lens culinaris*)	24.9	58.5	n.d.	Impact milling speed 8000 rpm; feed rate 2 rpm (0.5 kg/h); AC wheel 2,200–8,000 rpm; air flow rate 40–52 m^3/h; AC speed 5,000–10,000 rpm; AC feed rate 15 rpm (1.0 kg/h)	(Pelgrom, Boom et al., 2015; Pelgrom, Wang et al., 2015)

Continued

TABLE 4.3 Dry fractionation (by milling and air classification) of pulses, cereals, and by-products to obtain protein concentrates.—cont'd

Raw material	Initial protein content (% dm)	Protein content of fine fraction (% dm)	Mass yield (%)[a]	Conditions used	Refs.
Cereals					
Industrial barley endosperm fraction	8.3%	28.3%	22.1%	Aerosil 200F was used as flow aid; AC classifier wheel speed 4,000–21,500 rpm (optimum 8000 rpm); air flow rate 50 m^3/h	(Silventoinen et al., 2018)
Dehulled, non-defatted and defatted flaked oat	16.4% before and 17.2% after defatting with supercritical carbon dioxide extraction ($ScCO_2$)	73.0%	5.0%	Fine impact mill with pin disc grinders; milling twice at a rotor speed of 17,800 rpm, the feed rate of 10 kg/h; AC airflow rate of 220 m^3/h; AC feed rate 5 kg/h; AC wheel speed 3,000–7000 rpm	(Sibakov et al., 2011)
By-products					
Lentil by-products	24.8%	51.5%–53.0%	6.6–31.5	Milling speed 1900 rpm, feed rate 3.9×10^{-4} kg/s, AC speed 8036 rpm, air flow 46 m^3/h	(Afam-Mbah, Ketabi, Emami, Tabil, & Tyler, 2018)
Hard red winter wheat bran	14.9%	22.1%	14%	Milling feed rate 0.9–2.7 kg/h; AC wheel speed 3600/5860 rpm	(Ranhotra, Gelroth, Glaser, & Reddy, 1994)

AC, Air-classifier; *dm*, dry matter; *w.b.*, the wet basis.
[a]*percentage of the dry weight of protein fraction recovered from the total dry weight of the raw material.*

TABLE 4.4 Protein content of different microalgae.

Micro (alga)	Protein content (% dm)	Micro (alga)	Protein content (% dm)
Anabaena cylindrical	43–56	Pavlova sp.	24–29
Aphanizomenon flo-aquae	62	Phaeodactlylum tricornutum	35
Arthrospira maxima	60–71/56–77	Porphyridium aerugeneum	32
Arthrospira platensis	52–55	Porphyridium cruentum	28–39
Botryococcus braunni	39–40	Prymnesium parvum	28–45
Chatoceros calcitrans	34–38	Scenedesmus almeriensis	42
Chaetoceros gracilis	12	Scenedesmus dimorphus	8–18
Chaetoceros muelleri	59	Scenedesmus obliquus	48–56
Chlamydomonas rheinhardii	48	Scenedesmus quadricausa	47
Chlorela ellipsoidea	42	Skeletonema costatum	25
Chlorela ovalis	11	Spirulina maxima	56–71
Chlorella pyrenoidosa	57	Spirulina platensis	46–63
Chlorella spaerckii	7	Synechococcus sp.	63
Chlorela vulgaris	51–58	Spirogyra sp.	6–20
Dunaliella oculata	49	Tetraselmis sp.	36
Dunaliella primolecta	12	Tetraselmis chuii	31–46
Dunaliella salina	57	Tetraselmis maculate	52
Dunaliella tertiolecta	11	Thalassiosira pseudonana	34
Euglena gracilis	39–61		
Haematococcus pluvialis	51–53		

Continued

TABLE 4.4 Protein content of different microalgae.—cont'd

Micro (alga)	Protein content (% dm)	Micro (alga)	Protein content (% dm)
Isochrysis galbana	27–51		
Nannochloropsis granulata	18–34		
Nannochloropsis oculata	44–49		
Nitzschia closterium	26		

Adapted from Becker, E. W., (2007). Micro-algae as a source of protein. *Biotechnology Advances, 25,* 207–210. https://doi.org/10.1016/j.biotechadv.2006.11.002, Chiong, T., Acquah, C., Lau, S. Y., Khor, E. H., & Danquah, M. K., (2016). Microalgal-based protein by-products: Extraction, purification, and applications, *Protein byprod transform from environ burd into value-added prod.* 213–234. https://doi.org/10.1016/B978-0-12-802391-4.00012-4, Barkia, I., Saari, N., & Manning, S. R., (2019). Microalgae for high-value products towards human health and nutrition. *Marine Drugs, 17,* 1–29. https://doi.org/10.3390/md17050304, Barka, A., & Blecker, C., (2016). Microalgae as a potential source of single-cell proteins. A review. *Biotechnology, Agronomy and Society and Environment, 20,* 427–436. https://doi.org/10.25518/1780-4507.13132, Servaites, J. C., Faeth, J. L., & Sidhu, S. S., (2012). A dye binding method for measurement of total protein in microalgae. *Analytical Biochemistry, 421,* 75–80. https://doi.org/10.1016/j.ab.2011.10.047, Slocombe, S. P., Ross, M., Thomas, N., McNeill, S., & Stanley, M. S., (2013). A rapid and general method for measurement of protein in micro-algal biomass. *Bioresource Technology, 129,* 51–57. https://doi.org/10.1016/j.biortech.2012.10.163, Barbarino, E., & Lourenço, S. O., (2005) An evaluation of methods for extraction and quantification of protein from marine macro- and microalgae. *Journal of Applied Phycology, 17,* 447–460. https://doi.org/10.1007/s10811-005-1641-4, Schwenzfeier, A., Helbig, A., Wierenga, P. A., & Gruppen, H., (2013). Emulsion properties of algae soluble protein isolate from *Tetraselmis* sp. *Food Hydrocolloids, 30,* 258–263. https://doi.org/10.1016/j.foodhyd.2012.06.002, Romero García, J. M., Acién Fernández, F. G., & Fernández Sevilla, J. M., (2012). Development of a process for the production of l-amino-acids concentrates from microalgae by enzymatic hydrolysis. *Bioresource Technology, 112,* 164–170., González López, C. V., García M del, C. C., Fernández, F. G. A., Bustos, C. S., Chisti, Y., & Sevilla, J. M. F., (2010). Protein measurements of microalgal and cyanobacterial biomass. *Bioresource Technology, 101,* 7587–7591. https://doi.org/10.1016/j.biortech.2010.04.077.

The isolates, hydrolysates, concentrates, and bioactive peptides are obtained through extraction, concentration, and purification of proteins from microalgae cells (Soto-Sierra, Stoykova, & Nikolov, 2018). *Spirulina platensis* and *Chlorella Vulgaris* are the prevalent whole-cell microalgae used for consumption, due to their high protein content (46%–63% dm and 51%–58% dm, respectively) and favorable amino acids composition (Becker, 2007; Liestianty et al., 2019). Overall, in terms of protein quality, all the essential amino acids (branched-chain amino acids, lysine, valine, isoleucine, and leucine) are present in microalgae whose well-balanced profile are similar to egg albumin, lactoglobulin and soy (Bleakley & Hayes, 2017). Microalgae proteins are often presented and used in the form of capsules, liquids, and tablets (Pulz & Gross, 2004). Microalgae application in foods has been hampered by the existence of non-protein components, such as chlorophyll, which influence the color and taste of products. Furthermore, *Tetraselmis* and *Chlorella* are examples of microalgae that possess a rigid cell, lowering the extraction efficiency of intracellular proteins. Moreover, whole biomass applications can end in low digestibility (Barkia et al., 2019).

Protein extraction methods applied to microalgae include physical processes, chemical extraction, and enzymatic hydrolysis. Innovative methods, such as microwave-assisted extraction, pulsed electric field, and ultrasound-assisted extraction, have been used as disruption methods to liberate intracellular contents by the disintegration of microalgae cell walls and membranes (Show, Lee, Tay, Lee, & Chang, 2015).

4.3.1 Wet protein extraction methods for microalgae

The stiff cell walls of certain microalgae hinder the extraction of intact intracellular proteins. Therefore, cell lysis needs to be performed before the extraction of the protein. Therefore, disintegration methods are required for the recovery of proteins from microalgae. The relevant disintegration methods use ultrasounds, chemical or enzymatic treatments, mechanical action (bead mills and high-pressure homogenizers), thermal or osmotic shocks (freeze and thaw cycles) (Doucha & Lívanský, 2008). Schwenzfeier, Wierenga and Gruppen (2011) used bead milling followed by milling, centrifugation, and ion-exchange chromatography for cell disintegration to extract proteins from *Tetraselmis* sp. (Schwenzfeier et al., 2011). A protein isolate with a protein content of 64% dm was produced. Microalgae protein isolates must be free of color and taste, to increase their commercial value and potential applications in food. Therefore, the extract was also decolorized by precipitation at a pH of 3.5. Ursu et al. (2014) studied the extraction of proteins from *Chlorella vulgaris*. A high-pressure cell disrupter under pH seven or over pH 12 was used to solubilize proteins. The yield of solubilized proteins was 52% dm. Thereafter, precipitation in the acid media and concentration/fractionation by tangential ultrafiltration conducted to recover the solubilized proteins. Microalgae proteins from *Chlorella pyrenoidosa* extracted using the separation technique Three-Phase Partitioning (TPP) that has gain significance in the last few years (Waghmare, Salve, LeBlanc, & Arya, 2016). At optimal conditions, 78% protein concentration obtained from the dried biomass (the initial protein content was 45% dm). Teuling, Wierenga, Schrama and Gruppen (2017) studied the extraction and isolation of proteins from *Arthrospira maxima* (Spirulina), *Nannochloropsis gaditana*, *Tetraselmis impellucida* and *Scenedesmus dimorphus* (Teuling et al., 2017) using the method described previously by Schwenzfeier et al. (2011). Even though both the initial protein content of each microalga (27%−62% dm) and the protein extractability (17%−74% dm) were different, similar protein contents obtained in the final protein isolates (62%–77% dm). Pulsed Electric Field (PEF) has been used as a pre-treatment before enzymatic protein hydrolysis from *Scenedesmus almeriensis* to facilitate enzyme access to proteins. High-pressure homogenization (HPH) was also used to process some biomass samples (Akaberi et al., 2019). The initial protein content of the treated biomass was 50.8 ± 2.9% dm. Results showed that by using HPH treatment, 54% of dm protein was released into the suspension. However, by PEF treatment, only 1.15% dm of protein was released. Kulkarni and Nikolov (2018) studied alkaline extraction from wet, freeze-thawed *Chlorella vulgaris* and cell disruption using bead milling and high-pressure homogenization. Afterward, a two-stage membrane filtration process was used to concentrate and fractionate the proteins. The protein extraction method was effective and led to a final protein content of 76% (Kulkarni & Nikolov, 2018).

4.4 Insects

Entomophagy, or the act of eating insects, is a relatively new concept within the occidental culture. It arose from the need to find alternative and sustainable protein sources capable of compensating the increased demand for animal protein and the urgent need to lower the emission of greenhouse gases and ammonia (Gravel & Doyen, 2020; Jantzen da Silva Lucas, Menegon de Oliveira, da Rocha, & Prentice, 2020; Sogari, Liu, & Li, 2019; van Huis, 2015). Despite the struggle to be accepted by western consumers, insects have an excellent nutritional content (Table 4.5) and show a great potential to serve as a healthy, accessible and palatable food source (de Castro, Ohara, dos Santos Aguilar, & Domingues, 2018; Fasolin et al., 2019; Gravel & Doyen, 2020; Jantzen da Silva Lucas et al., 2020). In fact, according to Gravel and Doyen (2020), despite being slightly less digestible than animal-based proteins (beef and egg white – 100%), insect proteins appear to be more easily digested (76%–98%) than plant sources such as peanut and lentil proteins (52%) (Gravel & Doyen, 2020). Nonetheless, given that insects possess a hard exoskeleton rich in chitin, a protein particularly difficult to digest, its nutritional content is highly variable (Kim, Yong, Kim, Kim, & Choi, 2019).

Insects represent one of Earth's largest and most diverse groups of organisms, as shown by (Larsen et al., 2017), that estimated the existence of approximately six million insect species around the world (1 million formally described to date) (Stork, 2018; Larsen, Miller, Rhodes, & Wiens, 2017). Given the size and diversity of this enormous group, some authors predict that close to 2000 species are edible in a variety of forms (raw or processed) and life stages (eggs, larvae, pupae, or adults). Beetles (*Coleoptera* 31%); caterpillars (*Lepidoptera* 18%); ants, wasps, and bees (*Hymenoptera* 14%); locusts, crickets, and grasshoppers (*Orthoptera* 13%); scale insects, leafhoppers, true bugs, cicadas, and planthoppers (*Hemiptera* 10%); dragonflies (*Odonata* 3%);

TABLE 4.5 Protein Content (% dm) of different insects.

Order/Species	Protein content (% dm)[a]	Order/Species	Protein content (% dm)[a]
Blattodea (cockroaches, termites)	**57.3**	**Hymenoptera** (bees, ants, wasps)	**13.0–77.0**
Blaptica dubia (dubia cockroach)	19.3	*Apis mellifera* (honey bee brood)	40.5
Periplaneta americana (american cockroach)	49.4–65.6	*Oecophylla smaragdina* (green tree ant)	53.5
Coleoptera (adult beetles and grubs)	**23.0–66.0**	**Lepidoptera** (butterflies, moths)	**14.0–68.0**
Alphitobius diaperinus (lesser mealworm)	20.7	*Bombxy mori* (silkworm)	48.7–69.8
Rhynchophorus phoenicis (african palm weevil)	8.9–41.7	*Cirina forda* (shea defoliator)	20.2–74.4
Tenebrio molitor (mealworm)	47.1–65.6	*Galleria mellonella* (greater wax moth)	33.9–41.3
Zophobas morio (superworm)	20.6–46.8	*Samia cynthia ricini* (eri silkworm)	54.7
Hemiptera (true bugs)	**42.0–74.0**	**Odonata** (dragonflies, damselflies)	**46.0–65.0**
Belostoma sp. (giant water bugs)	70.9	*Aeschna multicolor* (blue-eyed darner)	54.2
Hoplophorion monograma ('Periquito del aguacate', Mexico)	59.6–64.0	*Anax* sp. (dragonfly)	26.2–56.2
Pachilis gigas ('Xamues', Mexico)	63.0–65.4	**Orthoptera** (crickets, grasshoppers, locusts)	**23.0–65.0**
Proarna sp. ('Cigarra', Mexico)	72.0	*Acheta domesticus* (house cricket)	64.4–70.7
Diptera (flies)	**49.5**	*Gryllodes sigillatus* (cricket)	56.8
Drosophila melanogaster (Fruit fly)	56.3	*Ruspolia differens* ('Nseenene' - Uganda grasshopper)	44.3–44.6

Continued

TABLE 4.5 Protein Content (% dm) of different insects.—cont'd

Order/Species	Protein content (% dm)[a]	Order/Species	Protein content (% dm)[a]
Musca domestica (Housefly)	63.1–64.0	*Schistocerca* sp. (grasshoppers)	61.0
		Sphenarium purpurascens (Cornfield grasshopper)	52.6–71.5

[a]*Information regarding protein may be ambiguous due to the use of Kjeldahl or Dumas based methods with a conversion factor of 6.25 overestimates protein content. Chitin content accounts for at least 10% of the whole dried insects and its amino group will also give rise false positive protein values when calculated from the crude nitrogen content. The words in bold correspond to the insect order.*

Adapted from Loveday, S. M., (2019). Food proteins: Technological, nutritional, and sustainability attributes of traditional and emerging proteins. *Annual Review of Food Science and Technology, 10,* 311–339. https://doi.org/10.1146/annurev-food-032818-121128, Fasolin, L. H., Pereira, R. N., Pinheiro, A. C., Martins, J. T., Andrade, C. C. P., Ramos, O. L., et al. (2019). Emergent food proteins – towards sustainability, health and innovation. *Food Research International, 125,* 108586. https://doi.org/10.1016/j.foodres.2019.108586, Jantzen da Silva Lucas, A., Menegon de Oliveira, L., da Rocha, M., & Prentice, C., (2020). Edible insects: An alternative of nutritional, functional and bioactive compounds. *Food Chemistry, 311,* 126022. https://doi.org/10.1016/j.foodchem.2019.126022, Kim, T. K., Yong, H.I., Kim, Y. B., Kim, H. W., & Choi, Y. S., (2019). Edible insects as a protein source: A review of public perception, processing technology, and research trends. *Food Science of Animal Resources, 39,* 521–540. https://doi.org/10.5851/kosfa.2019.e53, van Huis, A., Van Itterbeeck, J., Klunder, H., Mertens, E., Halloran, A., Muir, G., et al. (2013). *Edible insects: Future prospects for food and feed security,* Rumpold, B. A., & Schlüter, O. K., (2013). Nutritional composition and safety aspects of edible insects. *Molecular Nutrition and Food Research, 57,* 802–823. https://doi.org/10.1002/mnfr.201200735 and Gómez, B., Munekata, P. E. S., Zhu, Z., Barba, F. J., Toldrá, F., Putnik, P., et al. (2019). Challenges and opportunities regarding the use of alternative protein sources: aquaculture and insects. *Advances in Food and Nutrition Research, 89,* 259–295. https://doi.org/10.1016/bs.afnr.2019.03.003.

termites (*Blattodea* 3%) and flies (*Diptera* 2%) are among the mainly consumed species worldwide by order of percentage of preference (Imathiu, 2020; Sogari et al., 2019). The protein content of these species and others are summarized in Table 4.5. In Europe, *Tenebrio molitor* (mealworm larvae) and *Acheta domesticus* (crickets) – considered the most promising species – are already being included in food and feed industries (Regulation EU 2017/893, 2017). Worth noting that processing, through roasting, frying, boiling, drying, extrusion, among others, is directly connected with insects' protein quality and content, since chitin is degraded (de Castro et al., 2018; Imathiu, 2020; Kim, Yong, Chun, Lee, Kim, & Choi, 2020; van Huis et al., 2013).

Therefore, the commercial interest in edible insects relies on the fact of being an excellent source not only of protein (40%–70% dry matter - Table 4.5) but also of fat (predominantly polyunsaturated fatty acids), minerals (like phosphorous, magnesium, copper, iron, selenium, manganese and zinc) and vitamins (riboflavin, biotin, pantothenic acid and in some cases folic acid). Also, the wide range of functional properties, such as foaming, gelling, and structuring ability, as well as water and lipid retention, thickening and emulsifying capacity draws the attention towards the use of insect's proteins in different industries (Gravel & Doyen, 2020). For instance, the wide applicability of insect peptides is extended to its action as an antihypertensive, antimicrobial, and antioxidant agent (de Castro et al., 2018).

However, in spite of the tremendous potential of this alternative source, insect industry still faces many challenges and throwbacks: (1) doubts and concerning about food safety; (2) regulation; (3) consumer acceptability (4) upscaling of insect cultivation and their appropriate feed and (5) insect's shelf-life.

The lack of information regarding insects' consumption and farming rises undoubtedly a giant barrier to the growth and evolution of this industry, especially in the strictly regulated countries across Europe. To allow the safe and fair production and trade of this kind of product, the European Commission, through the European Food Safety Authority (EFSA), must assess all the underlying risks and hazards. This includes antinutritional factors, heavy metals, pesticide residues, pathogenic microorganisms, and parasites, as well as mycotoxins and allergens. The presence of such substances/organisms is powerfully determined by insects' production method, species, stage of harvest, and the substrate used (including sources) in the rearing process (Imathiu, 2020; Rumpold & Schlüter, 2013; van Huis, 2015).

In this way, following the Regulation (EU) 2015/2283 (2015), insect-based foods meant for human consumption must be submitted as Novel Food ("defined as food that had not been used for human consumption to a significant degree" within the EU before May 15, 1997, when the first regulation on novel food came into force) (Regulation (EU) 2015/2283. Regulation (Eu) 2015/2283 on novel foods, 2015).

Another barrier for the use of insects as foods, is as previously mentioned, consumer acceptability. Several studies conducted in some European countries reported food neophobia (defined as the tendency to refuse new or unfamiliar foods) and disgust, as two of the strongest factors leading western consumers to reject insects' consumption (Sogari et al., 2019). A step forward to solve this problem relies on the transformation of these ingredients, as suggested by Caparros Megido et al. (2014) and Melgar-Lalanne, Hernández-Álvarez and Salinas-Castro (2019) who observed that people are more willing to consume insects when presented in products like cookies, energy bars, burgers or sandwiches (Caparros Megido et al., 2014; Melgar-Lalanne et al., 2019). A different approach is trying to clarify and change consumers' perceptions about entomophagy. In this regard, van Huis (2015) interestingly demonstrated that in countries like Austria, Belgium, Netherlands, and France, farming and consumption of edible insects is somehow tolerated (Imathiu, 2020; van Huis, 2015).

In the future, while thinking about trade and distribution chains, shelf-life shall also be considered. Ssepuuya, Aringo, Mukisa and Nakimbugwe (2016) shows that wildly harvested insects are seasonal and usually have a very limited shelf-life, one or 2 days for a freshly harvested *Ruspolia nitidula*, an edible grasshopper (Ssepuuya et al., 2016).

Again, the idea of keeping a diversified diet and conjugate different protein sources is reinforced by van Huis et al. (2013). Sandra G. F. Bukkens gives the example of the people in Papua New Guinea, who compensate for the lack of lysine and leucine in some tubers, through the ingestion of palm weevil larvae (*Rhynchophorus ferrugineus*) (Bukkens, 2005). Conversely, tubers provide tryptophan and aromatic amino acids, which are limited in palm weevils. This fact is a good approach since insects' proteins, and amino acid content do not always match the daily intake levels recommended by WHO (World Health Organization, 2007). For example, two different species of termite – *Macrotermes bellicosus* and *Macrotermes subhyalinus* – have distinct amino acid compositions, being the first one rich in tryptophan and lysine, whereas in the other is quite the opposite (Sogbesan & Ugwumba, 2008).

4.4.1 Wet protein extraction methods for insects

As for plant-based products, protein extraction from edible insects also depends on the raw material features and composition. Likewise, different pretreatments (i.e., blanching and defatting) may be required, along with further downstream processing to recover, purify, and dry protein fractions (Gravel & Doyen, 2020; Melgar-Lalanne et al., 2019). Similarly, since insect matrixes possess a fiber-rich layer, the first pre-processing steps commonly include drying (by freeze-drying, oven-drying, microwave or spray drying), grinding and sieving, to obtain a fine powder (insect's flour) that enables protein solubilization (Bußler, Rumpold, Jander, Rawel, & Schlüter, 2016).

It's worth noting that for insects with high levels of fat, defatting becomes a crucial stage. The presence of high amounts of lipids limit insect's processability due to the loss of protein solubility caused by protein-lipid interactions. According to Bußler et al. (2016), defatting with hexane improved extraction yields from *Tenebrio molitor* flour and increased its crude protein content to 68%. Hence, in the food industry, hexane extraction is the most conventional way to remove fats, despite the associated health, safety, environmental and economic concerns (Gravel & Doyen, 2020).

To overcome these issues, new and more sustainable techniques, such as supercritical carbon dioxide extraction ($ScCO_2$), are being explored. For instance, Purschke, Brüggen, Scheibelberger and Jäger (2018), Purschke, Meinlschmidt, Horn, Rieder and Jäger (2018), Purschke, Tanzmeister et al. (2018) achieved a maximal defatting of 95% from *T. Molitor* larvae (105 min at 400/250 bar and 45°C), fostering new scale-up studies aiming to assess the cost-effective advantages of these methods (Purschke, Brüggen et al., 2018;

Purschke, Meinlschmidt et al., 2018; Purschke, Tanzmeister et al., 2018). Additionally, the successful application of cold atmospheric pressure plasma (CAPP) processing over *T. Molitor* L. flour (15 min exposure) led to an increase in protein oil binding capacity and to a decrease up to three log units in the microbiological loads, when compared with thermal treatments (20–140°C). According to the results obtained in this study, Bußler et al. (2016) recognized the efficient use of this innovative treatment to selectively modify proteins' structure and functional properties.

Nowadays, the extraction of proteins from edible insects is frequently done by conventional wet fractionation procedures, generally using water, chemicals, and enzymes to facilitate industrial processes (Kim et al., 2019; Melgar-Lalanne et al., 2019). The most widely used is alkaline-IEP extraction, that resulted in recovery and purity yields of 51.7% and 82.3% (on a dry basis), respectively, for a *Locusta migratoria* protein concentrate (Purschke, Brüggen et al., 2018; Purschke, Meinlschmidt et al., 2018; Purschke, Tanzmeister et al., 2018).

In a comparative study carried by Gresiana, Muzi Marpaung and Sutanto (2015) cricket powder (*Gryllus mitratus*) was extracted using two different solvents: water and sodium hydroxide, at extraction temperatures ranging from 30 to 50°C for 30, 60 and 90 min. Moreover, three precipitation methods (ammonium sulfate, isoelectric point, and acetone precipitation) were used to isolate the insect's protein fraction. Water at 30°C for 30 min was reported as the best extraction condition, resulting in protein content of 51.98 mg/mL. In contrast, ammonium sulfate precipitation allowed the highest recovery yield (up to 75.03%), comparatively with acetone (64.94%) and isoelectric precipitation (14.26%). Nevertheless, the highest protein purity was registered when acetone was added to the solution (up to 99.19%), while the purity of protein obtained from isoelectric precipitation was 96.44% and only 31.01% from ammonium sulfate precipitation (Gresiana et al., 2015).

Through a different approach, Yi et al. (2013) tested the aqueous solubilization of proteins extracted from five different species (*Acheta domesticus*, *Tenebrio molitor*, *Blaptica dubia, Zophobas morio,* and *Alphitobius diaperinus*), achieving recovery and purity yields ranging from 17% to 23% and 50%–75% (in a wet basis), respectively (Yi et al., 2013).

Independently of the separation method chosen to transform edible insects, proteins' properties – including techno-functional features (water and oil binding, foaming, gelling and emulsifying capacity), thermal stability, solubility, and amino acid profile – must be considered throughout the process. In this way, Bußler et al. (2016) emphasize the need of extensive knowledge about these properties, especially when these ingredients are used to supplement foodstuffs (Bußler, Steins, Ehlbeck, & Schlüter, 2015).

Bußler et al. disclosed evidence of the impact caused by processing methods over insect-based products, after assessing the composition and properties of the protein fractions recovered from *Tenebrio molitor* and

Hermetia illucens flours under varying extraction conditions (temperature, pH, and ionic strength). The authors determined that processing affected not only product's composition, appearance (alkaline conditions led to a dark-brown color of protein extracts) and microbial load, but also the protein and techno-functional properties of the isolated insect products (Bußler et al., 2016).

Through a different approach, Mishyna et al. carried out analysis of the techno-functional properties of insect's proteins (Mishyna, Martinez, Chen, & Benjamin, 2019). After the defatting and alkaline extraction of edible grasshopper (*S. gregaria*) and honey bee brood (*A. mellifera*) flours, the application of ultrasound-assisted extractions showed that protein content increased up to 57.5% and 55.2%, respectively. An improvement of foaming and emulsifying abilities was also reported. Likewise, using enzymatic hydrolysis to extract proteins from migratory locust (*Locusta migratoria* L.) flour, Purschke et al. showed an improvement in: (i) protein solubility (up to 55%, at pH 9), (ii) hydrolysates' emulsifying activity (54%, at pH 7), (iii) foaming ability (326%, at pH 3) and (iv) an advanced oil binding capacity. Through this study, the authors support and endorse the application of combined enzymatic hydrolysis to produce customized insect-based ingredients for food applications (Purschke, Brüggen et al., 2018; Purschke, Meinlschmidt et al., 2018; Purschke, Tanzmeister et al., 2018).

An example of the validation of such products is presented by Adam Mariod (2013), after preparing an ice cream using an insect's gelatine as a stabilizing agent. This product was further evaluated by a panel as acceptable, having no significant differences by the general preferences, when compared with an ice cream made with the same proportion of commercial gelatin (Mariod, 2013). According to Mariod et al. (2011), insect's gelatin was extracted from two defatted, dried, and ground Sudanese beetles *Aspongubus viduatus* and *Agonoscelis pubescens*, by an alkaline/acid pre-treatment, followed by hot water extraction. Besides promoting the removal of non-collagenous proteins, alkaline pre-treatment gave the best gelatin recovery yields, around 3.0% for both species (Mariod et al., 2011).

However, despite the recent advances in edible insects' processing techniques and the promising results exposed above, scale-up is still compromised. Additionally, in temperate countries, insect products keep facing tremendous challenges, namely consumer acceptance and legislative frameworks (Melgar-Lalanne et al., 2019; van Huis et al., 2013).

4.4.2 Dry protein extraction methods for insects

Dry fractionation (by milling and air-classification) of insects has been studied by Sipponen, Mäkinen, Rommi and Liisa (2018). Freeze-dried yellow mealworm larvae (*Tenebrio molitor*) and house crickets (*Acheta domesticus*) were defatted by $ScCO_2$ followed by milling and air classification. The initial protein content was 48.6% and 39.6% dm for the crickets and mealworm larvae, respectively. After defatting and dry fractionation, fine fractions containing 57.7% and 49.4% dm protein were obtained. The coarse fraction (starch-rich) contained more chitin than the fine fractions (protein-rich) from both insects and can lead to better consumer acceptance. Moreover, results showed that dry fractionation influenced the texture (coarseness), but its impact on flavor aspects was not evident.

Purschke et al. also reported dry fractionation (by milling and sieving), who successfully obtained protein-enriched fractions (up to 5.4% dry basis) of mealworm larvae (*T. molitor*), to a protein content of 58% (0−500 μm, for samples submitted to fluidized bed drying) and 66% (355−500 μm, for supercritical carbon dioxide defatted samples). Besides, physicochemical characteristics of the dried larvae were observed to be greatly affected by pre-processing techniques. After completing this study, the authors also concluded that dry fractionation could be improved, using alternative fractionation techniques such as air classification or triboelectric separation, respectively, to enhance protein concentration and chitin depletion (Purschke, Stegmann, Schreiner, & Jäger, 2017).

4.5 Disruptive application in the food industry: use of plant protein-rich ingredients to produce meat alternatives

Plant-based substitutes contribute to a more sustainable future food supply in the replacement of animal protein (Pelgrom, Berghout et al., 2014; Pelgrom, Boom et al., 2014). Consequently, the search for sustainable protein alternatives to partially substitute animal protein became inevitable (Palanisamy, Franke, Berger, Heinz, & Töpfl, 2019; Palanisamy, Töpfl, Berger, & Hertel, 2019).

Meat analogs are a food category possessing meat-like characteristics without including animal-derived protein. To structure plant proteins, different techniques exist, being extrusion the most commonly used, which consist of a continuous, thermomechanical process (Dekkers, Boom, & Van Der Goot, 2018; Philipp, Oey, Silcock, Beck, & Buckow, 2017). Basically, powdered protein-rich material and water are fed into the extruder barrel, hydrated, transported, and intermeshed by one or two co-rotating screws. Increasing temperature, pressure, and shear during extrusion processing create high mechanical and thermal energy, leading to endothermic physicochemical transformations of the constituents (Caporgno et al., 2020). Among extrusion, it is also possible to separate two classes of structuring according to the feed moisture content. It is called low-moisture extrusion (LME) when the feed moisture content is lower than 40% and high-moisture extrusion (HME) at feed moisture content higher than 40% (Akdogan, 1999). During LME, the protein concentrate is transformed in dried expanded products called texturized vegetable protein (TVP), which needs to be re-moisturized before cooking and consumption (Riaz, 2011). More interesting is the application of HME, through which it is possible to obtain non-expanded fibrous products to mimic the texture and mouthfeel of meat-products (Hood-Niefer, 2017). Fig. 4.3 illustrates both processes highlighting the main differences. Up to now, soybean has been the most common protein source used and technically excellent source, to produce meat analogs through LME and HME. Recently, issues related to food safety, environment, and social implications have contributed to holding back the use of soybean ingredients (Samard, Gu, & Ryu, 2019). Therefore, the research of alternative and more sustainable resources, such as pulses, cereals, microalgae, and insects, are seriously considered. A pea protein isolate (Osen, Toelstede, Wild, Eisner, & Schweiggert-Weisz, 2014), a lupin protein concentrate and isolate (Palanisamy, Franke et al., 2019; Palanisamy, Töpfl et al., 2019) have been explored for the production of meat analogs by HME.

Regarding cereals, only wheat gluten has been reported (Pietsch, Emin, & Schuchmann, 2017; Pietsch, Werner, Karbstein, & Emin, 2019; Samard et al., 2019). LME to produce meat analogs has been investigated only using wheat gluten (Samard et al., 2019). The proteins from the larvae *Alphitobius*

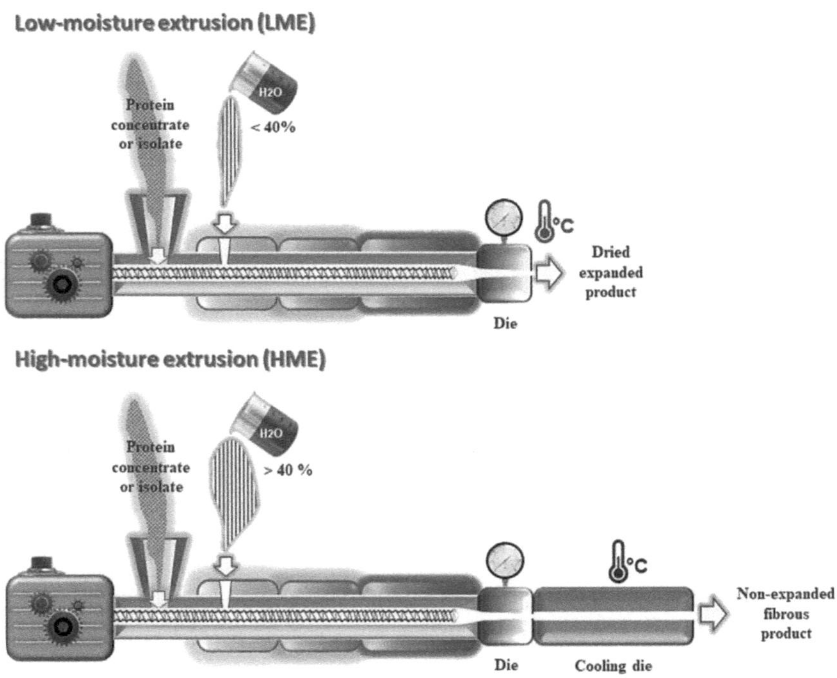

FIGURE 4.3 Low-moisture extrusion and high-moisture extrusion processes to produce meat analogues.

diaperinus and *Tenebrio molitor* were studied to produce meat analogs in combination with a soybean protein concentrate (Smetana, Pernutz, Toepfl, Heinz, & van Campenhout, 2018). Moreover, the microalgae *Auxenochlorela protothecoides* was also combined with soy protein concentrate to produce a meat analog by HME (Caporgno et al., 2020). *Spirulina platensis* and a lupin protein isolate and concentrate were also used and processed by HME (Palanisamy, Franke et al., 2019; Palanisamy, Töpfl et al., 2019). Table 4.6 summarizes the meat analog products obtained by HME and LME using pulses, cereals, insects, and microalgae protein rich-ingredients and the main outcomes related to their characterization in terms of protein quality, the content of bioactive compounds, texture properties, and functional analysis.

4.6 Conclusion and future perspectives

The global protein requirement is likely to grow in the next years due to the rising prosperity and world population. In this field, the market of alternative proteins is increasing due to the consumer demand for healthier and more sustainable products in addition to ethical considerations. Plant-based

TABLE 4.6 The use of insects, pulses, cereals and microalgae protein concentrates and isolates in the production of meat analogues.

Raw material	Processing technique	Main outcomes	References
Pulses			
Pea protein isolates	HME	• PPIs were texturized at a moisture content of 55% leading to a highly fibrous structure • The texture properties were dependent on the cooking temperature • RVA viscosity analysis of protein slurries could simulate the extrusion process • PPIs with larger particle size were simpler to process than PPIs with fine particle size • Fiber formation was dependent on the cooking temperature	(Osen et al., 2014)

Continued

TABLE 4.6 The use of insects, pulses, cereals and microalgae protein concentrates and isolates in the production of meat analogues.—cont'd

Raw material	Processing technique	Main outcomes	References
Lupin protein concentrate and isolate (50:50)	HME	• Water feed rate has a significant effect on the extruder responses • Higher number of fibrous layers and a denser microstructure were obtained at higher screw speeds and temperatures and lower water feed rate • Increased barrel temperature led to a decrease in the *in-vitro* protein digestibility (IVPD); increased water feed rate led to higher protein digestibility	(Palanisamy, Franke et al., 2019; Palanisamy, Töpfl et al., 2019)
Cereals			
Wheat gluten	HME	• The formation of a fibrous texture (anisotropic structure) correlates with the increase in gluten polymerization and with an increase in the hardness and Young's modulus • Wheat gluten polymerization was dependent on the thermal treatment (temperature) during extrusion	(Pietsch et al., 2019)
Wheat gluten	HME	• Pressure and specific mechanical energy (SME) had no impact on the polymerization behavior of wheat gluten • Extrusion temperature had a significant impact on the polymerization of wheat gluten (formation of different anisotropic structures)	(Pietsch et al., 2017)
Soy protein isolate (SPI),	LME and HME	• The formula containing gluten contained: SPI, WG and	(Samard et al., 2019)

TABLE 4.6 The use of insects, pulses, cereals and microalgae protein concentrates and isolates in the production of meat analogues.—cont'd

Raw material	Processing technique	Main outcomes	References
wheat gluten (WG) and corn starch (CS)		CS at the ratio of 50:40:10; moisture content was 64%; protein content was 84.6% • This formula was compared with another mixture without addition of gluten • LME without addition of gluten led to an expanded product with lower integrity and texture stability than the mixture containing gluten • LME of the mixture containing gluten led to the formation of a spongy structure • HME of the mixture containing gluten led to a distinctive meat-like texture, fibrous and dense	
Microalgae			
Auxenochlorela protothecoides and SPC	HME	• The fibrous structure formed was influenced by the mixture of SPC and microalgae protein possibly due to the greater fat content and rigid cell wall of the microalgae • Low water content promoted the formation a fibrous structure • Promising texture properties were achieved at a microalgae incorporation of 30% and a moisture content of 60% • The visual appearance was pleasant • The nutritional profile (vitamins B and E) of the meat analogues were improved by the microalgae incorporation and around 95% was retained after extrusion	(Caporgno et al., 2020)

Continued

TABLE 4.6 The use of insects, pulses, cereals and microalgae protein concentrates and isolates in the production of meat analogues.—cont'd

Raw material	Processing technique	Main outcomes	References
Spirulina platensis (15%–50%) and lupin protein isolate and concentrate	HME	• The addition of spirulina increased the TPC, TFC and the TEAC of the meat analogues • The addition of 30% spirulina decreased the IVPD of the meat analogues • Compared to the raw extrusion mixtures, the TPC, TFC, TEAC, and IVPD were increased by the extrusion processing • As the amount of spirulina was increased, the formation of a fibrous structure decreased	(Palanisamy, Franke et al., 2019; Palanisamy, Töpfl et al., 2019)
Insects			
40% Alphitobius diaperinus proteins and *Tenebrio molitor and 60% SPC dm*	HME	• Fibrous meat analogues were obtained, presenting hardness texture and protein composition comparable to meat • Best results were obtained for the mixture 40% *Alphitobius diaperinus* and 60% SPC dm • Texture improvement was obtained with addition of 5%–10% soy fiber	(Smetana et al., 2018)

HME, High-moisture extrusion; *IVPD*, in-vitro protein digestibility; *LME*, Low-moisture extrusion; *PPIs*, Pea protein isolates; *SPC*, Soy-protein concentrate; *SPI*, Soy protein isolate; *TEAC*, Trolox equivalent antioxidant activity; *TFC*, Total flavonoid content; *TPC*, Total phenolic content.

products, such as pulses and cereals, are candidate sources of high quantity and quality proteins that can be widely applied in the protein market. Insects and microalgae are also recognized to present a high nutritional value and high protein content. Still, they face some challenges regarding safety concerns, consumers' acceptance, and low protein extraction efficiency. Among all technologies, wet fractionation (conventional) methods are recognized to supply the highest protein content products, while dry fractionation (non-chemical) methods allow the development of protein concentrates with high functionality. Emerging technologies, such as dry fractionation (using air-classification or triboelectric separation) and supercritical fluid technology, are expected to expand in the protein industry. The further development and improvement of meat analogs are actuality one of the main challenges for the food industry and, despite the use of alternative proteins is already being investigated, future studies are needed to evaluate their acceptability and functional/sensorial properties.

Acknowledgments

The authors are thankful to António Gonçalves for the support of the artwork of this chapter. The authors acknowledge the financial support received from i) Norwegian Research Council through *FoodProFuture* Project "Innovative and Sustainable Exploitation of Plant Proteins in Future Foods" Nr 267858 and ii) EEA Grants 2014–2021 through the Bilateral Fund. The EEA Grants represent the contribution of Iceland, Liechtenstein, and Norway toward a green, competitive and inclusive Europe. There are two overall objectives: reduction of economic and social disparities in Europe, and to strengthen bilateral relations between the donor countries and 15 EU countries in Central and Southern Europe and the Baltics. The three donor countries cooperate closely with the EU through the Agreement on the European Economic Area (EEA). The donors have provided €3.3 billion through consecutive grant schemes between 1994 and 2014. For the period 2014–2021, the EEA Grants amount to €1.55 billion.

References

Afam-Mbah, C., Ketabi, L., Emami, S., Tabil, L., & Tyler, R. T. (2018). *Air classification of by-product of lentil dehulling into protein-rich and starch-rich fractions* (pp. 2–14). https://doi.org/10.13031/aim.201801275.

Akaberi, S., Gusbeth, C., Silve, A., Senthilnathan, D. S., Navarro-López, E., Molina-Grima, E., et al. (2019). Effect of pulsed electric field treatment on enzymatic hydrolysis of proteins of *Scenedesmus almeriensis*. *Algal Research, 43*, 101656. https://doi.org/10.1016/j.algal.2019.101656.

Akdogan, H. (1999). High moisture food extrusion. *International Journal of Food Science and Technology, 34*, 195–207. https://doi.org/10.1046/j.1365-2621.1999.00256.x.

Alzuwaid, N. T., Sissons, M., Laddomada, B., & Fellows, C. M. (2020). Nutritional and functional properties of durum wheat bran protein concentrate. *Cereal Chemistry, 97*, 304–315. https://doi.org/10.1002/cche.10246.

Arntfield, S. D., & Maskus, H. D. (2011). Peas and other legume proteins. In *Handbook of food proteins* (pp. 233–266). https://doi.org/10.1016/B978-1-84569-758-7.50009-5.

Barbarino, E., & Lourenço, S. O. (2005). An evaluation of methods for extraction and quantification of protein from marine macro- and microalgae. *Journal of Applied Phycology, 17*, 447–460. https://doi.org/10.1007/s10811-005-1641-4.

Barka, A., & Blecker, C. (2016). Microalgae as a potential source of single-cell proteins. A review. *Biotechnology, Agronomy and Society and Environment, 20*, 427–436. https://doi.org/10.25518/1780-4507.13132.

Barkia, I., Saari, N., & Manning, S. R. (2019). Microalgae for high-value products towards human health and nutrition. *Marine Drugs, 17*, 1–29. https://doi.org/10.3390/md17050304.

Becker, E. W. (2007). Micro-algae as a source of protein. *Biotechnology Advances, 25*, 207–210. https://doi.org/10.1016/j.biotechadv.2006.11.002.

Bessada, S. M. F., Barreira, J. C. M., & Oliveira, M. B. P. P. (2019). Pulses and food security: Dietary protein, digestibility, bioactive and functional properties. *Trends in Food Science and Technology, 93*, 53–68. https://doi.org/10.1016/j.tifs.2019.08.022.

Bleakley, S., & Hayes, M. (2017). Algal proteins: Extraction, application, and challenges concerning production. *Foods, 6*, 33. https://doi.org/10.3390/foods6050033.

Boye, J. I., Aksay, S., Roufik, S., Ribéreau, S., Mondor, M., Farnworth, E., et al. (2010). Comparison of the functional properties of pea, chickpea and lentil protein concentrates processed using ultrafiltration and isoelectric precipitation techniques. *Food Research International, 43*, 537–546. https://doi.org/10.1016/j.foodres.2009.07.021.

Boye, J., Zare, F., & Pletch, A. (2010). Pulse proteins: Processing, characterization, functional properties and applications in food and feed. *Food Research International, 43*, 414–431. https://doi.org/10.1016/j.foodres.2009.09.003.

Brouns, F., Hemery, Y., Price, R., & Anson, N. M. (2012). Wheat aleurone: Separation, composition, health aspects. *Critical Reviews in Food Science and Nutrition*. https://doi.org/10.1080/10408398.2011.589540.

Brückner-Gühmann, M., Vasil'eva, E., Culetu, A., Duta, D., Sozer, N., & Drusch, S. (2019). Oat protein concentrate as alternative ingredient for non-dairy yoghurt-type product. *Journal of the Science of Food and Agriculture, 99*, 5852–5857. https://doi.org/10.1002/jsfa.9858.

Bukken, S. G. F. (2005). Insects in the human diet: Nutritional aspects. *Ecological Implications of minilivestock*, 545–577.

Bußler, S., Rumpold, B. A., Jander, E., Rawel, H. M., & Schlüter, O. K. (2016). Recovery and techno-functionality of flours and proteins from two edible insect species: Meal worm (*Tenebrio molitor*) and black soldier fly (*Hermetia illucens*) larvae. *Heliyon, 2*. https://doi.org/10.1016/j.heliyon.2016.e00218. e00218–e00218.

Bußler, S., Steins, V., Ehlbeck, J., & Schlüter, O. (2015). Impact of thermal treatment versus cold atmospheric plasma processing on the techno-functional protein properties from *Pisum sativum* "Salamanca.". *Journal of Food Engineering, 167*, 166–174. https://doi.org/10.1016/j.jfoodeng.2015.05.036.

Caparros Megido, R., Sablon, L., Geuens, M., Brostaux, Y., Alabi, T., Blecker, C., et al. (2014). Edible insects acceptance by belgian consumers: Promising attitude for entomophagy development. *Journal of Sensory Studies, 29*, 14–20. https://doi.org/10.1111/joss.12077.

Caporgno, M. P., Böcker, L., Müssner, C., Stirnemann, E., Haberkorn, I., Adelmann, H., et al. (2020). Extruded meat analogues based on yellow, heterotrophically cultivated *Auxenochlorella* protothecoides microalgae. *Innovative Food Science and Emerging Technologies, 59*, 102275. https://doi.org/10.1016/j.ifset.2019.102275.

do Carmo, C. S., Serra, A. T., & Duarte, C. M. M. (2017). Recovery technologies for lipophilic bioactives. In Y. H. Roos, & Y. D. Livney (Eds.), *Eng. foods bioact. stab. deliv.* (pp. 1–49). New York, NY: Springer New York. https://doi.org/10.1007/978-1-4939-6595-3_1.

de Castro, R. J. S., Ohara, A., Dos Santos Aguilar, J. G., & Domingues, M. A. F. (2018). Nutritional, functional and biological properties of insect proteins: Processes for obtaining, consumption and future challenges. *Trends in Food Science & Technology, 76*, 82–89. https://doi.org/10.1016/j.tifs.2018.04.006.

Chiong, T., Acquah, C., Lau, S. Y., Khor, E. H., & Danquah, M. K. (2016). Microalgal-based protein by-products: Extraction, purification, and applications. In *Protein byprod transform from environ burd into value-added prod* (pp. 213–234). https://doi.org/10.1016/B978-0-12-802391-4.00012-4.

Contreras, M. del M., Lama-Muñoz, A., Manuel Gutiérrez-Pérez, J., Espínola, F., Moya, M., & Castro, E. (2019). Protein extraction from agri-food residues for integration in biorefinery: Potential techniques and current status. *Bioresource Technology, 280*, 459–477. https://doi.org/10.1016/j.biortech.2019.02.040.

Dapčević-Hadnadev, T., Hadnadev, M., Pojić, M., & Balandrán-Quintana, R. R. (2018). *The healthy components of cereal by-products and their functional properties*. https://doi.org/10.1016/B978-0-08-102162-0.00005-8.

Day, L. (2013). Proteins from land plants - potential resources for human nutrition and food security. *Trends in Food Science and Technology, 32*, 25–42. https://doi.org/10.1016/j.tifs.2013.05.005.

Day, L. (2016). Protein: Food sources. In *Encyclopedia of healing foods* (pp. 530–537). https://doi.org/10.1016/B978-0-12-384947-2.00576-6.

Dekkers, B. L., Boom, R. M., & Van Der Goot, A. J. (2018). Structuring processes for meat analogues. *Trends in Food Science and Technology, 81*, 25–36. https://doi.org/10.1016/j.tifs.2018.08.011.

Dhillon, G. S. (2016). *Protein by products - transformation from environmental burden into value-added products*. Nikki Levy. https://doi.org/10.1016/B978-0-12-802391-4/00018-5.

Doucha, J., & Lívanský, K. (2008). Influence of processing parameters on disintegration of chlorella cells in various types of homogenizers. *Applied Microbiology and Biotechnology, 81*, 431–440. https://doi.org/10.1007/s00253-008-1660-6.

Elkowicz, K., & Sosulski, F. W. (1982). Antinutritive factors in eleven legumes and their air-classified protein and starch fractions. *Journal of Food Science, 47*, 1301–1304. https://doi.org/10.1111/j.1365-2621.1982.tb07673.x.

Fasolin, L. H., Pereira, R. N., Pinheiro, A. C., Martins, J. T., Andrade, C. C. P., Ramos, O. L., et al. (2019). Emergent food proteins – towards sustainability, health and innovation. *Food Research International, 125*, 108586. https://doi.org/10.1016/j.foodres.2019.108586.

Flambeau, M., Redl, A., & Respondek, F. (2017). Proteins from wheat: Sustainable production and new developments in nutrition-based and functional applications. *Sustainable Protein Sources*, 67–78.

Gerzhova, A., Mondor, M., Benali, M., & Aider, M. (2015). A comparative study between the electro-activation technique and conventional extraction method on the extractability, composition and physicochemical properties of canola protein concentrates and isolates. *Food Bioscience, 11*, 56–71. https://doi.org/10.1016/j.fbio.2015.04.005.

Gómez, B., Munekata, P. E. S., Zhu, Z., Barba, F. J., Toldrá, F., Putnik, P., et al. (2019). Challenges and opportunities regarding the use of alternative protein sources: aquaculture and insects. *Advances in Food and Nutrition Research, 89*, 259–295. https://doi.org/10.1016/bs.afnr.2019.03.003.

González López, C. V., García, M. del CC., Fernández, F. G. A., Bustos, C. S., Chisti, Y., & Sevilla, J. M. F. (2010). Protein measurements of microalgal and cyanobacterial biomass. *Bioresource Technology, 101*, 7587–7591. https://doi.org/10.1016/j.biortech.2010.04.077.

Gravel, A., & Doyen, A. (2020). The use of edible insect proteins in food: Challenges and issues related to their functional properties. *Innovative Food Science and Emerging Technologies, 59*, 102272. https://doi.org/10.1016/j.ifset.2019.102272.

Gresiana, F., Muzi Marpaung, A., & Sutanto, H. (2015). Protein isolation from cricket (Gryllus-mitratus). *Proceedings of International Conference on Innovation, Enterprenuership and Technology*, 214–221.

Heusala, H., Sinkko, T., Sözer, N., Hytönen, E., Mogensen, L., & Knudsen, M. T. (2020). Carbon footprint and land use of oat and faba bean protein concentrates using a life cycle assessment approach. *Journal of Cleaner Production, 242*, 1–9. https://doi.org/10.1016/j.jclepro.2019.118376.

Holtekjølen, A. K., Uhlen, A. K., & Knutsen, S. H. (2008). Barley carbohydrate composition varies with genetic and abiotic factors. *Acta Agriculturae Scandinavica Section B Soil and Plant Science, 58*, 27–34. https://doi.org/10.1080/09064710601160037.

Hood-Niefer, S. (2017). Insights into extrusion of protein. *Cereal Foods World, 62*, 148–150. https://doi.org/10.1094/CFW-62-4-0148.

Houde, M., Khodaei, N., Benkerroum, N., & Karboune, S. (2018). Barley protein concentrates: Extraction, structural and functional properties. *Food Chemistry, 254*, 367–376. https://doi.org/10.1016/j.foodchem.2018.01.156.

van Huis, A. (2015). Edible insects contributing to food security? *Agriculture and Food Security, 4*. https://doi.org/10.1186/s40066-015-0041-5.

van Huis, A., Van Itterbeeck, J., Klunder, H., Mertens, E., Halloran, A., Muir, G., et al. (2013). *Edible insects: Future prospects for food and feed security.*

Imathiu, S. (2020). Benefits and food safety concerns associated with consumption of edible insects. *NFS Journal, 18*, 1–11. https://doi.org/10.1016/j.nfs.2019.11.002.

Jantzen da Silva Lucas, A., Menegon de Oliveira, L., da Rocha, M., & Prentice, C. (2020). Edible insects: An alternative of nutritional, functional and bioactive compounds. *Food Chemistry, 311*, 126022. https://doi.org/10.1016/j.foodchem.2019.126022.

Kim, T. K., Yong, H. I., Chun, H. H., Lee, M. A., Kim, Y. B., & Choi, Y. S. (2020). Changes of amino acid composition and protein technical functionality of edible insects by extracting steps. *Journal of Asia-Pacific Entomology, 23*, 298–305. https://doi.org/10.1016/j.aspen.2019.12.017.

Kim, T. K., Yong, H. I., Kim, Y. B., Kim, H. W., & Choi, Y. S. (2019). Edible insects as a protein source: A review of public perception, processing technology, and research trends. *Food Science of Animal Resources, 39*, 521–540. https://doi.org/10.5851/kosfa.2019.e53.

Klamczynska, B., & Mooney, W. D. (2017). Heterotrophic microalgae: A scalable and sustainable protein source. *Sustainable Protein Sources*, 327–339.

Kong, W., Baeyens, J., Qin, P., Zhang, H., & Tan, T. (2018). Towards an energy-friendly and cleaner solvent-extraction of vegetable oil. *Journal of Environmental Management, 217*, 196–206. https://doi.org/10.1016/J.JENVMAN.2018.03.061.

Krimpen, MM van, Bikker, P., Meer, IM van der, Peet-Schwering, CMC van der, & Vereijken, J. M. (2013). *Cultivation, processing and nutritional aspects for pigs and poultry of European protein sources as alternatives for imported soybean products* (Vol. 361). LR - Diervoeding: Wageningen UR Livestock Research.

Kulkarni, S., & Nikolov, Z. (2018). Process for selective extraction of pigments and functional proteins from *Chlorella vulgaris*. *Algal Research, 35*, 185–193. https://doi.org/10.1016/j.algal.2018.08.024.

Larsen, B. B., Miller, E. C., Rhodes, M. K., & Wiens, J. J. (2017). Inordinate fondness multiplied and redistributed: The number of species on earth and the new pie of life. *The Quarterly Review of Biology, 92*, 229–265. https://doi.org/10.1086/693564.

Lásztity, R. (1998). Oat grain—a wonderful reservoir of natural nutrients and biologically active substances. *Food Reviews International, 14*, 99–119. https://doi.org/10.1080/87559129809541150.

Leitner, W., & Poliakoff, M. (2008). Supercritical fluids in green chemistry. *Green Chemistry, 10*, 730–735.

Liestianty, D., Rodianawati, I., Arfah, R. A., Assa, A., Patimah, S., et al. (2019). Nutritional analysis of spirulina sp to promote as superfood candidate. In , *509*. *IOP Conference Series: Materials Science and Engineering*. https://doi.org/10.1088/1757-899X/509/1/012031.

Liu, K. (2009). Some factors affecting sieving performance and efficiency. *Powder Technology, 193*, 208–213.

Liu, K. (2014). Fractionation of oats into products enriched with protein, beta-glucan, starch, or other carbohydrates. *Journal of Cereal Science, 60*, 317–322. https://doi.org/10.1016/j.jcs.2014.06.002.

Liu, S., Li, Y., Obadi, M., Jiang, Y., Chen, Z., Jiang, S., et al. (2019). Effect of steaming and defatting treatments of oats on the processing and eating quality of noodles with a high oat flour content. *Journal of Cereal Science, 89*, 102794. https://doi.org/10.1016/j.jcs.2019.102794.

Loveday, S. M. (2019). Food proteins: Technological, nutritional, and sustainability attributes of traditional and emerging proteins. *Annual Review of Food Science and Technology, 10*, 311–339. https://doi.org/10.1146/annurev-food-032818-121128.

Ma, C. Y. (1983). Preparation, composition and functional properties of oat protein isolates. *Canadian Institute of Food Science and Technology Journal, 16*, 201–205. https://doi.org/10.1016/S0315-5463(83)72208-X.

Mäkinen, O. E., Sozer, N., Ercili-Cura, D., & Poutanen, K. (2017). Protein from oat: Structure, processes, functionality, and nutrition. *Sustainable Protein Sources, 105*. https://doi.org/10.1016/B978-0-12-802778-3.00006-8.

Mariod, A. A. (2013). Insect oil and protein: Biochemistry, food and other uses: Review. *Agricultural Sciences, 4*.

Mariod, A. A., Abdelwahab, S. I., Ibrahim, M. Y., Mohan, S., Elgadir, M. A., & Ain, N. M. (2011). Preparation and characterization of gelatins from two sudanese edible insects. *Journal of Food Science and Engineering, 1*, 45.

Mccurdy, S. M., & Knipfel, J. E. (1990). Investigation of faba bean protein recovery and application to pilot scale processing. *Journal of Food Science, 55*, 1093–1094. https://doi.org/10.1111/j.1365-2621.1990.tb01606.x.

Melgar-Lalanne, G., Hernández-Álvarez, A. J., & Salinas-Castro, A. (2019). Edible insects processing: Traditional and innovative technologies. *Comprehensive Reviews in Food Science and Food Safety*. https://doi.org/10.1111/1541-4337.12463.

Mishyna, M., Martinez, J.-J. I., Chen, J., & Benjamin, O. (2019). Extraction, characterization and functional properties of soluble proteins from edible grasshopper (*Schistocerca gregaria*) and honey bee (*Apis mellifera*). *Food Research International, 116*, 697–706. https://doi.org/10.1016/j.foodres.2018.08.098.

Ochoa-Rivas, A., Nava-Valdez, Y., Serna-Saldívar, S. O., & Chuck-Hernández, C. (2017). Microwave and ultrasound to enhance protein extraction from peanut flour under alkaline conditions: Effects in yield and functional properties of protein isolates. *Food and Bioprocess Technology, 10*, 543–555. https://doi.org/10.1007/s11947-016-1838-3.

Osborne, T. B. (1924). *The vegetable proteins*. Longmans, Green and Company.

Osen, R., Toelstede, S., Wild, F., Eisner, P., & Schweiggert-Weisz, U. (2014). High moisture extrusion cooking of pea protein isolates: Raw material characteristics, extruder responses, and texture properties. *Journal of Food Engineering, 127*, 67–74. https://doi.org/10.1016/j.jfoodeng.2013.11.023.

Palanisamy, M., Franke, K., Berger, R. G., Heinz, V., & Töpfl, S. (2019a). High moisture extrusion of lupin protein: Influence of extrusion parameters on extruder responses and product properties. *Journal of the Science of Food and Agriculture, 99*, 2175–2185. https://doi.org/10.1002/jsfa.9410.

Palanisamy, M., Töpfl, S., Berger, R. G., & Hertel, C. (2019). Physico-chemical and nutritional properties of meat analogues based on spirulina/lupin protein mixtures. *European Food Research and Technology, 245*, 1889–1898. https://doi.org/10.1007/s00217-019-03298-w.

Pelgrom, P. J. M., Berghout, J. A. M., van der Goot, A. J., Boom, R. M., & Schutyser, M. A. I. (2014). Preparation of functional lupine protein fractions by dry separation. *Lebensmittel-Wissenschaft und -Technologie- Food Science and Technology, 59*, 680–688. https://doi.org/10.1016/j.lwt.2014.06.007.

Pelgrom, P. J. M., Boom, R. M., & Schutyser, M. A. I. (2014). Functional analysis of mildly refined fractions from yellow pea. *Food Hydrocolloids, 44*, 12–22. https://doi.org/10.1016/j.foodhyd.2014.09.001.

Pelgrom, P. J. M., Boom, R. M., & Schutyser, M. A. I. (2015). Method development to increase protein enrichment during dry fractionation of starch-rich legumes. *Food and Bioprocess Technology, 8*, 1495–1502. https://doi.org/10.1007/s11947-015-1513-0.

Pelgrom, P. J. M., Schutyser, M. A. I., & Boom, R. M. (2013). Thermomechanical morphology of peas and its relation to fracture behaviour. *Food and Bioprocess Technology, 6*, 3317–3325. https://doi.org/10.1007/s11947-012-1031-2.

Pelgrom, P. J. M., Vissers, A. M., Boom, R. M., & Schutyser, M. A. I. (2013b). Dry fractionation for production of functional pea protein concentrates. *Food Research International, 53*, 232–239. https://doi.org/10.1016/j.foodres.2013.05.004.

Pelgrom, P. J. M., Wang, J., Boom, R. M., & Schutyser, M. A. I. (2015). Pre- and post-treatment enhance the protein enrichment from milling and air classification of legumes. *Journal of Food Engineering, 155*, 53–61. https://doi.org/10.1016/j.jfoodeng.2015.01.005.

Petterson, D. M. (2011). Storage proteins. In St Paul (Ed.), *Oats chemistry and technology*.

Philipp, C., Oey, I., Silcock, P., Beck, S. M., & Buckow, R. (2017). Impact of protein content on physical and microstructural properties of extruded rice starch-pea protein snacks. *Journal of Food Engineering, 212*, 165–173. https://doi.org/10.1016/j.jfoodeng.2017.05.024.

Pietsch, V. L., Emin, M. A., & Schuchmann, H. P. (2017). Process conditions influencing wheat gluten polymerization during high moisture extrusion of meat analog products. *Journal of Food Engineering, 198*, 28–35. https://doi.org/10.1016/j.jfoodeng.2016.10.027.

Pietsch, V. L., Werner, R., Karbstein, H. P., & Emin, M. A. (2019). High moisture extrusion of wheat gluten: Relationship between process parameters, protein polymerization, and final product characteristics. *Journal of Food Engineering, 259*, 3–11. https://doi.org/10.1016/j.jfoodeng.2019.04.006.

Poel, TFB Van Der, & Aarts, H. L. M. (1990). *Air classification of bean flour - Effects on protein, antinutritional factors and the effect of a fines fraction on cultured explants of small intestinal mucosa.*

Pojić, M., Mišan, A., & Tiwari, B. (2018). Eco-innovative technologies for extraction of proteins for human consumption from renewable protein sources of plant origin. *Trends in Food Science and Technology, 75*, 93–104. https://doi.org/10.1016/j.tifs.2018.03.010.

Pulz, O., & Gross, W. (2004). Valuable products from biotechnology of microalgae. *Applied Microbiology and Biotechnology, 65*, 635–648. https://doi.org/10.1007/s00253-004-1647-x.

Purschke, B., Brüggen, H., Scheibelberger, R., & Jäger, H. (2018). Effect of pre - treatment and drying method on physico - chemical properties and dry fractionation behaviour of mealworm larvae (*Tenebrio molitor* L .). *European Food Research and Technology, 244*, 269–280. https://doi.org/10.1007/s00217-017-2953-8.

Purschke, B., Meinlschmidt, P., Horn, C., Rieder, O., & Jäger, H. (2018). Improvement of techno-functional properties of edible insect protein from migratory locust by enzymatic hydrolysis. *European Food Research and Technology, 244*, 999–1013. https://doi.org/10.1007/s00217-017-3017-9.

Purschke, B., Stegmann, T., Schreiner, M., & Jäger, H. (2017). Pilot-scale supercritical CO_2 extraction of edible insect oil from *Tenebrio molitor* L. larvae – influence of extraction conditions on kinetics, defatting performance and compositional properties. *European Journal of Lipid Science and Technology, 119*, 1600134. https://doi.org/10.1002/ejlt.201600134.

Purschke, B., Tanzmeister, H., Meinlschmidt, P., Baumgartner, S., Lauter, K., & Jäger, H. (2018). Recovery of soluble proteins from migratory locust (*Locusta migratoria*) and characterisation of their compositional and techno-functional properties. *Food Research International, 106*, 271–279. https://doi.org/10.1016/j.foodres.2017.12.067.

Ramachandran, S., Singh, S. K., Larroche, C., Soccol, C. R., & Pandey, A. (2007). Oil cakes and their biotechnological applications – a review. *Bioresource Technology, 98*, 2000–2009. https://doi.org/10.1016/j.biortech.2006.08.002.

Ranhotra, G. S., Gelroth, J. A., Glaser, B. K., & Reddy, P. V. (1994). Nutritional profile of a fraction from air-classified bran obtained from a hard red wheat. *Cereal Chemistry, 71*, 321–324.

Rebello, C. J., Greenway, F. L., & Finley, J. W. (2014). Whole grains and pulses: A comparison of the nutritional and health benefits. *Journal of Agricultural and Food Chemistry, 62*, 7029–7049. https://doi.org/10.1021/jf500932z.

Regulation (EU) 2015/2283. (2015). Regulation (Eu) 2015/2283 on novel foods. *Official Journal of the European Union, 327*, 1–22.

Regulation (EU) 2017/893. (2017). Commission regulation (EU) 2017/1151. *Official Journal of the European Union 2017*, 92–116.

Rempel, C., Geng, X., & Zhang, Y. (2019). Industrial scale preparation of pea flour fractions with enhanced nutritive composition by dry fractionation. *Food Chemistry, 276*, 119–128. https://doi.org/10.1016/j.foodchem.2018.10.003.

Riaz, M. N. (2011). Texturized vegetable proteins. In *Handbook of Food Proteins* (pp. 395–418). https://doi.org/10.1016/B978-1-84569-758-7.50015-0.

Romero García, J. M., Acién Fernández, F. G., & Fernández Sevilla, J. M. (2012). Development of a process for the production of l-amino-acids concentrates from microalgae by enzymatic hydrolysis. *Bioresource Technology, 112*, 164–170. https://doi.org/10.1016/j.biortech.2012.02.094.

Roselló-Soto, E., Barba, F. J., Parniakov, O., Galanakis, C. M., Lebovka, N., Grimi, N., et al. (2015). High voltage electrical discharges, pulsed electric field, and ultrasound assisted extraction of protein and phenolic compounds from olive kernel. *Food and Bioprocess Technology, 8*, 885–894. https://doi.org/10.1007/s11947-014-1456-x.

Roth, M., Jekle, M., & Becker, T. (2019). Opportunities for upcycling cereal byproducts with special focus on Distiller's grains. *Trends in Food Science and Technology, 91*, 282–293. https://doi.org/10.1016/j.tifs.2019.07.041.

Roy, F., Boye, J. I., & Simpson, B. K. (2010). Bioactive proteins and peptides in pulse crops: Pea, chickpea and lentil. *Food Research International, 43*, 432–442. https://doi.org/10.1016/j.foodres.2009.09.002.

Rumpold, B. A., & Schlüter, O. K. (2013). Nutritional composition and safety aspects of edible insects. *Molecular Nutrition and Food Research, 57*, 802–823. https://doi.org/10.1002/mnfr.201200735.

Saldanha do Carmo, C., Silventoinen, P., Nordgård, C. T., Dessev, T., Zobel, H., Holtekjølen, A., et al. (2020). Is dehulling of peas and faba beans necessary prior to dry fractionation for the production of protein- and starch-rich fractions? Impact on physical properties, chemical composition and techno-functional properties. *Journal of Food Engineering*, 109937. https://doi.org/10.1016/j.jfoodeng.2020.109937.

Samard, S., Gu, B. Y., & Ryu, G. H. (2019). Effects of extrusion types, screw speed and addition of wheat gluten on physicochemical characteristics and cooking stability of meat analogues. *Journal of the Science of Food and Agriculture, 99*, 4922–4931. https://doi.org/10.1002/jsfa.9722.

Sá, A. G. A., Moreno, Y. M. F., & Carciofi, B. A. M. (2020). Plant proteins as high-quality nutritional source for human diet. *Trends in Food Science and Technology, 97*, 170–184. https://doi.org/10.1016/j.tifs.2020.01.011.

Sarkar, A., & Kaul, P. (2014). Evaluation of tomato processing by-products: A comparative study in a pilot scale setup. *Journal of Food Process Engineering, 37*, 299–307. https://doi.org/10.1111/jfpe.12086.

Schutyser, M. A. I., Pelgrom, P. J. M., van der Goot, A. J., & Boom, R. M. (2015). Dry fractionation for sustainable production of functional legume protein concentrates. *Trends in Food Science and Technology, 45*, 327–335. https://doi.org/10.1016/j.tifs.2015.04.013.

Schutyser, M. A. I., & van der Goot, A. J. (2011). The potential of dry fractionation processes for sustainable plant protein production. *Trends in Food Science and Technology, 22*, 154–164. https://doi.org/10.1016/j.tifs.2010.11.006.

Schwenzfeier, A., Helbig, A., Wierenga, P. A., & Gruppen, H. (2013). Emulsion properties of algae soluble protein isolate from *Tetraselmis* sp. *Food Hydrocolloids, 30*, 258–263. https://doi.org/10.1016/j.foodhyd.2012.06.002.

Schwenzfeier, A., Wierenga, P. A., & Gruppen, H. (2011). Isolation and characterization of soluble protein from the green microalgae *Tetraselmis* sp. *Bioresource Technology, 102*, 9121–9127. https://doi.org/10.1016/j.biortech.2011.07.046.

Servaites, J. C., Faeth, J. L., & Sidhu, S. S. (2012). A dye binding method for measurement of total protein in microalgae. *Analytical Biochemistry, 421*, 75–80. https://doi.org/10.1016/j.ab.2011.10.047.

Sharma, P., & Gujral, H. S. (2013). Extrusion of hulled barley affecting β-glucan and properties of extrudates. *Food and Bioprocess Technology, 6*, 1374–1389. https://doi.org/10.1007/s11947-011-0777-2.

Show, K. Y., Lee, D. J., Tay, J. H., Lee, T. M., & Chang, J. S. (2015). Microalgal drying and cell disruption - recent advances. *Bioresource Technology, 184*, 258–266. https://doi.org/10.1016/j.biortech.2014.10.139.

Sibakov, J., Myllymäki, O., Holopainen, U., Kaukovirta-Norja, A., Hietaniemi, V., Pihlava, J. M., et al. (2011). Lipid removal enhances separation of oat grain cell wall material from starch and protein. *Journal of Cereal Science, 54*, 104–109. https://doi.org/10.1016/j.jcs.2011.04.003.

Silventoinen, P., Sipponen, M. H., Holopainen-Mantila, U., Poutanen, K., & Sozer, N. (2018). Use of air classification technology to produce protein-enriched barley ingredients. *Journal of Food Engineering, 222*, 169–177. https://doi.org/10.1016/j.jfoodeng.2017.11.016.

Singhal, A., Karaca, A. C., Tyler, R., & Nickerson, M. (2016). Pulse proteins: From processing to structure-function relationships. In A. K. Goyal (Ed.), *Grain legum*. Rijeka: IntechOpen. https://doi.org/10.5772/64020.

Singh, P., Kumar, R., Sabapathy, S. N., & Bawa, A. S. (2008). Functional and edible uses of soy protein products. *Comprehensive Reviews in Food Science and Food Safety, 7*, 14–28.

Sipponen, M. H., Mäkinen, O. E., Rommi, K., & Liisa, R. (2018). Biochemical and sensory characteristics of the cricket and mealworm fractions from supercritical carbon dioxide extraction and air classification. *European Food Research and Technology, 244*, 19–29. https://doi.org/10.1007/s00217-017-2931-1.

Slocombe, S. P., Ross, M., Thomas, N., McNeill, S., & Stanley, M. S. (2013). A rapid and general method for measurement of protein in micro-algal biomass. *Bioresource Technology, 129*, 51–57. https://doi.org/10.1016/j.biortech.2012.10.163.

Smetana, S., Pernutz, C., Toepfl, S., Heinz, V., & Campenhout, L Van (2018). High-moisture extrusion with insect and soy protein concentrates: Cutting properties. *Journal of Insects as Food and Feed, 1*, 1–6. https://doi.org/10.3920/JIFF2017.0066.

Sogari, G., Liu, A., & Li, J. (2019). Understanding edible insects as food in western and eastern societies. Environ. Heal. Bus. Oppor. New Meat Altern. Mark. *IGI Global Business Science Reference*, 166–181. https://doi.org/10.4018/978-1-5225-7350-0.ch009.

Sogbesan, A. O., & Ugwumba, A. A. A. (2008). Nutritional evaluation of termite (*Macrotermes subhyalinus*) meal as animal protein supplements in the diets of *Heterobranchus longifilis* (Valenciennes, 1840) fingerlings. *Turkish Journal of Fisheries and Aquatic Sciences, 157*, 149–157.

Sosulski, F. W., Elkowicz, L., & Reichert, R. D. (1982). Oligosaccharides in eleven legumes and their air-classified protein and starch fractions. *Journal of Food Science, 47*, 498–502. https://doi.org/10.1111/j.1365-2621.1982.tb10111.x.

Sosulski, F., & Youngs, C. G. (1979). Yield and functional properties of air-classified protein and starch fractions from eight legume flours. *Journal of the American Oil Chemists Society, 56*, 292–295. https://doi.org/10.1007/BF02671477.

Soto-Sierra, L., Stoykova, P., & Nikolov, Z. L. (2018). Extraction and fractionation of microalgae-based protein products. *Algal Research, 36*, 175–192. https://doi.org/10.1016/j.algal.2018.10.023.

Ssepuuya, G., Aringo, R. O., Mukisa, I. M., & Nakimbugwe, D. (2016). Effect of processing, packaging and storage-temperature based hurdles on the shelf stability of sautéed ready-to-eat *Ruspolia nitidula*. *Journal of Insects as Food and Feed, 2*, 245–253. https://doi.org/10.3920/JIFF2016.0006.

Stork, N. E. (2018). How many species of insects and other terrestrial arthropods are there on earth? *Annual Review of Entomology, 63*, 31–45. https://doi.org/10.1146/annurev-ento-020117-043348.

Sunil, L., Appaiah, P., Prasanth Kumar, P. K., & Gopala Krishna, A. G. (2015). Preparation of food supplements from oilseed cakes. *Journal of Food Science and Technology, 52*, 2998–3005. https://doi.org/10.1007/s13197-014-1386-7.

Sun, X. H., Zhu, K. X., & Zhou, H. M. (2009). Optimization of a novel backward extraction of defatted wheat germ protein from reverse micelles. *Innovative Food Science and Emerging Technologies, 10*, 328–333. https://doi.org/10.1016/j.ifset.2009.01.006.

Teuling, E., Wierenga, P. A., Schrama, J. W., & Gruppen, H. (2017). Comparison of protein extracts from various unicellular green sources. *Journal of Agricultural and Food Chemistry, 65*, 7989–8002. https://doi.org/10.1021/acs.jafc.7b01788.

Ursu, A. V., Marcati, A., Sayd, T., Sante-Lhoutellier, V., Djelveh, G., & Michaud, P. (2014). Extraction, fractionation and functional properties of proteins from the microalgae *Chlorella vulgaris*. *Bioresource Technology, 157*, 134–139. https://doi.org/10.1016/j.biortech.2014.01.071.

Vogelsang-O'Dwyer, M., Petersen, I. L., Joehnke, M. S., Sørensen, J. C., Bez, J., Detzel, A., et al. (2020). Comparison of faba bean protein ingredients produced using dry fractionation and isoelectric precipitation: Techno-functional, nutritional and environmental performance. *Foods, 9*, 322. https://doi.org/10.3390/foods9030322.

Vose, J. R., Basterrechea, M. J., Gorin, P. A. J., Finlayson, A. J., & Youngs, C. G. (1976). Air classification of field peas and horsebean flours: Chemical studies of starch and protein fractions. *Cereal Chemistry*, 928–936.

Waghmare, A. G., Salve, M. K., LeBlanc, J. G., & Arya, S. S. (2016). Concentration and characterization of microalgae proteins from *Chlorella pyrenoidosa*. *Bioresour Bioprocess, 3*. https://doi.org/10.1186/s40643-016-0094-8.

Wang, R., Tian, Z., & Chen, L. (2011). A novel process for microencapsulation of fish oil with barley protein. *Food Research International, 44*, 2735–2741. https://doi.org/10.1016/j.foodres.2011.06.013.

Wilson, S. (2009). Grasses and grassland ecology. *Annals of Botany, 104*. https://doi.org/10.1093/aob/mcp219. ix–ix.

World Health Organization. (2007). *Protein and amino acid requirements in human nutrition* (Vol. 935).

Wu, Y. V., & Bagby, M. O. (1987). Recovery of protein-rich by-products from sweet potato stillage following alcohol distillation. *Journal of Agricultural and Food Chemistry, 35*, 321–325.

Wu, Y. V., & Nichols, N. N. (2005). Fine grinding and air classification of field pea. *Cereal Chemistry, 82*, 341–344.

Wu, Y., Qian, Y., Pan, Y., Li, P., Yang, J., Ye, X., et al. (2015). Association between dietary fiber intake and risk of coronary heart disease: A meta-analysis. *Clinical Nutrition, 34*, 603–611. https://doi.org/10.1016/J.CLNU.2014.05.009.

wu, Y. V., Sexson, K. R., Cluskey, J. E., & Inglett, G. E. (1977). Protein isolate from high-protein oats: Preparation, composition and properties. *Journal of Food Science, 42*, 1383–1386. https://doi.org/10.1111/j.1365-2621.1977.tb14504.x.

Xing, Q., Dekker, S., Kyriakopoulou, K., Boom, R. M., Smid, E. J., & Schutyser, M. A. I. (2020). Enhanced nutritional value of chickpea protein concentrate by dry separation and solid state fermentation. *Innovative Food Science and Emerging Technologies, 59*, 102269. https://doi.org/10.1016/j.ifset.2019.102269.

Yi, L., Lakemond, C. M. M., Sagis, L. M. C., Eisner-schadler, V., Huis, A Van, & Boekel, MAJS Van (2013). Extraction and characterisation of protein fractions from five insect species. *Food Chemistry, 141*, 3341–3348. https://doi.org/10.1016/j.foodchem.2013.05.115.

Zhu, K.-X., Sun, X.-H., & Zhou, H.-M. (2009). Optimization of ultrasound-assisted extraction of defatted wheat germ proteins by reverse micelles. *Journal of Cereal Science, 50*, 266–271. https://doi.org/10.1016/j.jcs.2009.06.006.

Chapter 5

IoT technologies in the food supply chain

Sandeep Jagtap[1], Linh Duong[2], Hana Trollman[3], Farah Bader[3], Guillermo Garcia-Garcia[4], George Skouteris[5], Jie Li[6], Pankaj Pathare[7], Wayne Martindale[2], Mark Swainson[2], Shahin Rahimifard[3]

[1]*Sustainable Manufacturing Systems Centre, School of Aerospace, Transport & Manufacturing, Cranfield University, United Kingdom;* [2]*The National Centre for Food Manufacturing, University of Lincoln, Holbeach, United Kingdom;* [3]*Wolfson School of Mechanical, Electrical and Manufacturing Engineering, Loughborough University, Loughborough, United Kingdom;* [4]*Department of Chemical & Biological Engineering, The University of Sheffield, Sheffield, United Kingdom;* [5]*Institute of Fluid Dynamics, Helmholtz-Zentrum Dresden-Rossendorf, Dresden, Germany;* [6]*College of Mechanical Engineering, Donghua University, Shanghai, China;* [7]*Department of Soils, Water & Agricultural Engineering, College of Agricultural & Marine Sciences, Sultan Qaboos University, Muscat, Oman*

5.1 Overview of the Internet of Things

FSCs are under tremendous pressure to improve not only their revenues but also their overall sustainability, as well as supply chain efficiencies. At the same time, they aim to keep their costs low – a difficult task if investments to improve their sustainability performance and efficiencies are needed. However, the advent of digitization and its related technologies is helping businesses to overcome this difficult task. Particularly, FSCs have seen the coming together of information and operational technology thanks to developments and synergies between corresponding areas, which has resulted in the IoT.

The term IoT was coined back in 1999 by the MIT Auto-ID Lab with a special mention to Kevin Ashton (Sundmaeker, Guillemin, Friess, & Woelfflé, 2010). The Global Standards Initiative on IoT (IoT-GSI) has defined IoT "as a global infrastructure for the information society, enabling advanced services by interconnecting (physical and virtual) things based on existing and evolving interoperable information and communication technologies" (ITU-T, 2012). Lee (2013) described IoT as a complicated cyber-physical system, which incorporates all kinds of sensing, identification, communication, networking and informatics devices and systems, and impeccably connects people and things based on interests, so that anyone, at any time and any place can more

Food Technology Disruptions. https://doi.org/10.1016/B978-0-12-821470-1.00009-4

efficiently access the information of any object and service through devices and media. Some features of IoT include ubiquitous: identification, sensing, computing, and intelligence (Da Xu, He, & Li, 2014). Therefore, it is possible to consider the concept of IoT as providing a solution that gathers and consolidates huge amounts of data generated from supply chains. Applications built on IoT platforms enable the collection, analysis, and faster and better decision-making based on data to enhance operational efficiency.

Gartner, Inc. predicted that 20.4 billion connected things would be in use worldwide by 2020 (van der Meulen, 2017). IoT services have reached many different players and are now focused on new applications by end-user organizations and vendors, thereby gaining further recognition. Apart from smart cars, homes and industries, public safety, energy, and environmental conservation, agriculture, and other consumer uses will continue to account for the largest number of connected things, while enterprise will account for the biggest spending. In accordance with this development, the UK Government allocated £40 million toward research into IoT in their 2015 budget (GOV.UK, 2015). The British Ex-Chancellor of the Exchequer, George Osborne, highlighted that the IoT is the next stage of the information revolution, connecting everything from urban transport to medical devices to household appliances (GOV.UK, 2015).

IoT is a relatively new manufacturing concept and is one of the nine principle technologies in Industry 4.0. It consists of innovative information technology (IT) infrastructure for data collection and distribution, which significantly influences the efficiency and performance of manufacturing systems. IoT supports manufacturing through real-time data acquisition and sharing between several resources such as equipment, employees, materials, jobs, etc (Bi, Da Xu, & Wang, 2014). The real-time data collection is usually based on sensors, radio frequency identification (RFID), and wireless communication standards. By using these technologies, crucial information related to manufacturing, movement of materials or people, and various other information can be tracked and monitored in real-time, giving full visibility and traceability of manufacturing operations (Lu, Bateman, & Cheng, 2006; Zhong et al., 2013). This allows management to make faster and better decisions based on real-time information and thereby improve the effectiveness and efficiency of manufacturing operations.

Daugherty, Banerjee, Negm, and Alter (2015) stated that some companies have started considering IoT as a method to improve operational efficiency and its potential benefits by using it as a tool for finding growth in unexpected opportunities. Furthermore, it accelerated the pace of IoT development by coining new terms within the IoT concept, such as Industrial IoT (IIoT) and adding additional components to it. They predicted that, in the future, successful businesses would use IIoT to increase their revenues by boosting production, building new hybrid business models, exploiting intelligent technologies to energize innovation, and transform their workforce. Rapid advancements in IoT that are already developing, and which can be further enhanced by combining it with other related technologies such as cloud computing, future internet, big data, robotics, and semantic technologies (Vermesan & Friess, 2013).

However, Vermesan and Friess (2013) argued that IoT implementation faces several challenges, such as:

- ➢ Not enough work completed on semantic interoperability for exchanging sensor data in heterogeneous environments
- ➢ Privacy, trust, security, and ownership of data
- ➢ Hurdles in pursuing businesses to adopt IoT fully
- ➢ Absence of testing and learning environments to stimulate innovation
- ➢ Partial deployment in supply chains
- ➢ Missing technical availability of robust network connectivity

Overcoming these difficulties would result in better utilization of the IoT. IIoT is often used in manufacturing industries to focus on industrial applications. It represents the industrial subclass of the IoT, which has already started transforming the manufacturing sector resulting in a fourth industrial revolution, termed Industry 4.0. Industry 4.0 aims to build intelligent factories where manufacturing technologies are improved and transformed using cyber-physical systems (CPSs), IoT, and cloud computing. Industry 4.0 incorporates advanced manufacturing techniques with the IoT to create manufacturing systems that are interconnected, with the ability to communicate and analyze the collected information to make better decisions (Zhong, Xu, Klotz, & Newman, 2017). The 2014 Accenture technology vision stated that every business is a digital business, and the digital-physical focus is turning industrial companies into customer-service companies (Accenture, 2014). These businesses have recognized the huge potential of IIoT, as it supports businesses with new-growth by adding digital services and innovations to their product mix. Businesses already spent $20 billion in 2012, and it is expected they will spend $500 billion by 2020 (Floyer, 2013). One of the important aspects of IIoT is operational efficiency, which is achieved by implementing automation and agile production techniques such as predictive maintenance to avoid downtime and plant and facility shutdowns (Daugherty et al., 2015). Companies are starting to realize the huge potential of IIoT: the universe of intelligent industrial products, processes, and services that connect with each other and with people over a worldwide network thus boosting new growth opportunities by adding digital services and innovation to their product mix (Daugherty et al., 2015).

IoT has already gained entrance within a number of FSC activities such as food traceability, warehouse management, transport, and logistics. Dunbrack, Ellis, Hand, Knickle, and Turner (2016) stated that 33% of all industry leaders would be disrupted by digitally enabled competitors. They further stated that 58% of companies think IoT is strategic, and 24% of all organizations see IoT as transformational; furthermore, the young population will accelerate IoT adoption (Dunbrack et al., 2016). Therefore, it can be assumed that a key enabler to improve productivity and growth in the short-term future will be IoT, which will drive and transform most FSCs.

Based on the above prediction, Fig. 5.1 depicts an IoT-powered food-manufacturing plant. This figure shows how IoT can empower the stakeholders and factory management by the ability to monitor food production flow in real-time, allowing them to monitor and manage the connected equipment. It further allows them to identify quality issues and address them immediately, giving more transparency and visibility in inventory management and flexibility to perform predictive maintenance. Fig. 5.1 also illustrates how managers, research, and development professionals, and the plant floor staff can achieve benefits through the application of IoT.

5.2 The architecture of IoT in the food supply chain

IoT for FSCs is designed to connect different machines, equipment, and other things over the networks easily. Therefore, an IoT architecture is required to collect the data seamlessly and transfer it securely for further analysis.

The most basic IoT architecture consisted of three-layers: perception, network, and application layers (Kumar & Mallick, 2018; Sethi & Sarangi, 2017), which are described below:

- Perception layer – is a physical layer consisting of sensors and actuators for sensing and collecting data on physical parameters, as well as identifying other smart objects from its surroundings.
- Network layer – is responsible for communicating with other smart objects, networking devices, and servers and also transfers and processes the sensor data.
- Application layer – This layer delivers specific application services to the users and, at the same time, defines several applications in which IoT can be installed, such as smart health, smart home, and smart city.

However, for FSCs, the most suitable IoT architecture consists of four layers: the sensing layer, the network layer, the service layer, and the application layer (see Fig. 5.2), as described below:

- The sensing layer – The data such as time, temperature, location, machine, etc. in the whole FSC are collected using sensors, RFID, camera, etc. All of the collected data, i.e., from the point of raw material sourcing until the end of life, is then preprocessed in this layer.
- The network layer – transports the data collected in the sensing layer to the service layer via various networking technologies such as Wireless Sensor Network (WSN), Bluetooth, WiFi router, etc.
- The Service layer – consists of a wide range of analytic engines and services where data can be analyzed or stored.

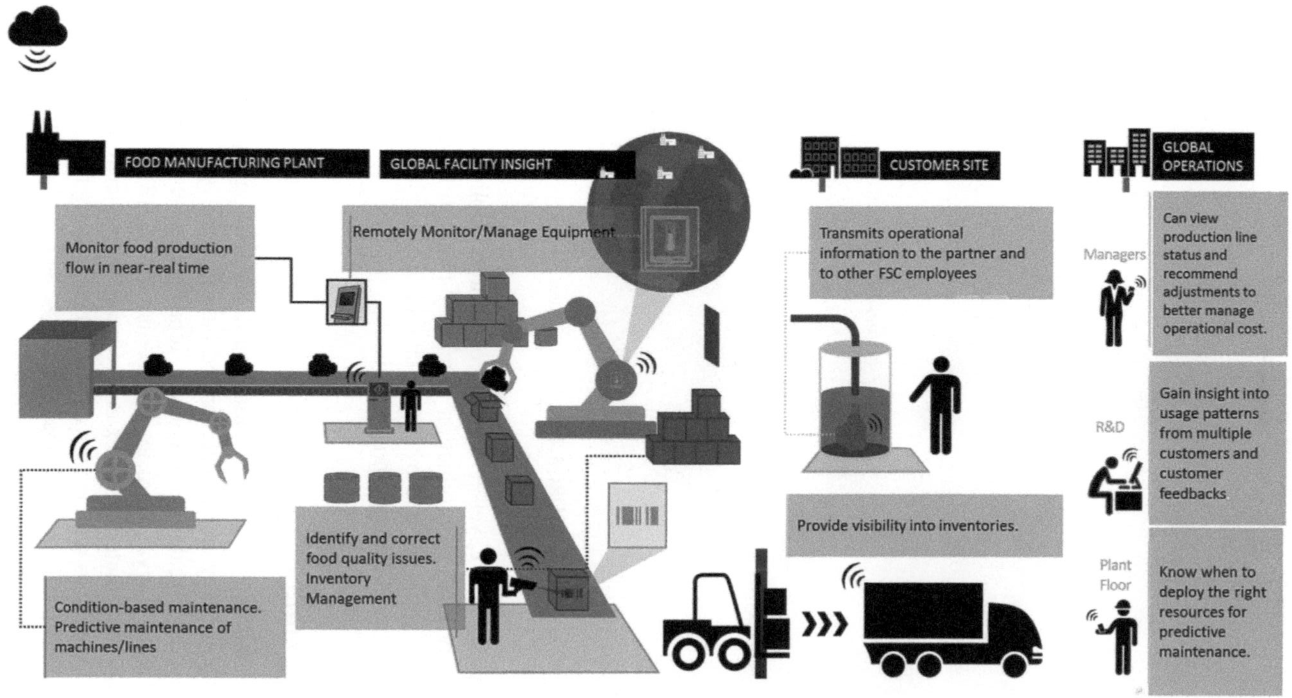

FIGURE 5.1 Applications of IoT in food manufacturing.

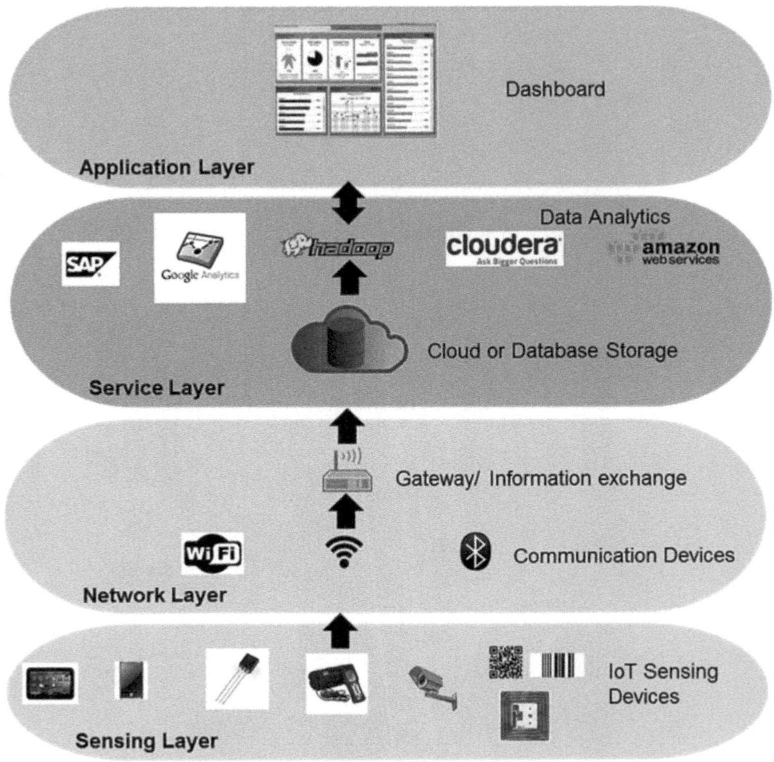

FIGURE 5.2 IoT architecture for FSC.

- The Application Layer – consists of various applications such as food traceability module, production efficiency module, food quality module, etc., which could be used by supply chain actors to view the information in real-time and make appropriate decisions.

5.3 Application of IoT in the food supply chain

Currently, numerous IoT applications are depending on the type of industry that uses them. They can be categorized based on the type of network availability, coverage, scale, heterogeneity, repeatability, user involvement, and impact (Gubbi, Buyya, Marusic, & Palaniswami, 2013). FSCs are uniquely complex and distributed, have extensive geographical coverage, complicated operational processes with numerous stakeholders along the chain, and present issues for food quality, operational efficiency, and food safety (Pang, 2013).

FSCs' ultimate objectives are to eliminate food shortages and ensure the availability of nutritious food to all is above any economic, social and environmental developments. The increasing population estimated to reach 9 billion by 2050, which will completely change the world, as well as adds tremendous pressure on FSCs. However, developing and evolving IoT technology is likely to offer encouraging solutions (Sundmaeker, Verdouw, Wolfert, & Pérez Freire, 2016).

IoT can play a role in the functioning of every actor within FSCs, ranging from farms to food production, processing, storage, distribution, and consuming. IoT can address traceability, visibility, and transparency and controllability challenges. Safety, efficiency, transparency, and sustainability are some of the features needed in FSCs.

5.3.1 IoT application in food transportation logistics

During food transportation, the product that is being transported can face various challenges, such as temperature control, hygiene and pest control, traceability, goods management (goods handling, damage, rejection, and safety), vehicle/container preventive maintenance, and employee management (goods handling, personal hygiene, safety, policies, and training). With the application of IoT, it is possible to track all the activities involving the food movement. Radio Frequency Identification (RFID) technology is one of the efficient and cheapest enablers of the IoT for tracking food products within FSC. RFID tags can hold specific but important information with regards to food products that are being transported, and which can be easily communicated over a wireless network. In case of food recalls or food safety issues, alerts can be sent in real-time throughout the supply chain, and the affected product can be quarantined immediately. Low cost wireless remote system enables the creation of wireless networking in food transport vehicles allowing to monitor food safety during transportation. For example, with the help of IoT systems integrated into the Hazard analysis critical control points (HACCP) processes, the FSC actors can monitor and document the temperatures and other conditions in real-time, ensuring effective cold chain management, as well as complying with global and local regulations. The food transport vehicles can be fitted with simple to complex wireless systems providing connectivity and continuous access to information in real-time. Ferreira et al. (2012) discuss the use of several applications related to vehicular sensing, as well as connectivity issues related to the mobility and limited wireless range of an infrastructure-less network based only on vehicular nodes. Furthermore, Chen, Tian, Fortino, Zhang, and Humar (2018) proposed the Cognitive Internet of Vehicle (CIoV) concept to enhance transportation safety and network security by mining effective information from both physical and network data space. The MovingNet – a Vehicular Ad-Hoc Sensor Network consisting of multiple sensors was installed on the public transport system to detect the production of spurious alcohol (Ramesh & Das, 2011). Zhang, Chen, and Lu (2012) in order to improve transport efficiency, proposed a design of an intelligent monitoring system based on IoT, including RFID, sensor and wireless communication technology to monitor temperature and humidity inside a refrigerator truck, as well as the cargo identification and tracking location in real-time. Ma, Li, Yin, and Ji (2012) demonstrated the integration of IoT into the Enterprise Resource Planning (ERP) system of a pork supply chain to generate an early warning system to inform about the pork quality and reduce the logistics costs and improve the circulation efficiency.

One of the widely used IoT applications in the food sector is in transportation, as shown in Fig. 5.3. The food haulage truck can be fitted with sensors to track vehicle location, monitor temperature, control stock management, and protection, assess needs for maintenance of the vehicle, inform about route changes in case of traffic jams, etc. These sensors can be connected and controlled through IoT (Jhingran, 2016).

Lacey, Lisachuk, Giannopoulos, and Ogura (2015) categorized IoT applications in transportation into demand and supply. Storage facilities, transport network (i.e., roads/sea/air) and mode of transportation (e.g., truck, ship, plane) come under the "supply" category, as shown in Table 5.1, and the goods to be transported, and the intended consumer who shall receive goods comes under the "demand" category (Table 5.2).

5.3.2 IoT application in food production

The data collected from the factory floors (machinery, staff, vehicle, materials, etc.) can be used to automate work procedures and processes to optimize food production systems without human interference. With the support of real-time information, algorithms, and actuators, the specially designed control software can make self-decisions and initiate actuators to minimize any deviation from the plan. It also helps in generating optimal decisions and imparting autonomy

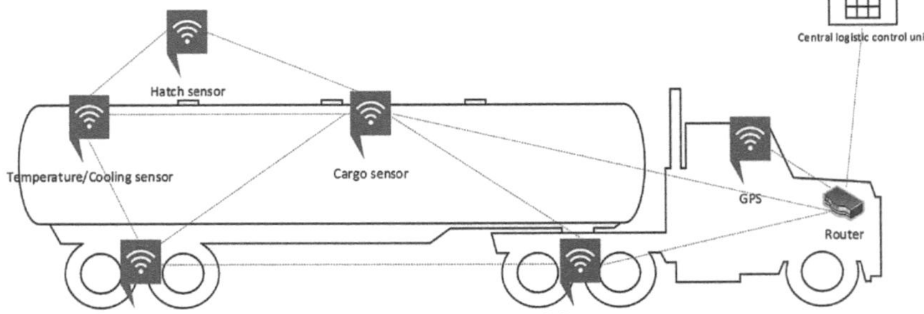

FIGURE 5.3 IoT scenario – food haulage truck (Jhingran, 2016).

TABLE 5.1 Common applications of IoT for logistics supply (Lacey et al., 2015).

Capacity sensing	Planning and reporting	Route optimization	Energy management	Fault detection and resolution
Systems that can detect and communicate open spaces in a warehouse, port or parking lot	Systems that can detect and analyze events such as traffic accidents within a delivery network, allowing for more accurate delivery dates	Tools that can map the shortest or most fuel-efficient route, e.g., delivery vehicles	Tools that monitor and enable decision making about the use of fuel, lighting, and heating/cooling within vehicle fleets and facilities	Systems that can monitor fleets of vehicles, aircraft, or ships for faults and maintenance needs, improving uptime for the fleet

TABLE 5.2 Common applications of IoT for logistics demand (Lacey et al., 2015).

Environment monitoring and management	Threat detection and prevention	Real-time traceability
Systems that can monitor and adjust the temperature at which a package is maintained	Tools that can help detect unauthorized openings of shipping containers, helping to prevent and reduce theft	Systems that can track not only vehicles or shipments but individual items

by controlling the food production processes and addressing any disruptions. IoT can also help to continuously monitor and compare the energy usage and FSC activities in real-time by embedding sensors at various energy hotspots. IoT makes FSC actors become energy-aware and go beyond a single process/machine optimization to adopt a more multidomain approach (e.g., machinery, people, materials, vehicles, etc.) to improve overall energy-efficiency of FSC. Some food manufacturers have understood the importance of predictive/proactive maintenance, which enables early diagnosis and repairs based on the condition and performance of the machines in order to reduce the impact of unplanned downtime and unwanted breakdowns. Low-cost sensors with wireless connectivity and powered by big data tools can provide and analyze useful data with regards to machine performances. Historical and real-time data obtained from the machines can be modeled, compared, analyzed, and displayed to predict machine degradation. IoT connected FSCs enables all the actors to view real-time information with regards to production, logistics, maintenance, resources, inventory, and sales, etc. This action makes effective lean implementation as all the actors are aware of interdependencies, the flow of materials, and the identification of any potential issues. All the information can be viewed by all the actors in real-time, which will eliminate the issue of the laggard level of information.

IoT can be utilized in a wide range of applications in food production (Table 5.3). Lee, Bagheri, and Kao (2015) differentiate between today's food factories and IoT-powered Industry 4.0 food factories.

Therefore, in food manufacturing sensors and networked devices can be connected to collect data such as temperature, quality, machine maintenance, sales, orders, and information on transportation, storage, environment during the whole lifecycle of the food product on a 24/7 basis. The huge amount of data can be processed into valuable information for the decision making of all intended users or stakeholders.

5.3.3 IoT application in resource/waste management

The supply chain practices within FSCs are considered to be unsustainable due to the continuous demand for various vital resources such as raw materials, energy, and water. This issue has led researchers and industry practitioners to develop resource-efficient sustainable FSCs without compromising supply chain productivity. However, managing resource efficiency in FSCs could be a difficult task due to the nature of the finished food product (i.e., shelf-life, storage temperature, handling, etc.). Furthermore, the complexity originating from the range of resources used across several processes with each process having different usage levels impacts resource management. In order to improve sustainability across FSCs, there is a need for better communication between the various actors of FSCs. However, the lack of accurate and readily accessible data is one of the major roadblocks in attaining resource efficiency in FSCs. Moreover, FSC actors are completely unaware of its resource usage, e.g., they are aware of total water intake and water discharge in the form of effluent but are generally unaware of water usage at individual process level or actor level. The traditional methods of collecting data using pen and paper are inefficient, tedious, and laborious.

TABLE 5.3 Today's food factory versus Industry 4.0 food factory (Lee et al., 2015).

	Data source	Today's food factory		Industry 4.0 food factory	
		Attributes	**Technologies**	**Attributes**	**Technologies**
Component	Sensor	Precision	Smart sensors and fault detection	Self-aware Self-predict	Degradation monitoring and remaining useful life prediction
Machine	Controller	Producibility and performance	Condition-based monitoring and diagnostics	Self-aware Self-predict Self-compare	Uptime with predictive health monitoring
Production system	Networked system	Productivity and overall equipment effectiveness	Lean operations, work and waste reduction	Self-configure Self-maintain Self-organize	Worry-free productivity

Rahimifard et al. (2017) highlighted the need for producing more food with fewer resources in order to achieve sustainable FSCs. Elaine Weidman-Grunewald, the vice president of sustainability and corporate responsibility at global ICT Company Ericsson in 2016, said that "2016 is going to be the year we realize the huge range of applications of the IoT, and that's a huge opportunity from a sustainability point of view, and the possibilities are just endless" (Edie, 2016). IoT's main goal within food manufacturing can be described as sustainable when it serves the needs for producing, manufacturing, transporting and supplying food products with minimal economic, environmental, and societal impacts (Kumar, Dash, & Singh, 2018, pp. 68–72). IoT can be used to reduce global emissions and wastes through interconnected devices and objects (Maksimovic, 2017). Lyons Hardcastle (2013) predicted that by 2020, IoT applications would reduce world emissions by seven billion metric tons in energy, transportation, and the built environment and agricultural sectors. Kyriazis, Varvarigou, White, Rossi, and Cooper (2013) described how to optimize energy usage in commercial and residential areas by deploying heat and electricity meters and cruise controlling public transportation through the utilization of environmental and traffic sensors to provide recommendations to enhance eco-efficiency.

The usage of internal and external sensors and simulation systems can provide complete monitoring of energy in manufacturing systems as per ISO 14955 and ISO 50001 standards (Gontarz, Hampl, Weiss, & Wegener, 2015). Petersen et al. (2007) showed that real-time resource responsive systems motivated students to decrease resource consumption in dormitories (electricity by 32% and water by 3%). Henningsson, Hyde, Smith, and Campbell (2004) demonstrated the value of resource efficiency in the food industry by reducing the resource consumption at source, at the end of the process, and showed the importance of reducing electricity and water consumption, as well as improving effluent quality, which can lead to significant savings. Hence, a more sustainable production and consumption system can be achieved by improving the communication between all the links (producer, manufacturer, retailer, and consumer) within FSC and embracing changing technologies to reduce labor costs, materials, and utilities (Henningsson et al., 2004). Comprehensive information is required with regard to resource consumption patterns and behaviors to plan and execute a resource-efficient production system (Weinert, Chiotellis, & Seliger, 2011), as well as incorporating resource efficiency in production management (Bunse, Vodicka, Schönsleben, Brülhart, & Ernst, 2011). Hence, to achieve these targets, consumers require more real-time information on resource production and consumption costs (Chetty, Tran, & Grinter, 2008, pp. 242–251). Stich, Brandenburg, and Kropp (2011, pp. 390–395) underlined that Information and Communication Technologies (ICT) allow resource-efficient manufacturing by incorporating real-time information from the shop floor to planning systems by focusing on idle times on the shop floor, which can be planned as per the order demand.

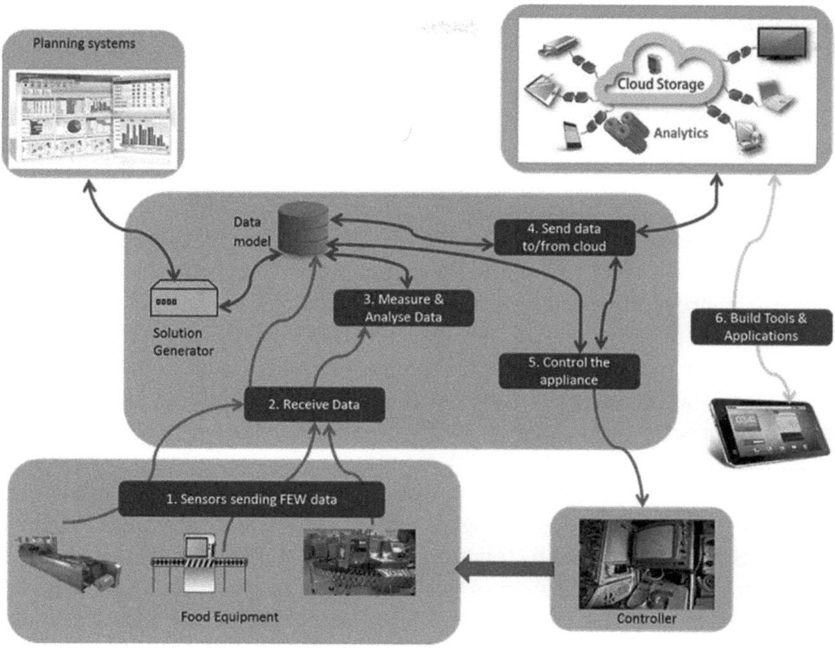

FIGURE 5.4 Resource efficiency monitoring in a food factory.

This situation can be effectively managed by IoT wherein product manufacturing can be optimized by following a resource-efficient route (Jagtap & Rahimifard, 2017, 2019b; Stich et al., 2011), as shown in Fig. 5.4. Therefore, in order to reduce resource consumption and wastage, it is important to have access to accurate, meaningful, and real-time data. Hence, in this effort to improve the resource efficiency of FSCs, IoT-based applications that have the capability to address the mentioned issues becomes an absolute necessity.

5.3.3.1 Relevant IoT-research for reducing food waste generation

Food waste management within FSCs is currently managed by the supply chain actors by placing food waste bins next to the point of waste generation. The food waste generated is typically used as animal feed or sent to anaerobic digestion for energy generation rather than human consumption due to stringent regulations for food redistribution. This is a massive environmental issue since huge amounts of resources are used to produce food that ends up being wasted. Moreover, FSC actors consider food waste production as a part of the supply chain processes, and each actor uses a different measuring method.

Garcia-Garcia, Woolley, and Rahimifard (2015) discussed the magnitude of food waste in FSCs of both developing and developed countries and options to manage it and then proposed a systematic methodology to identify the most sustainable solution to manage it (Garcia-Garcia, Woolley, Rahimifard, Colwill, White, & Needham, 2017). Sheppard, Garcia-Garcia, Stone, and Rahimifard (2020) presented a knowledge-based management system to enhance the effectiveness and productivity of existing tools for food waste management and valorization, and therefore, enhance their sustainability performance. Wang and Yue (2017) developed a framework supported by IoT to provide a warning in the event of food safety risks and avoid food waste within FSCs. Hong et al. (2014) proposed an IoT-based garbage system for managing food waste to improve the effectiveness and efficiency of food waste management in Seoul. The proposed system not only reduced food waste by 33% but also resulted in energy savings of 16%. Jagtap and Rahimifard (2019a) demonstrated a digital food tracking system based on IoT, which reduced food waste in a ready meal food factory by 60.7%. It was achieved by better real-time visibility of the food waste hotspots, reasons for food waste generation, collection of more reliable data, operational improvements, and behavioral transformation in employees (Jagtap, Bhatt, Thik, & Rahimifard, 2019a). Ostojić, Stankovski, Tegeltija, Đukić, and Tejić (2017) stated that food waste is a result of inefficiencies within FSCs and lack of information, and to address this issue, developed a Kinetic Arrhenius model based on IoT system to predict shelf life. All the researches have demonstrated that measuring accurate real-time food waste data with reasons for its generation can lead to behavioral changes, process redesign, and investment into types of machinery, which would eventually reduce food waste generation. Therefore, IoT-based food waste management systems that allow accurate weight measurements with reasons for food waste generation have a high impact on food waste reduction. These IoT-based food waste management systems can be easily integrated with ERP systems and will enable all the actors within FSC to view the effect of last-minute order changes on food waste generation. These food waste management systems are more reliable, mobile (easy installation), continuous and convenient as compared to the traditional food waste bins.

5.3.3.2 IoT application for reducing energy consumption

Energy is fundamental for many activities within FSCs. Increasing energy prices, stricter environmental legislation, added with energy security concerns, are driving FSCs to adopt energy-efficient practices to improve profitability and competitiveness. In other words, energy efficiency is defined as achieving more with less energy. FSCs consume a considerable amount of energy, and especially the meat industry has one of the largest carbon footprints (Pathare, Roskilly, & Jagtap, 2019).

Energy efficiency is often measured in terms of the ratio between the output of performance, goods or services, and the input of energy. In FSCs, there are mainly two ways to enhance the energy efficiency, first, by deploying energy-efficient equipment/or machines, and second, by having energy-efficient production planning/or scheduling. From the energy-efficient equipment/or machines viewpoint, installing new equipment and machines is deployed to minimize power and energy consumptions of those components. However, this option of upgrading to energy-efficient components could be rather expensive for small and medium scale FSC actors. From the viewpoint of energy-efficient production planning/or scheduling, energy consumption forms the basis of planning/or scheduling models and is considered as an easy alternative to the reduction of energy in FSCs. For example, during food production, the food manufacturers usually start the production with a product with no or low-level allergen product and eventually finishes production with a product having high allergen level rather than starting with high allergen product and finishing with low allergen level product. This practice enables manufacturers to save on time, energy, and water required for cleaning the line as the line contaminated with allergens needs thorough cleaning and inspection. Furthermore, in the production planning/or scheduling stage, activities are assigned on the basis of historical information and the current status of the machines. As a result, these activities may not be necessarily assigned to the machines at an optimum level and results in energy inefficiency. Moreover, in FSCs, the lack of assessing and analyzing energy data in real-time can result in making energy wastage decisions and difficulty for supply chain actors to explore energy-saving opportunities.

Currently, IoT applications supported by RFID technology, Bluetooth, etc. have the ability to address these issues. IoT plays an important role in monitoring energy usage (Haller, Karnouskos, & Schroth, 2009). With the support of these novel technologies, the IoT-based smart meters can communicate with other computing systems within FSC and control the usage of energy. By analyzing the data obtained from the smart energy meters in real-time, energy usage patterns and status can be monitored, and the correlation between various supply chain activities and their energy consumption could be understood. IoT-based technology (e.g., smart energy meters and energy sensors) provide energy consumption patterns by continuously collecting energy consumption data in real-time. Shrouf and Miragliotta (2015) illustrated how management in a factory could approach the IoT implementation in a profitable way by easily collecting and analyzing energy consumption, and thereby, improving their energy-aware decision-making process. Jagtap and Rahimifard (2018) and Jagtap, Rahimifard, & Duong, 2019 demonstrated the implementation of IoT-centred technology based on embodied product energy modeling, which led to better energy monitoring and management with substantial cost savings. With this additional information available from IoT-based technologies monitoring energy, the inefficient energy consumption practices can be reduced or eliminated.

5.3.3.3 IoT application for reducing water consumption

Water management in FSCs involves water usage planning, appropriate water distribution, and managing the efficient use of water resources. The water is used in every stage of FSC, including food production, cleaning, as a constituent of finished food products, washing, etc. The lack of water information within FSC leads to its overconsumption and wastage.

Robles et al. (2015) proposed a smart water management model based on linking IoT technologies with operational processes and decision support systems. Researchers from Brazil built an automated solution based on IoT technology to minimize water wastage, and at the same time, improve energy required in pumping and heating systems (De Freitas Melo, De Souza lage, Vagner Rocha, & De Jesus Cardoso, 2017). Wang et al. (2018) developed a real-time smart water Long-range IoT system, which allowed operators to detect water leakage problems, and thereby, improve operational efficiency. Drenoyanis, Raad, Wady, and Krogh (2019) presented a novel IoT system based on the Low Range Wide Area Network to provide a complete review of wastewater monitoring networks. Another example is the improvement in the efficiency of the use of water by deploying IoT-based real-time systems in a beverage factory and reducing the annual consumption by 6.7% (Jagtap, Skouteris, Choudhari, & Rahimifard, 2018).

IoT-based smart water system allows us to monitor the water usage at various hotspots within FSC, and any deviation from standard daily usages such as leakages and overconsumption triggers an alert to the respective actor. This ability to send alerts in real-time ensures timely actions from the actors and reduces overall water consumption.

5.3.4 IoT application in improving food safety

Current FSCs are long and complex, with increased safety risk and continuous pressure from consumers to deliver high-quality and safe food products. All actors within FSCs contribute to food safety information, which could result in unpredictable risks caused due to incorrect data sharing or latency. However, with the development of IoT, which enables effective collection and sharing, as well as an opportunity for analyzing the data can highlight any shortcomings and generate prewarning food safety alerts throughout the supply chain. This results in developing a system that can predict potential food safety risks and generate an advanced warning in case of any breach in food safety and quality of the food product. Thus, each actor within FSCs can contribute to the reduction of deviation in quality and resource waste and thereby avoid any food safety-related incidents. Maintaining and providing quality food with an advanced warning on food safety is the most important aspect of achieving sustainability in FSC. Food safety advance warning continuously poses a challenge for supply chain actors. Most food products pass through several supply chain actors before reaching the end consumer, and during this movement, the complexity with regards to product quality can be heavily compromised. And it becomes more severe when more actors join the FSC. Hence, in this complicated and intensely competitive environment, the stakeholders within supply chain actors fail to see the knock-on effect it can cause throughout their FSC activities. In the worst case, the unsafe food product may result in product recall, and that can cause great ordeal for the supply chain actors. Another factor that impacts the effectiveness of food product quality guarantee is low visibility in the FSC. The phenomenal rise in food product recalls, informs that the multi-layered FSCs with low transparency are prone to food safety risk. However, the consumers expect their food products to be of high quality and managed through better supply chain processes, thus, forcing FSC actors to enhance quality. Therefore, these days the competition among various FSCs is based on quality. Moreover, the demand for the freshness of food products and the extent of transparency is gradually increasing with more focus on temperature monitoring, microbiological information, and food quality specifications. Therefore, in order to achieve these goals, it becomes necessary to research in improving FSC sustainability through information sharing and food safety advance warning.

Information sharing is an important aspect to enhance communication and manage the effective collaboration between actors of FSC, as well as to improve the sustainability of FSC. IoT is capable of achieving the information sharing tasks and enables real-time process monitoring, collecting and transmitting the crucial information to all the FSC actors, and has the ability to address the issue of FSC transparency. Furthermore, IoT allows FSCs to utilize virtualizations effectively in supply chain activities, which supports FSCs in dealing with short shelf-life products, uncertain supply fluctuations, and stricter food safety and sustainability standards. The application of IoT in the FSC has significantly reduced the risk of food fraud, the use of counterfeit products, and food-related illnesses. Various sensors are used for tracking product movement and environment conditions (e.g., temperature). The application of real-time sensors enables numerous traceability opportunities, such as tracking the location of products, production, and delivery schedules, ingredient sources, handling, and usage history. It also allows tracking of temperatures at crucial food safety hotspots within cold FSCs. With the support of IoT, it would be swifter and easier to comply with global and local regulations. For example, the use of digitized and automated HACCP, already common practice in a number of FSCs, can result in authentic and robust data leading to the designing of food safety solutions (Bader & Jagtap, 2020).

Currently, there is a lot of attention to the food safety issue, and numerous tasks are undertaken to face this challenge. One of the tasks is an advance warning system that can detect and alert the actors about the food safety issue before it becomes a major crisis. The advance warning system is implemented in conjugation with HACCP-based traceability systems. Current FSCs consist of several actors that makes it hard to regulate, track, monitor, and control the food product trade. Hence, the majority of food safety events are triggered by an inadequate amount of supervision, and thus there is a need for an advance warning system to control the food product trade in an efficient and automatic manner.

5.3.5 IoT application in maintaining food quality

Various image-processing technologies and sensors can help in maintaining the quality and specifications of raw materials and finished products (Jagtap et al., 2019a). These sensors can continuously monitor the product quality, and any deviation from the set standards can be immediately notified and rectified. The other benefits of these sensors are product tracking, tracking employee activities, and performing real-time production analysis for efficiency. Overall, this would result in the optimization of FSC activities.

As earlier stated, that IoT is making a significant influence in the areas associated with food safety and consumer safety. The gas sensor is one of the important IoT applications, which is gaining attention from both industrial and academic individuals. Recently developed Paper-based electrical gas sensors (PEGS) can detect spoilage gases like ammonia and trimethylamine in meat and fish products and warn users whether food is safe to process/or consume (Barandun et al., 2019). Electronic nose systems contain many kinds of electronic chemical gas sensors to identify complex odors (Schaller, Bosset, & Escher, 1998). It was successfully used for quality control, process monitoring, freshness evaluation, shelf-life investigation and authenticity assessment on meat, grains, coffee, mushrooms, cheese, sugar, fish, beer, and other beverages, as well as on the odor quality evaluation of food packaging material (Matindoust, Baghaei-Nejad, Abadi, Zou, & Zheng, 2016).

The embedding of gas sensors in food packaging has provided the consumer with smart packaging solutions. These developments have resulted in enhancing food quality, long shelf-life, and usability. The emergence of nanosensors would be the future of smart packaging developments with a focus primarily on food safety (microbiological growth detection, oxidation, and tamper visibility), food safety (volatile flavors and odor detection), shelf-life monitoring, verification, convenience and the sustainability of food products (Kuswandi, Wicaksono, Abdullah, Heng, & Ahmad, 2011). It is proving important to make devices embedded with inexpensive commercial gas sensors capable of sensing and determining the concentration of the gases emitted through various food products.

5.3.6 IoT application in improving food supply chain transparency

The end consumers/buyers require transparency from the manufacturers in order to have full visibility of how, where, and when their food products are sourced from and processed. Having full traceability and visibility of the whole supply chain will aid food manufacturers in growing their business by gaining customer trust and loyalty. Although current FSCs are often very long and complex, IoT technology could make it easier for all FSC actors to track products. Furthermore, being transparent can be advantageous for food manufacturers as it would result in better stock management, labor management, reduced costs, and shorter lead times. These benefits could be achieved by addressing the inefficiencies within the supply chains, meeting and going beyond minimum food safety requirements, and giving full visibility to customers.

Skilton and Robinson (2009) have acknowledged that transparency and traceability are interrelated. They also defined traceability as the process to recognize and validate the components and the timeline of events throughout the supply chain, whereas transparency is about having all the information readily available. They further added that the relationship between transparency and traceability is not the same since having more transparency may lead to increased traceability, but not vice-versa as increased traceability does not mean increased transparency if some of the supply chain actors are loosely affiliated to the supply chain. Traceability can be compromised when material information is not accurate, insufficient, incomplete, or totally missing. It can be further impacted due to the complexity involved in the supply chain network. The complexity of the supply chain is due to the fact that it consists of multiple numbers of actors, hidden elements, and poses a challenge for effective and secure food product monitoring. Supply chain experts have the expertise to resolve these issues due to their holistic view of the value chain (Markman & Krause, 2014).

In the FSC context, transparency can be defined as the information available to all the actors operating and involved within a supply chain network. Transparency within FSC can demonstrate the product traceability from farm to fork, i.e., where the raw material is sourced/or originated from, how it was processed, and delivered to consumers. Blockchain technology can provide improved transparency throughout FSC. RFID based system was suggested for tracing beef from farm to slaughter (Shanahan et al., 2009). This RFID based traceability system involved all the supply chain actors along the beef supply chain in order to enhance consumer confidence in beef products by ensuring that traceability data is available to the consumers. Similarly, Abad et al. (2009) utilized RFID smart tag for real-time traceability and cold chain monitoring along an intercontinental fresh fish logistic chain. They further demonstrated that their system has many more benefits than conventional traceability tools. Mattoli, Mazzolai, Mondini, Zampolli and Dario (2010) designed, developed, and tested a Flexible Tag Datalogger (FTD) for improving the logistics during transport, storage, and vending of wine bottles, as well as evaluate their safety status.

5.4 Pros and cons for IoT implementation in food supply chains

Tracking and monitoring almost everything within FSC, from raw materials to machines and staff, could save a large amount of time and money. However, at the same time, Visich, Li, Khumawala, and Reyes (2009), Visich, Powers, and Roethlein (2009) identified significant challenges for IoT applications, such as the high investment needed, concerns about return on investment (ROI), complexity, trust, and privacy/security concerns. They further stated that technical issues such as the gap between technology development and business innovation, i.e., the alignment of enabling technology and practical business requirements could pose further challenges.

These challenges have also been identified by Gontarz et al. (2015). For instance, delayed RFID implementation in Walmart's operation raised doubts on the applicability of the RFID technology (Visich Li et al., 2009a; Visich, Powers, & Roethlein, 2009). There are several significant hurdles for the implementation of IoT in FSC, for example:

1. Nonbeneficial value proposition - It is an obstacle for large-scale implementation of IoT in FSCs. For instance, the food traceability system in which RFID tags are used to monitor and record the operations and time taken within the supply chains is one of the most common IoT applications in the FSCs. Although RFID can reduce labor cost and processing times in manufacturing, distribution and retailing, this added value is not convincing enough to implement it throughout the entire FSC: "The suppliers were reluctant to adopt the RFID because their initial investment cost, required by the third-party logistics firm, has produced the minimum level benefits for themselves, which, in turn, has a cascading effect on the minimum level business benefits realized by the third-party logistics firm" (Wamba & Chatfield, 2011). Visich Li et al. (2009) also showed that most of the profitable RFID applications in today's world are derived from the intended targets when the technology was first implemented. Therefore, more benefits could be delivered, and new functionalities and capacities may be developed.
2. Absence of device and service integration - Many valued technologies have been developed in recent years for FSC applications. Nevertheless, there is a vast amount of separated technologies and devices, but there are few integrated services (Ruiz-Garcia, Lunadei, Barreiro, & Robla, 2009; Ruiz-Garcia & Lunadei, 2011). Hence, there is a need for a holistic design to successfully incorporate the vast number of devices and technologies into more value-added services.
3. Training for the workforce – The FSCs works on low-profit margins, need a high degree of flexibility in terms of production and most of the food

products have shorter product lifecycles; therefore, the supply chain actors may not be willing to invest in IoT-related activities such as training of workforce or hiring of IT specialists. Also, FSCs experience high staff turnover due to low wages and seasonality, which would add to skill shortages associated with IoT (Trollman & Trollman, 2019). The other important consideration needed is to have an interaction of food industry experts and IT specialists to develop an IoT application. Furthermore, FSCs would need staff to maintain the IoT network and correct functioning of sensors, which would be an additional cost (Jagtap, 2019).

4. Trust/privacy/security concerns - The CERP-IoT report 2010 (Vermesan et al., 2010), has stressed security and privacy as major challenges, including:
   - ➢ Event-driven agents to enable an intelligent/self-aware behavior of networked devices
   - ➢ Privacy-preserving technology for heterogeneous sets of devices
   - ➢ Models for decentralized authentication and trust
   - ➢ Energy-efficient encryption and data protection technologies
   - ➢ Technologies for objects and network authentication
   - ➢ Anonymity mechanisms
   - ➢ Security and trust for cloud computing
   - ➢ Data ownership

Perera, Ranjan, Wang, Khan, and Zomaya (2015) described how data collected through sensors or other devices pose a challenge and opportunity for research and innovation with regard to privacy in IoT. The above discussion leads to both advantages and disadvantages of implementing IoT in FSC, as summarized in Table 5.4.

TABLE 5.4 Advantages and disadvantages of IoT.

Advantages	Disadvantages
Real-time information – more information collected and made available supporting in making better decisions.	**Compatibility** – numerous equipment embedded with sensors will require unique internet protocol addresses, and there is no standard way for the tagging and monitoring of equipment.
Monitoring – monitoring capability will be improved and be easier.	**Complexity** – building secure and large networks will be a cumbersome job.
Time and money – IoT may help to make quicker decisions and more efficient use of systems.	**Privacy/Security/Safety** – hacking is a potential threat and losing privacy is a major concern.

5.5 Future trends

The IoT technology has a tremendous potential to adapt and integrate with the majority of routine operations within FSCs. However, the speed at which the IoT is evolving daily, it will be hard to predict its full potential. There are several potentials and opportunities based on the IoT technology are being made available or being promised to people in forthcoming days. Some examples of its application could be in the elements of Industry 4.0, blockchain and intelligent packaging.

5.5.1 Industry 4.0

Industry 4.0 concept consists of digitizing the manufacturing by connecting the product, processes, and people within a factory environment to allow greater visibility of the entire production line and sometimes make decisions on its own. It fosters the creation of smart factories wherein novel technological developments will result in an increased level of communication between machines, sensor data processing at the machine level leading to greater flexibility to new production necessities. Furthermore, it also involves linking of information systems and communicating crucial data across the whole supply chain in order to enhance the operational efficiency. The ready availability of cheap sensor and information technologies have provided FSC actors with a more significant opportunity to bank on IoT powered applications. The actors who are connected via the internet are effective, productive, swifter, and considered to be smarter than their competitors who are not connected to the internet. Food and drink supply chains will be more benefitted as this will reduce the number of product recalls, make it more resource-efficient, and cater to broader customer demands. Industry 4.0 will change supply chain actors from being reactive to become more proactive, which will improve the overall sustainability of FSC by saving time, money, and vital resources. An additional advantage of Industry 4.0 is predictive maintenance feature wherein sensors installed on various machines in factories, transport vehicles, etc. could continuously gather and analyze the data on the current status of that equipment and predict its potential breakdown. This feature of predictive maintenance can be of significant help to FSC actors is seamless and timely production, transport, and cutting massively on waste and breakdowns, thereby increasing the efficiency of the supply chain. Furthermore, this will give an added advantage to the supply chain actors in this age of intensive competition, and this is all possible because of Industry 4.0 concept.

Generally, there are nine components of Industry 4.0, as shown in Fig. 5.5 that includes robotics, simulation, system integration, IoT, cybersecurity, cloud computing, additive printing, augmented reality, and big data. The combination of all these components can create significant opportunities, as well as new services or products.

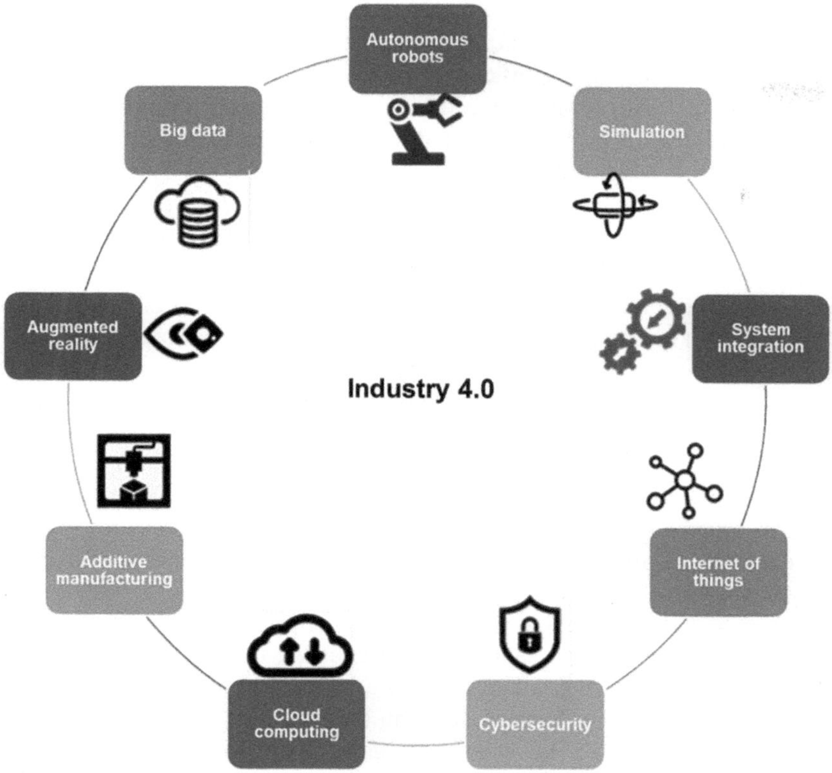

FIGURE 5.5 Nine components of Industry 4.0.

5.5.1.1 Robotics and automation

Due to consumer needs for personalized foods, political pressure on food handlers to improve on food safety and quality, shortage of workforce, and business competitiveness has resulted in supply chain actors exploring the usage of robotics and automation (Bader & Rahimifard, 2018, p. 37). These authors identified three motives for using robots in the food sector. First, due to their efficient performance that allows higher repeatability with speed and accuracy. Second, due to their durability, which will enable them to perform tedious and unwanted tasks in adverse conditions. Finally, they can be flexible, adaptable, and reconfigurable to produce a variety of products under an array of processes. These characteristics of robots make it ideal for its application in the food sector.

At Boomf, a marshmallow producer, the simple job of cutting marshmallow in squares by staff took 5 min resulting in a bottleneck in production, however, after the adoption of a robot this task was reduced to mere 17 s leading to increase in efficiency by 600% (Pendrous, 2016). Scott Technology has developed a lamb processing slaughterhouse that uses robots to process 600 carcasses per hour (Garfield, 2016). While KUKA Robotics has developed robots that can work 100% efficiently in adverse conditions such as −30°C (KUKA, 2018), ABB Robotics developed a robot named Flexipicker that can perform up to 200 picks/minute if fitted with an appropriate gripper depending on the foodstuff being handled (ABB, 2019).

5.5.1.2 Big data

Smart manufacturing involves creating manufacturing intelligence by collecting real-time information to provide timely and better decision making. The amount of information generated throughout the FSC with the adoption of IoT is expected to create an enormous volume of data, also termed as Big Data. Big data analytics is a process of analyzing big and varied data sets using techniques such as Statistical Process Monitoring to unearth unseen patterns, unidentified relationships, market trends and consumer preferences that can support in making informed business decisions. It will result in the development of smart manufacturing. As per McKinsey's report, the examination of big data will be the most significant factor for competitiveness, efficiency, innovation, growth, and development.

Big data complements the FSCs with product pricing, product promotion, new product development, and demand forecasting (Edwards, 2017). Jagtap and Duong (2019) demonstrated through a case study of a beverage manufacturer that how a new food product can be developed and launched in the shortest possible time. Harvesting big data opens new avenues such as actionable consumer insights leading to new product design, developing new products, and improving existing product lines (Zhan, Tan, Li, & Tse, 2018).

5.5.1.3 Augmented reality

Augmented reality is defined as a real-time direct or indirect experience of a real-world environment that is enhanced by computer-generated information. The user of augmented reality receives an interactive experience with the real world or product (Azuma et al., 2001). In today's highly competitive food sector, the application of augmented reality could provide an innovative simulative solution that supports in improving the supply chain processes. Apart from supply chain processes, it could also enhance the product/or process development resulting in reducing lead times, cost, and improving the quality.

Adopting Augmented reality within FSCs could provide valuable information about food or food products by scanning the interactive labels. It helps to attract customers' attention and to get ahead in the competitive market space. Also, it provides more detailed information compared to traditional labels. It allows customers to feel the product, enhancing their experience, and at the same time, gaining their loyalty (Visuar, 2012). Food manufacturers could use the augmented technology to provide consumers with nutritional information and composition of food products (Gallagher, 2017), as well as changing the color of bland-looking foods to make them more appealing (Mangiaracina, 2017).

5.5.1.4 Additive manufacturing

Additive manufacturing or 3D printing is an evolving technology that is estimated to reach £23.21 billion by 2022, and McKinsey's forecast that it will have around £422.78 billion annual economic impacts by 2025 (Markets & Markets, 2019). This technology has the potential to support the sustainability of FSCs through reduced consumption of resources, reduced carbon emissions, efficient resource utilization, better product life cycles, and aiding in the digitization of the supply chain (Nyman & Sarlin, 2014, pp. 4190–4199). Today, numerous companies have developed 3D printer to address such as FoodJet (De Grood Innovations), Chefjet and Cocojet (3D systems), Food form (RIG), Foodini (Natural Machines), Replicator (Makerbot), Choc Creator (Choc Edge) and Imagine3D (Essential Dynamics) (Sun et al., 2015). The advantage of using additive manufacturing is its ability to produce complex product shapes faster as compared to traditional methods of going through various unit operations to make a final product (Lipton, Cutler, Nigl, Cohen, & Lipson, 2015). Lipton et al. (2015) further demonstrated that products made with additive manufacturing had acceptable outcomes, and with some pre-processing and little modifications in the material supply can quickly meet consumer expectations.

5.5.1.5 Cloud technology

Earlier, FSC actors had to buy software, hardware, or seek services of IT professionals to manage software systems and expecting that it all supported and worked for them. However, this process was expensive and did not help the supply chain actors with low margins. But cloud computing could support them by providing third-party solutions to access its software on monthly subscription charges with no or minimum fees for hardware (Nessen & Cowan, 2012). The primary benefit of cloud computing, apart from lower upfront costs, is to have access to the internet connection, which is the only requirement for actors. The other benefits for supply chain actors include no hassle of upgrading the system, managing databases, virus patches, integrating systems, and archiving or recovery of data. Cloud computing also offers greater flexibility whenever actors' requirements change and gives a competitive advantage in a low margin but highly competitive marketplace.

5.5.1.6 Cybersecurity

As per Trustwave's (2019) Global Security Report, 7% of all reported data breaches occurred in the food and beverage sector (Trustwave, 2019).

Obsolete firewalls, insecure remote access, operating systems defects, weak security configurations, flawed security policies, lack of employee training, carelessness, and inadequate change control procedures are some of the security issues that affect FSC actors (Straka, 2014). The security risk to FSCs could be theft of data, the revelation of data to the public, loss of data, data corruption, and data fabrication (Woehl, 2020). In 2017, the Cadbury factory in Tasmania was affected by ransomware that resulted in the halting of production and growth of the company dropped 3% in the second quarter (Guardian, 2017). As the food sector works on low margins, the threat of data theft is very low, but if it occurs, then it could cause significant financial loss. For example, incapability to show a HACCP record could cost millions in product recall or disposal (Woehl, 2020). It could also result in delivery disruptions, amendment of recipe formulations, hacking into confidential information, and tampering threats (James, 2018). Therefore, with the increasing risk of cyber-attacks, protecting the data has become crucial. It has resulted in assessing the threat or chances of occurrences and developing solutions to avoid such cyber-attacks.

5.5.1.7 System integration

By having a universal, standardized data network system allows various entities or actors within FSC to be integrated and connected. It enables seamless communication and the possibility of the automated supply chain to be a reality. Epicor's Tropos Enterprise Resource Planning (ERP) tool is linked to various departments and functions within a factory and provides traceability, production planning, stock control, finance, engineering, purchasing, etc (Knott, 2018). Having integrated all the functions within the supply chain makes it possible to access information by all supply chain actors in real-time, resulting in savings and better decision-making (Hasnan & Yusoff, 2018, pp. 1–6).

5.5.1.8 Simulation

The simulation software allows FSC actors to use real-time data to model the physical supply chain system. It will enable the actors to test, analyze, and optimize the parameters virtually former to any real production/or changeover takes place (Rostkowska, 2014; Rüßmann et al., 2015). Various simulation software, such as ARENA Simulation Software, TrackSYS, Tecnomatix, and FlexSim, are currently available in the market. A case study was conducted to design a new brewery with all the supply chain functions within the brewery that were simulated to understand, analyze, and evaluate various strategies and scenarios beforehand (Siemens, 2015). This action led to better decision-making in the early start-up stages with regards to cost planning, minimizing downtimes, and any other production issues.

5.5.2 Blockchain

A blockchain ensures that all the participating parties can view all transactions at the same time and in real-time. In FSCs, the end consumer/or retailer would be able to see with whom their manufacturing partner or suppliers had conducted transactions. Since all these transactions are not stored in any single database, it is almost challenging to breach the information.

Global FSCs are often nontransparent and lack trust due to their complexity and confidentiality. Food crises such as the 2013 horse meat scandal in Europe (BBC, 2013) and the 2008 Chinese milk scandal (Branigan, 2008) has made transparency an ultimate criterion for every FSC. The IBM blockchain technology platform, also called Food Trust, has demonstrated how Walmart can trace mangoes to their source in approximately 2 s, which earlier took a week (IBM, n.d.). It would reduce a significant amount of time in case of a product safety recall and confine/or restrict the product from being sold to consumers.

Blockchain's ability to deliver every supply chain actor with end-to-end visibility allows for tracking the movement of food through the supply chain, providing food history, as well as source location of food products (Martindale, Hollands, Swainson, & Keogh, 2018). Due to the tamper-proof nature of blockchain technology, it is not possible for any actor within the supply chain to change, delete, or amend any data without the approval of others in the network. This feature allows to reduce the risk of food frauds, reduce food transportation times, enhances efficiency, improve inventory management, and minimizes wastage and cost. Apart from transparency, resource efficiency, and information accuracy, Martindale et al. (2018) highlighted that food recall and processes, data inspection and food certification process as the areas, which could benefit most from blockchain technology.

5.5.3 Intelligent packaging

Food packaging has various functions, such as physical protection (shock, compression, vibration, bacteria, temperature), barrier protection (oxygen, dust, water vapor), information transmission (transport, nutritional information, how to use, recycle, or dispose), convenience, security, containment, marketing, and portion control. The packaging material needs to be inert to the food product packed inside.

Ongoing IoT research and technological advances have resulted in the development of interactive, aware, and intelligent food packaging systems (Vanderroost, Ragaert, Devlieghere, & De Meulenaer, 2014). This packaging system can help in continuous monitoring of food for its quality and safety through sensing, detecting, or recording any changes in the food product, its packaging, or the environment it is stored in. Other than providing information on food quality and safety, it can also offer health and nutritional information, product sourcing, processing, its movement in the supply chain, preparing instructions, packaging date, batch, weight, and any other information that might be beneficial to consumer (Fang, Zhao, Warner, & Johnson, 2017). Smart or intelligent packaging utilizes data carriers, indicators, and sensors to monitor food quality and safety from farm to fork (Müller & Schmid, 2019). They further collated the information of the companies and currently available intelligent packaging systems in the market, as shown in Table 5.5.

Smart or intelligent packaging is defined as a subset of IoT (Lingle, 2015). It can leverage both IoT and Big data to create an interface, as well as establish

TABLE 5.5 Commercially available intelligent packaging systems (Müller & Schmid, 2019).

Technologies	Type	Name	Company
Data carriers	Radiofrequency identification technologies	Easy2log Intelligent Box CS8304 Temptrip	CAEN RFID Srl Mondi Pic Convergence Systems Ltd. Temptrip LLC
Indicators	Time-temperature indicators	CheckPoint Fresh-Check On Vu MonitorMark eO Keep-it Cook-Chex Timestrip Colour-Therm TopCryo	Vitsab LifeLines Ciba Speciality Chemical and FreshPoint 3M, Minnesota CRYOLOG S.A. Keep-it Technologies Pymah Corp. Timestrip Plc Colour-Therm TRACEO
	Freshness indicators	Fresh Tag SensorQ RipeSense Food fresh	COX Technologies DSM NV and Food Quality Sensor RIpSense and ort Research Vanprob
Sensors	Biosensors	Toxin Guard Food Sentinel	Toxin Alert SIRA Technologies
	Gas sensors	Tell-Tab Ageless Eye O2Sense	IMPAK Corporation Mitsubishi Gas Chemical Inc. FreshPoint Lab

communication with sensors on the packaging, including RFID, Bluetooth, Near-Field communications (NFC), and smart labels. With the support of these technologies, supply chain actors can track and monitor various environmental conditions throughout the FSC. Application of IoT-based packaging can enhance better decision making, minimizes the risk of product recalls, transparency, and increased accountability. However, there are some barriers to the acceptance of smart or intelligent packaging such as extra cost, the unreliability of indicating devices, food safety, and regulatory issues (e.g., migration of materials into the product), recycling conditions, and environmental regulations (AZoM, 2003).

5.5.4 Artificial intelligence

Artificial intelligence (AI) is an area of computer science that deals with intelligent machines and deploys two of the most common algorithms, i.e., deep learning and machine learning. These algorithmic models learn from data and are used by various supply chain actors to make predictions and offer the most appropriate solutions to any problem. It has attracted many businesses, including FSC actors, to deal with the complicated and varied data. As per McKinsey's report, the food processing and handling industry is worth around $100 billion and will continue to grow at a Compound annual growth rate (CAGR) of 5% until 2021 (McKinsey, 2018) and to support this growth AI would play a major role.

TOMRA, a leader in food sorting technology, uses machine learning to sort foods for their optimized use. For example, apart from sorting good and bad potatoes, it uses machine learning to identify the potatoes that are best suited for french fries or potato chips production with minimum waste generation (TOMRA, 2015). Similarly, Kewpie Corporation, a baby food production company, used AI to pick out the defective diced potato from the ingredients at their Tosu factory (Ogino, 2017). KanKan uses AI trained cameras in a food production facility to ensure that staff is wearing masks and hairnets, and any deviations can be identified and corrected immediately (Remark Holdings, 2020). SOCIP system developed by Martec uses AI to autonomously optimize the cleaning process for food manufacturing equipment. It uses multisensor technology to monitor food and microbial debris and controls the cleaning system until this debris is completely removed (Martec, 2020). Coca-Cola is using AI to develop new products. For example, they collected data from self-service machines to analyze customer preference and taste and developed Cherry Sprite (Purdy, 2017). Kellogg company teamed up with IBM's Chef Watson to launch Bear Naked customized granola using AI to suggest to people the ingredients they could add to their granola and whether the ingredients they added would taste good or not (Kellogg's, 2016).

5.6 Conclusion

IoT technology has been around for a few years now and has reached a mature stage. We believe that a large number of benefits of implementing the IoT in the FSC outweigh the disadvantages. Since the costs associated with digitization of supply chains have dropped considerably with the improvement in IoT technologies, more food manufacturers are ready to apply such technologies. Leading the adoption of this technology will result in reaping benefits for the longer term and gaining significant market share before everyone follows suit.

Adopting IoT technologies in every aspect of FSCs would help to optimize and increase the sustainability of the whole supply chain. Benefits such as better transparency, monitoring, and control over various food operational activities would enable IoT to enhance production, alert and avoid emerging issues and automate multiple functions within FSCs.

5.7 Chapter summary

This chapter has presented an overview of the IoT and its potential usage within the food sector. It has also shown current IoT applications in FSCs and described the pros and cons of IoT implementation. IoT presents more benefits than disadvantages, and therefore, it is recommended that more supply chain actors should adopt IoT technologies. This chapter has also emphasized the future trends and the importance of elements of industry 4.0, as well as how these elements will support the generation of business opportunities and new product developments.

On the other hand, this chapter has also demonstrated that supply chain activities suffer due to a lack of awareness or availability of accurate data in real-time. Monitoring and analyzing the activity of each actor is an essential step toward attaining efficiency and transparency. The availability of authentic information is fundamental for bringing changes across the various activities within FSCs. By collecting the data in real-time, and analyzing that data, decisions can be made with consideration to not only economic profits but also environmental impacts, and thereby leading to an optimized and improved food supply chain.

References

Abad, E., Palacio, F., Nuin, M., De Zarate, A. G., Juarros, A., Gómez, J. M., et al. (2009). RFID smart tag for traceability and cold chain monitoring of foods: Demonstration in an intercontinental fresh fish logistic chain. *Journal of Food Engineering, 93*(4), 394–399.

ABB. (2019). *Robotics product specification IRB 360.* [Online] Available at https://library.e.abb.com/public/0a5e13aa301e4325bc270a76e4636676/3HAC029963%20PS%20IRB%20360-en.pdf. (Accessed 1 January 2020).

Accenture. (2014). *Accenture technology vision 2014 - every business is a digital business: From digitally disrupted to digital disrupter.* [Online] Available at https://www.accenture.com/t20160316T224443__w__/fr-fr/_acnmedia/PDF-10/Accenture-Technology-Vision_2014.pdf. (Accessed 12 March 2019).

AZoM. (2003). *Smart packaging - intelligent packaging for food, beverages, pharmaceuticals and household products* [Online] Available at: https://www.azom.com/article.aspx?ArticleID=2152. (Accessed 3 January 2020).

Azuma, R., Baillot, Y., Behringer, R., Feiner, S., Julier, S., & MacIntyre, B. (2001). Recent advances in augmented reality. *IEEE Computer Graphics and Applications, 21*(6), 34–47.

Bader, F., & Jagtap, S. (2020). Internet of things-linked wearable devices for managing food safety in the healthcare sector. In N. Dey, A. S. Ashour, & C. Bhatt (Eds.), *Wearable and implantable medical devices - applications and challenges. s.l.* (pp. 229–253). Academic Press.

Bader, F., & Rahimifard, S. (2018). *Challenges for industrial robot applications in food manufacturing. s.l.* ACM.

Barandun, G., Soprani, M., Naficy, S., Grell, M., Kasimatis, M., Chiu, K. L., et al. (2019). Cellulose fibers enable near-zero-cost electrical sensing of water-soluble gases. *ACS Sensors, 4*(6), 1662–1669.

BBC. (2013). *Findus beef lasagne contained up to 100% horsemeat, FSA says.* [Online] Available at https://www.bbc.co.uk/news/uk-21375594. (Accessed 11 January 2020).

Bi, Z., Da Xu, L., & Wang, C. (2014). Internet of things for enterprise systems of modern manufacturing. *IEEE Transactions on Industrial Informatics, 10*(2), 1537–1546.

Branigan, T. (2008). *Chinese figures show fivefold rise in babies sick from contaminated milk* [Online] Available at: https://www.theguardian.com/world/2008/dec/02/china. (Accessed 11 January 2020).

Bunse, K., Vodicka, M., Schönsleben, P., Brülhart, M., & Ernst, F. O. (2011). Integrating energy efficiency performance in production management—gap analysis between industrial needs and scientific literature. *Journal of Cleaner Production, 19*(6–7), 667–679.

Chen, M., Tian, Y., Fortino, G., Zhang, J., & Humar, I. (2018). Cognitive internet of vehicles. *Computer Communications, 120*, 58–70.

Chetty, M., Tran, D., & Grinter, R. E. (2008). *Getting to green: Understanding resource consumption in the home.* Seoul: ACM.

Da Xu, L., He, W., & Li, S. (2014). Internet of things in industries: A survey. *IEEE Transactions on Industrial Informatics, 10*(4), 2233–2243.

Daugherty, P., Banerjee, P., Negm, W., & Alter, A. E. (2015). *Driving unconventional growth through the industrial internet of things.* [Online] Available at https://www.accenture.com/ph-en/_acnmedia/Accenture/next-gen/reassembling-industry/pdf/Accenture-Driving-Unconventional-Growth-through-IIoT.pdf. (Accessed 2 May 2019).

De Freitas Melo, D., De Souza lage, E., Vagner Rocha, A., & De Jesus Cardoso, B. (2017). *Improving the consumption and water heating efficiency in smart buildings.* IEEE: Stony Brook.

Drenoyanis, A., Raad, R., Wady, I., & Krogh, C. (2019). Implementation of an IoT based radar sensor network for wastewater management. *Sensors, 19*(2), 254.

Dunbrack, L., Ellis, S., Hand, L., Knickle, K., & Turner, V. (2016). *IoT and digital transformation: A tale of four industries.* IDC White Paper.

Edie. (2016). *How the Internet of Things is transforming sustainability.* [Online] Available at https://www.edie.net/library/How-the-Internet-of-Things-is-transforming-sustainability/6679. (Accessed 30 January 2019).

Edwards, J. (2017). *Developing a taste for big data.* [Online] Available at https://www.hpe.com/us/en/insights/articles/developing-a-taste-for-big-data-1710.html. (Accessed 14 December 2019).

Fang, Z., Zhao, Y., Warner, R. D., & Johnson, S. K. (2017). Active and intelligent packaging in meat industry. *Trends in Food Science and Technology, 61*, 60–71.

Ferreira, M., Fernandes, R., Conceiçao, H., Gomes, P., d'Orey, P. M., Moreira-Matias, L., et al. (June 2012). Vehicular sensing: Emergence of a massive urban scanner. In *International conference on sensor systems and software* (pp. 1–14). Berlin, Heidelberg: Springer.

Floyer, D. (2013). *Defining and sizing the industrial internet.* [Online] Available at http://wikibon.org/wiki/v/Defining_and_Sizing_the_Industrial_Internet. (Accessed 6 April 2019).

Gallagher, J. (2017). *Learn what your food is made of using augmented reality.* [Online] Available at https://mobile-ar.reality.news/news/apple-ar-learn-what-your-food-is-made-using-augmented-reality-0178874/. (Accessed 20 January 2020).

Garcia-Garcia, G., Woolley, E., & Rahimifard, S. (2015). A framework for a more efficient approach to food waste. *International Journal of Food Engineering, 1*(1), 65–72.

Garcia-Garcia, G., Woolley, E., Rahimifard, S., Colwill, J., White, R., & Needham, L. (2017). A methodology for sustainable management of food waste. *Waste and Biomass Valorization, 8*(6), 2209–2227.

Garfield, L. (2016). *The world's biggest meat producer is planning to test out robot butchers.* [Online] Available at https://www.businessinsider.com/jbs-meatpacking-testing-robot-butchers-2016-10?r=US&IR=T. (Accessed 2 January 2020).

Gontarz, A. M., Hampl, D., Weiss, L., & Wegener, K. (2015). Resource consumption monitoring in manufacturing environments. *Procedia Cirp, 26*, 264–269.

Guardian. (2017). *Petya cyber-attack: Cadbury factory hit as ransomware spreads to Australian businesses.* [Online] Available at https://www.theguardian.com/technology/2017/jun/28/petya-cyber-attack-cadbury-chocolate-factory-in-hobart-hit-by-ransomware. (Accessed 11 January 2020).

Gubbi, J., Buyya, R., Marusic, S., & Palaniswami, M. (2013). Internet of things (IoT): A vision, architectural elements, and future directions. *Future Generation Computer Systems, 29*(7), 1645–1660.

Haller, S., Karnouskos, S., & Schroth, C. (2009). The internet of things in an enterprise context. In J. Domingue, D. Fensel, & P. Traverso (Eds.), *Future internet – FIS 2008* (pp. 14–28). Berlin: Springer.

Hardcastle, J. L. (2013). *'Internet of Things' can cut emissions 19%, report finds* [Online] Available at: https://www.environmentalleader.com/2013/02/internet-of-things-can-cut-emissions-19-report-finds/. (Accessed 1 January 2020).

Hasnan, N. Z. N., & Yusoff, Y. M. (2018). *Short review: Application areas of industry 4.0 technologies in food processing sector. s.l.* IEEE.

Henningsson, S., Hyde, K., Smith, A., & Campbell, M. (2004). The value of resource efficiency in the food industry: A waste minimisation project in East Anglia, UK. *Journal of Cleaner Production, 12*(5), 505–512.

Hong, I., Park, S., Lee, B., Lee, J., Jeong, D., & Park, S. (2014). IoT-based smart garbage system for efficient food waste management. *The Scientific World Journal, 2014*, 13.

IBM, n.d. *IBM Food Trust. A new era for the world's food supply.* [Online] Available at https://www.ibm.com/uk-en/blockchain/solutions/food-trust[Accessed 12 January 2020].

ITU-T. (2012). *Overview of the internet of things.* [Online] Available at https://www.itu.int/ITU-T/recommendations/rec.aspx?rec=y. (Accessed 9 March 2019), 2060.

Jagtap, S. (2019). *Utilising the Internet of Things concepts to improve the resource efficiency of food manufacturing.* Loughborough: Loughborough University.

Jagtap, S., Bhatt, C., Thik, J., & Rahimifard, S. (2019a). Monitoring potato waste in food manufacturing using image processing and internet of things approach. *Sustainability, 11*(11), 3173.

Jagtap, S., & Duong, L. N. K. (2019). Improving the new product development using big data: A case study of a food company. *British Food Journal, 121*(11), 2835–2848.

Jagtap, S., & Rahimifard, S. (2017). Utilisation of internet of things to improve resource efficiency of food supply chains. *Chania, Greece, CEUR Workshop Proceedings*, 8–19.

Jagtap, S., & Rahimifard, S. (2018). Real-time data collection to improve energy efficiency in food manufacturing. In *Castelo branco, +AGRO 2018 – international congress on organizational management*. Energy Efficiency and Occupational Health.

Jagtap, S., & Rahimifard, S. (2019a). The digitisation of food manufacturing to reduce waste–case study of a ready meal factory. *Waste Management, 87*, 387–397.

Jagtap, S., & Rahimifard, S. (2019b). Unlocking the potential of the internet of things to improve resource efficiency in food supply chains. In A. Theodoridis, A. Ragkos, & M. Salampasis (Eds.), *Innovative approaches and applications for sustainable rural development* (pp. 287–301). Cham: Springer.

Jagtap, S., Rahimifard, S., & Duong, L. N. K. (2019b). Real-time data collection to improve energy efficiency: A case study of food manufacturer. *Journal of Food Processing and Preservation*. p.e14338. https://doi.org/10.1111/jfpp.14338

Jagtap, S., Skouteris, G., Choudhari, V., & Rahimifard, S. (2018). *Improving water efficiency in the beverage industry with the internet of things*. Changa, s.n.

James, N. C. K. (2018). *Cyberterrorism: How food companies are planning for threat of cyber-security risks*. [Online] Available at https://www.foodqualityandsafety.com/article/cyberterrorism-food-industry-cybersecurity-risks/?singlepage=1. (Accessed 11 January 2020).

Jhingran, P. (2016). *Reducing food spoilage using IoT*. [Online] Available at https://www.linkedin.com/pulse/reducing-food-spoilage-using-iot-prashant-jhingran. (Accessed 11 January 2018).

Kellogg, 's (2016). *Bear Naked® Introduces Customized Granola, Powered by IBM Chef Watson* [Online] Available at: http://newsroom.kelloggs.com/2016-04-19-Bear-Naked-Introduces-Customized-Granola-Powered-by-IBM-Chef-Watson. (Accessed 11 January 2020).

Knott, M. (2018). *How Industry 4.0 offers a smarter future*. [Online] Available at https://www.foodmanufacture.co.uk/Article/2018/07/26/Food-firms-begin-to-embrace-the-power-of-Industry-4.0. (Accessed 15 January 2020).

KUKA. (2018). *KUKA robots for the food industry*. [Online] Available at https://www.kuka.com/-/media/kuka-downloads/imported/9cb8e311bfd744b4b0eab25ca883f6d3/kuka-robots-for-food-industry.pdf?rev=391363ca61de4a8cbc334c089e39f8fc. (Accessed 2 January 2020).

Kumar, N. M., Dash, A., & Singh, N. K. (2018). *Internet of Things (IoT): An opportunity for energy-food-water Nexus. s.l.* IEEE.

Kumar, N. M., & Mallick, P. K. (2018). The Internet of Things: Insights into the building blocks, component interactions, and architecture layers. *Procedia computer science, 132*, 109–117.

Kuswandi, B., Wicaksono, Y., Abdullah, A., Heng, L. Y., & Ahmad, M. (2011). Smart packaging: Sensors for monitoring of food quality and safety. *Sensing and Instrumentation for Food Quality and Safety, 5*(3–4), 137–146.

Kyriazis, D., Varvarigou, T., White, D., Rossi, A., & Cooper, J. (2013). *Sustainable smart city IoT applications: Heat and electricity management & Eco-conscious cruise control for public transportation. s.l.* IEEE.

Lacey, M., Lisachuk, H., Giannopoulos, A., & Ogura, A. (2015). *Shipping smarter: IoT opportunities in transport and logistics*. [Online] Available at https://www2.deloitte.com/content/dam/insights/us/articles/iot-in-shipping-industry/DUP1271_IoT_Transportation-and-Logistics_MASTER.pdf. (Accessed 7 January 2019).

Lee, E. A. (2013). *Cyber-physical systems, A fundamental intellectual challenge*. [Online] Available at https://ptolemy.berkeley.edu/projects/chess/pubs/1045.html. (Accessed 26 December 2019).

Lee, J., Bagheri, B., & Kao, H. A. (2015). A cyber-physical systems architecture for industry 4.0-based manufacturing systems. *Manufacturing Letters, 3*, 18–23.

Lingle, R. (2015). *Smart packaging experts see applications and opportunity*. [Online] Available at https://www.packagingdigest.com/smart-packaging/smart-packaging-experts-see-applications-opportunity1507. (Accessed 5 January 2020).

Lipton, J. I., Cutler, M., Nigl, F., Cohen, D., & Lipson, H. (2015). Additive manufacturing for the food industry. *Trends in Food Science and Technology, 43*(1), 114–123.

Lu, B. H., Bateman, R. J., & Cheng, K. (2006). RFID enabled manufacturing: Fundamentals, methodology and applications. *International Journal of Agile Systems and Management, 1*(1), 73–92.

Maksimovic, M. (2017). The role of green internet of things (G-IoT) and big data in making cities smarter, safer and more sustainable. *International Journal of Computing and Digital Systems, 6*(04), 175–184.

Ma, C., Li, Y., Yin, G., & Ji, J. (July 2012). The monitoring and information management system of pig breeding process based on internet of things. In *2012 fifth international conference on information and computing science* (pp. 103–106). IEEE.

Mangiaracina, B. (2017). *Augmented reality can make good-for-you food look more attractive, making it easier to eat well*. [Online] Available at https://augmented.reality.news/news/augmented-reality-can-make-good-for-you-food-look-more-attractive-making-easier-eat-well-0176795/. (Accessed 13 January 2020).

Markets, & Markets. (2019). *3D printing market by offering (printer, material, software, service), process (binder jetting, direct energy deposition, material extrusion, material jetting, powder bed fusion), application, vertical, technology, and geography - global forecast to 2024*. [Online] Available at https://www.marketsandmarkets.com/Market-Reports/3d-printing-market-1276.html. (Accessed 21 December 2019).

Markman, G., & Krause, D. (2014). Special topic forum on theory building surrounding sustainable supply chain management. *Journal of Supply Chain Management, 50*(4), III.

Martec. (2020). SELF-OPTIMISING CIP. [Online] Available at: http://www.marplug-technologies.com/SOCIP. (Accessed 9 January 2020).

Martindale, W., Hollands, T., Swainson, M., & Keogh, J. G. (2018). Blockchain or bust for the food industry? *Food Science and Technology, 33*(4).

Matindoust, S., Baghaei-Nejad, M., Abadi, M. H. S., Zou, Z., & Zheng, L. R. (2016). Food quality and safety monitoring using gas sensor array in intelligent packaging. *Sensor Review, 36*(2), 169–183.

Mattoli, V., Mazzolai, B., Mondini, A., Zampolli, S., & Dario, P. (2010). Flexible tag datalogger for food logistics. *Sensors and Actuators A: Physical, 162*(2), 316–323.

McKinsey. (2018). *Food Processing and Handling - Ripe for disruption?* [Online] Available at: https://www.mckinsey.com/~/media/mckinsey/industries/advanced%20electronics/our%20insights/whats%20ahead%20for%20food%20processing%20and%20handling/mckinsey-on-food-processing-and-handling-ripe-for-disruption.ashx. (Accessed 20 January 2020).

van der Meulen, R. (2017). *Gartner says 8.4 billion connected "things" will Be in use in 2017, up 31 percent from 2016.* [Online] Available at https://www.gartner.com/en/newsroom/press-releases/2017-02-07-gartner-says-8-billion-connected-things-will-be-in-use-in-2017-up-31-percent-from-2016. (Accessed 23 January 2019).

Müller, P., & Schmid, M. (2019). Intelligent packaging in the food sector: A brief overview. *Foods, 8*(1), 16.

Nessen, T., & Cowan, J. (2012). *Cloud computing maximizes usefulness of food safety system data.* [Online] Available at https://www.foodqualityandsafety.com/article/cloud-computing-maximizes-usefulness-of-food-safety-system-data/?singlepage=1. (Accessed 5 January 2020).

Nyman, H. J., & Sarlin, P. (2014). *From bits to atoms: 3D printing in the context of supply chain strategies. s.l.* IEEE.

Ogino, T. (2017). *How AI can help make safer baby food (and other products)* [Online] Available at: https://www.blog.google/topics/google-cloud/how-ai-can-help-make-safer-baby-food-and-other-products/. (Accessed 10 January 2020).

Ostojić, G., Stankovski, S., Tegeltija, S., Đukić, N., & Tejić, B. (2017). Implementation of IoT for food wastage minimisation. Novi Sad, Serbia. In *XVII international scientific conference on industrial systems.*

Pang, Z. (2013). *Technologies and architectures of the Internet-of-Things (IoT) for health and well-being, s.l.* KTH Royal Institute of Technology.

Pathare, P. B., Roskilly, A. P., & Jagtap, S. (2019). Energy efficiency in meat processing. In P. D. Gaspar, & P. D. da Silva (Eds.), *Novel technologies and systems for food preservation. s.l* (pp. 78–107). IGI Global.

Pendrous, R. (2016). *Marshmallow maker Boomf installs robotic ultrasonic cutter.* [Online] Available at https://www.foodmanufacture.co.uk/Article/2016/11/17/Marshmallow-maker-Boomf-installs-robotic-ultrasonic-cutter. (Accessed 15 December 2019).

Perera, C., Ranjan, R., Wang, L., Khan, S. U., & Zomaya, A. Y. (2015). Big data privacy in the internet of things era. *IT Professional, 17*(3), 32–39.

Petersen, J. E., et al. (2007). Dormitory residents reduce electricity consumption when exposed to real-time visual feedback and incentives. *International Journal of Sustainability in Higher Education, 8*(1), 16–33.

Purdy, C. (2017). *AI told Coca-Cola to make Cherry Sprite. So it did* [Online] Available at: https://qz.com/1088885/coca-cola-uses-ai-robots-to-invent-new-sodas-like-cherry-sprite/. (Accessed 7 January 2020).

Rahimifard, S., Woolley, E., Webb, D. P., Garcia-Garcia, G., Stone, J., Jellil, A., et al. (2017). *Forging new frontiers in sustainable food manufacturing. s.l.* Springer.

Ramesh, M. V., & Das, R. N. (2011). August. A public transport system based sensor network for fake alcohol detection. In *International conference on wireless communications and applications* (pp. 137–144). Berlin, Heidelberg: Springer.

Holdings, Remark (2020). *AI Simplified* [Online] Available at: https://www.remarkholdings.com/. (Accessed 9 January 2020).

Robles, T., Alcarria, R., de Andrés, D. M., de la Cruz, M. N., Calero, R., Iglesias, S., et al. (2015). An IoT based reference architecture for smart water management processes. *Journal of Wireless Mobile Networks Ubiquitous Computing, and Dependable Applications, 6*(1), 4–23.

Rostkowska, M. (2014). Simulation of production lines in the education of engineers: How to choose the right software? *Management and Production Engineering Review, 5*(4), 53–65.

Ruiz-Garcia, L., & Lunadei, L. (2011). The role of RFID in agriculture: Applications, limitations and challenges. *Computers and Electronics in Agriculture, 79*(1), 42–50.

Ruiz-Garcia, L., Lunadei, L., Barreiro, P., & Robla, I. (2009). A review of wireless sensor technologies and applications in agriculture and food industry: State of the art and current trends. *Sensors, 9*(6), 4728–4750.

Rüßmann, M., et al. (2015). Industry 4.0: The future of productivity and growth in manufacturing industries. *Boston Consulting Group, 9*(1), 54–89.

Schaller, E., Bosset, J. O., & Escher, F. (1998). 'Electronic noses' and their application to food. *Lebensmittel-Wissenschaft+ Technologie, 31*(4), 305–316.

Sethi, P., & Sarangi, S. R. (2017). Internet of things: Architectures, protocols, and applications. *Journal of Electrical and Computer Engineering, 2017*, 25.

Shanahan, C., Kernan, B., Ayalew, G., McDonnell, K., Butler, F., & Ward, S. (2009). A framework for beef traceability from farm to slaughter using global standards: An Irish perspective. *Computers and Electronics in Agriculture, 66*(1), 62–69.

Sheppard, P., Garcia-Garcia, G., Stone, J., & Rahimifard, S. (2020). A complete decision-support infrastructure for food waste valorisation. *Journal of Cleaner Production, 247*, 119608.

Shrouf, F., & Miragliotta, G. (2015). Energy management based on internet of things: Practices and framework for adoption in production management. *Journal of Cleaner Production, 100*(2015), 235–246.

Siemens. (2015). *Food & beverage:virtually brewing*. [Online] Available at https://webservices.siemens.com/adtree/newsdb/detail/html.aspx?filename=PN2015-01-04-Virtuelle_Braukunst.xml&language=en&design=1. (Accessed 11 January 2020).

Skilton, P. F., & Robinson, J. L. (2009). Traceability and normal accident theory: How does supply network complexity influence the traceability of adverse events? *Journal of Supply Chain Management, 45*(3), 40–53.

Stich, V., Brandenburg, U., & Kropp, S. (2011). *Benchmarking concept for energy efficiency in the manufacturing industry—A holistic energy efficiency model*. Berlin: Springer.

Straka, C. (2014). *Cybersecurity in food*. [Online] Available at https://www.foodqualityandsafety.com/article/cybersecurity-in-food-and-beverage-industry/. (Accessed 1 January 2020).

Sundmaeker, H., Guillemin, P., Friess, P., & Woelfflé, S. (2010). Vision and challenges for realising the internet of things. *Cluster of European Research Projects on the Internet of Things, European Commision, 3*(3), 34–36.

Sundmaeker, H., Verdouw, C., Wolfert, S., & Pérez Freire, L. (2016). Internet of food and farm 2020. In O. Vermesan, & P. Friess (Eds.), *Digitising the Industry-Internet of Things connecting physical, digital and virtual worlds. s.l* (pp. 129–151). River Publishers.

Sun, J., Peng, Z., Zhou, W., Fuh, J. Y., Hong, G. S., & Chiu, A. (2015). A review on 3D printing for customized food fabrication. *Procedia Manufacturing, 1*, 308–319.

TOMRA. (2015). *Artifical intelligence and food sorting* [Online] Available at: https://food.tomra.com/blog/en/sorting/food/food-news/2015/leading-the-resource-revolution. (Accessed 11 January 2020).

Trollman, H., & Trollman, F. (2019). A sustainability assessment of smart innovations for mass production, mass customisation and direct digital manufacturing. In *Mass production processes. s.l.* IntechOpen.

Trustwave. (2019). *2019 Trustwave global security report*. [Online] Available at https://securenation.net/wp-content/uploads/2018/12/Trustwave_GSR_2019_Final-.pdf. (Accessed 1 January 2020).

GOV.UK. (2015). *Budget 2015: Some of the things we've announced*. [Online] Available at https://www.gov.uk/government/news/budget-2015-some-of-the-things-weve-announced. (Accessed 18 January 2018).

Vanderroost, M., Ragaert, P., Devlieghere, F., & De Meulenaer, B. (2014). Intelligent food packaging: The next generation. *Trends in Food Science and Technology, 39*(1), 47–62.

Vermesan, O., & Friess, P. (2013). *Internet of things: Converging technologies for smart environments and integrated ecosystems. s.l.* River publishers.

Vermesan, O., Harrison, M., Vogt, H., Kalaboukas, K., Tomasella, M., Wouters, K., et al. (2010). *Vision and challenges for realising the Internet of Things*. Brussels: Cluster of European Research Projects on the Internet of Things & European Commission.

Visich, J. K., Li, S., Khumawala, B. M., & Reyes, P. M. (2009). Empirical evidence of RFID impacts on supply chain performance. *International Journal of Operations and Production Management, 29*(12), 1290–1315.

Visich, J. K., Powers, J. T., & Roethlein, C. J. (2009). Empirical applications of RFID in the manufacturing environment. *International Journal of Radio Frequency Identification Technology and Applications, 2*(3–4), 115–132.

Visuar. (2012). *Food industry augmented reality*. [Online] Available at http://www.visuar.es/eng/realidad_aumentada_alimentacion_visuar_eng.html. (Accessed 11 January 2020).

Wamba, S. F., & Chatfield, A. T. (2011). The impact of RFID technology on warehouse process innovation: A pilot project in the TPL industry. *Information Systems Frontiers, 13*(5), 693–706.

Wang., J, & Yue, H. (2017). Food safety pre-warning system based on data mining for a sustainable food supply chain. *Food Control, 73*, 223–229.

Wang, J. X., Liu, Y., Lei, Z. B., Wu, K. H., Zhao, X. Y., Feng, C., et al. (2018). *Smart water lora IoT system*. Nagoya, ACM.

Weinert, N., Chiotellis, S., & Seliger, G. (2011). Methodology for planning and operating energy-efficient production systems. *CIRP Annals, 60*(1), 41–44.

Woehl, R. (2020). *Cyber security threats to the food industry: Consider the cloud*. [Online] Available at https://globalfoodsafetyresource.com/cyber-security-threats-food-industry-consider-cloud/. (Accessed 15 January 2020).

Zhang, Y., Chen, B., & Lu, X. (2012). Intelligent monitoring system on refrigerator trucks based on the internet of things. In P. Sénac, M. Ott, & A. Seneviratne (Eds.), *Wireless communications and applications* (pp. 201–206). Berlin, Heidelberg: Springer.

Zhan, Y., Tan, K. H., Li, Y., & Tse, Y. K. (2018). Unlocking the power of big data in new product development. *Annals of Operations Research, 270*, 577–595.

Zhong, R. Y., Li, Z., Pang, L. Y., Pan, Y., Qu, T., & Huang, G. Q. (2013). RFID-enabled real-time advanced planning and scheduling shell for production decision making. *International Journal of Computer Integrated Manufacturing, 26*(7), 649–662.

Zhong, R. Y., Xu, X., Klotz, E., & Newman, S. T. (2017). Intelligent manufacturing in the context of industry 4.0: A review. *Engineering, 3*(5), 616–630.

Chapter 6

Innovative distribution and delivery of food

Arghya Ray[1], **Pradip Kumar Bala**[2]
[1]*Information Technology Area, FORE School of Management, New Delhi, India;* [2]*Indian Institute of Management Ranchi, Suchana Bhawan, Ranchi, Jharkhand, India*

6.1 Introduction

In recent years, the advancement of various technological innovations and the penetration of the internet among masses has seen a growth of various internet-based services or better known as e-services in almost every sector (Baroková, Kryvinska, & Strauss, 2016; Ray, Bala, & Dasgupta, 2019, 2020). Even in the food industry, providers are shifting from traditional ways of providing services to online-based ones. While some food delivery operators are mainly focused on the delivery of raw and fresh food items, some food delivery providers focus on the delivery of prepared food to customers. Past researchers have mostly used the term online food delivery services to refer to the ordering and delivery of food from various restaurants through a website or mobile application (Pigatto, Machado, Negreti, & Machado, 2017). Some researchers have also proposed the term food delivery apps for referring to online food delivery services. The difference between online food delivery services and the food delivery applications is that orders can be placed over the internet in case of online food delivery services, but for food delivery applications, the orders are placed using mobile applications. The service providers are responsible for accepting orders, tracking orders, and receiving payments, but they are not liable for the actual food preparation (Pigatto et al., 2017). The online food delivery service providers can be categorized as the restaurants itself (like Domino's, Pizza Hut, KFC) and multiple types of restaurant intermediaries (like Foodpanda, Zomato, and Uber Eats) (Sjahroeddin, 2018; Yeo, Goh, & Rezaei, 2017). In case of online food delivery services, the usual process that is followed is that users search for a restaurant on the website or mobile application from the service provider of their choice, choose from available food items, and provide their delivery address and place an order (Pigatto et al., 2017). Once the order is placed, it is up to the restaurant to prepare the ordered

Food Technology Disruptions. https://doi.org/10.1016/B978-0-12-821470-1.00004-5

item within a certain specified period, pack the prepared food item and hand over the packed, prepared food item to the delivery staff, who gets it delivered to the delivery address as mentioned by the customer. In case of delivery of raw food items, like, fish, meat, cereals, canned food, etc., to customers, the process is somewhat similar, the customer usually places an order over website or app., the store manager accepts the order and packs the items and hands it over to the delivery staff who then delivers it to the specified address.

The online food delivery services expressed a revenue of US $ 95.4 billion in 2018 alone (Statista Reports, 2018). The growing popularity of online food delivery services has seen a rise in the expected revenues, and the expected market volume stands at the US $137.6 bn by 2023 (Statista Reports, 2018). With the increasing popularity of online food delivery services and the growing number of customers, there is also an increase in the number of players or aggregators providing delivery services (Klein, 2019). With low entry and exit barriers and with the venture investment on food startups on the rise (Klein, 2019), there is constant competition among service-providers for gaining a strong customer base. However, it is easier said than done. The challenges faced by service providers are multifold. First, the online food delivery service providers face challenges related to the trade-off between expansion and losses (Tandon, 2018). The service providers know that for attracting customers, they need to provide discounts and offers (Klein, 2019). As service providers take measures for expansion using funds from investors (Hector, 2017) and targeting more customers by providing discounts and offers, they may suffer suffering heavy losses (Bhushan, 2019). However, the service providers need to decide how long they can provide discounts since providing deep discounts means losses for the company. Second, service providers face challenges related to recruiting delivery staff and improving their working conditions (Bhattacharjee, 2018; Tandon, 2018). Service providers need to recruit the correct staff with the right attitude. The staff behavior is important in the service industry. Customers also expect polite and friendly delivery staff (Elvandari, Sukartiko, & Nugrahini, 2017). Additionally, it is important for service providers to also focus on the safety of delivery staff (Maimaiti, Zhao, Jia, Ru, & Zhu, 2018). Third, the restaurants face challenges related to faster preparation time and inappropriate prices set by the service providers (Varma, 2018; Bhushan, 2019; Johari, 2019). Fourth, the online food delivery service providers face challenges due to high expectations from consumers related to food quality, food packaging, and delivery charges (Elvandari et al., 2017). Fifth, consumers want more offers, several foods, and restaurant options, free deals, ease-of-use, different payment options (Kundu & Chatterjee, 2018), and superior food quality (He et al., 2018). Sixth, service providers need to look into issues related to tampering of packaging or food by restaurants or delivery staff (Verma, 2018), since such issues can affect the brand image and customer's trust. Seventh, customers want faster preparation and delivery of ordered food items (Klein, 2019). Hence, service providers need to have a proper plan for maintaining the distribution network. It is also noted that off-late online food delivery services are suffering from not only increase resistance from consumers (Dua, 2017; Nandita, 2018) but also from the rising competition due to the increase in the number of online food delivery service providers (Mundy, 2018). To address these gaps, it is important for service providers to not only look at food quality and hygiene, food packaging, working conditions of the delivery staff, sustainability, and strategies for better delivery and meeting customer needs. This study also discusses the scenario concerning the appropriate strategies where the supply chain gets disrupted, and there are risks related to not only the health of delivery staff but also the customers. Also, it denotes some ethical considerations that service providers need to consider while formulating various strategies during emergencies.

6.2 Prior research works on food delivery services

In the past few years, although scholars have been working on understanding the factors that affect consumer behavior in the context of food delivery services, there is a paucity of studies that tries to understand the various aspects affecting user's intention to use food delivery services. Let us now explore the various scholarly works on food delivery services. Prior literature on online food delivery services in case of prepared food was reviewed (refer to Table 6.1) that have tried to capture the factors affecting the user's behavioral intention. Elvandari et al. (2017), in their study, noted that order conformity, politeness, and delivery staff friendliness affected user's intention to use the online food delivery service. Additionally, the researchers also noted that customers value cleanliness of packaged food, and the quality of food received while using food delivery services. The researchers also noted that service providers need to provide skill-related training to delivery staff and periodic evaluations of service performance for improving their delivery operations. Yeo et al. (2017) found that the customer's experience (both pre and post) with the food delivery service affects attitude and behavioral intentions. Hence in the services industry, service providers need to reduce service gaps and solve service issues that exist at the earliest possible. Pigatto et al. (2017) found that in this era of digitalization, social media platforms help in improving the company's presented brand and business visibility. Social media platforms also help service providers to engage with customers better. The researchers also found that content, functionality, and usability play important roles while designing websites for online food delivery services. See-Kwong, Soo-Ryue, Shiun-Yi, and Lily (2017) found that the desire for more revenue, increase customer reach, and improve customer base forces business owners to outsource the delivery services. Maimaiti et al. (2018) argued that online food delivery service providers need to take note of the safety of delivery staff because of the increase in the number of road accidents. Additionally, there is

TABLE 6.1 Reviews of important articles on online food delivery and food delivery apps. (between years 2000–2019).

Author (Year)	Variables used	What the study tries to find out?
Çavuşoğlu (2007)	–	Finding the online food delivery sector growth and problems faced (like food hygiene, lack of proper food standards, inadequate diffusion, etc.) in Turkey
Elvandari et al. (2017)	Satisfaction, service quality, technical requirements, service delivery, intentions to use online food delivery services	Finding satisfaction from the quality of service and the technical requirements for better service delivery.
Lee et al. (2017)	User and firm generated information, system quality, design quality, perceived usefulness, ease of use, attitude.	Influence of user/firm generated information and system/design quality on usefulness, ease of use and attitude.
Pigatto et al. (2017)	Content of the website, functionality of the website, and usability of the website, Performance of OFD companies based on the site's ability to conduct business transactions.	Examining the feasibility of the site based on content, functionality, and usability for conducting business transactions.
See-Kwong et al. (2017)	Opportunity for revenue increase, broader customer reach, better customer base, outsource intention.	Exploring factors that influence owners' decisions to outsource food delivery services (an increase of revenue, broader customer reach, better customer base).
Yeo et al. (2017)	Hedonic motivation, price saving, time-saving, online purchase experience, convenience, postusage usefulness, attitude, behavioral intention.	Examining the relationship between various factors like convenience motivation, post-usage usefulness, hedonic motivation, etc. on intention toward the usage of online food delivery services.
Kundu and Chatterjee (2018)	Price, variety, quality, website compatibility, proximity to home, offers/ discounts and payment option, intentions to use online food delivery services	Impact of digitalization on the online food delivery sector.

TABLE 6.1 Reviews of important articles on online food delivery and food delivery apps. (between years 2000–2019).—cont'd

Author (Year)	Variables used	What the study tries to find out?
Maimaiti et al. (2018)	Intentions to use, food shopping habits, the prevalence of overweight and obesity, diet-related noncommunicable diseases.	Food shopping habits, increasing prevalence of overweight and obesity, as well as diet-related noncommunicable diseases.
Sjahroeddin (2018)	Food quality, efficiency, fulfillment, system availability, and privacy, perceived value, user satisfaction.	Impact of efficiency, fulfillment, system availability, privacy, perceived value, food quality and user satisfaction in the online food delivery sector.
Suhartanto et al. (2019)	E-service quality, food quality, customer loyalty.	Quality of food and service, satisfaction, perceived value and consumer loyalty toward online food delivery services.
Yusra and Agus (2018)	Contact, responsiveness, efficiency, privacy, and fulfillment, customer satisfaction, and loyalty.	Association between mobile service quality or M-S-Qual (Contact, Responsiveness, Efficiency, Privacy, and Fulfillment), Customer Response (Personal Innovativeness, Customer Satisfaction, and Customer Loyalty) and Demographic Information.
Correa et al. (2018)	Traffic conditions, indicators of online food delivery services (cost of delivery, expected delivery time, minimum ordering, number of comments for the customer, delivery time fulfillment).	Influence of traffic conditions on factors influencing the adoption of online food delivery services.
Cho et al. (2018)	Convenience, design, trustworthiness, price and food choices, perceived value, attitude, intention to continue.	Examine the impact of quality attributes (convenience, design, trustworthiness, price, and food choices), attitude, and perceived value on continuance intention.
Roh and Park (2018)	Convenience orientation, subjective norm, moral obligation, compatibility, ease of use, usefulness, intention to use.	People's value systems, moral obligations, and influence on adoption decision.

Continued

TABLE 6.1 Reviews of important articles on online food delivery and food delivery apps. (between years 2000–2019).—cont'd

Author (Year)	Variables used	What the study tries to find out?
He et al. (2018)	Average waiting time, food quality, food preparation time, and accumulated order count, service experience.	Testing the agent-based online to offline food ordering model-based proposed model that consists of three main stakeholders, namely, the consumers, the restaurants, and the online food delivery platform.
Chai and Yat (2019)	Perceived ease of use, time-saving orientation, convenience motivation, privacy and security, behavioral intention.	To explore the influence of the factors (perceived ease of use, time-saving orientation, convenience, and privacy) on behavioral intention.
Ray et al. (2019)	Convenience, societal pressure, customer experience, delivery experience, a search of restaurants, quality control, listing, and ease-of-use, user behavioral intention.	To explore the influence of various gratifications on the intention to use the FDA's.
Alalwan (2020)	Performance expectancy, effort expectancy, online tracking, online rating, online reviews, social influence, hedonic motivation, perceived value, facilitating condition, e-satisfaction, habit, continuance intention.	To investigate the influence of the factors identified in the UTAUT2 model along with online reviews and ratings on the intention to continue using online food delivery services.

Note: Unified Theory of Acceptance and Use of Technology = UTAUT.

also a need to focus on food hygiene. Kundu and Chatterjee (2018) found that customer's behavioral intention is affected by factors like offers, variety, free deliveries, and ease of use. He et al. (2018) stated that the policies set by the service providers also affect consumer's decisions. Roh and Park (2018) found that moral obligations (married or single) also affect the user's decisions to adopt online food delivery services. Yusra and Agus (2018) found that customer satisfaction affects both the service quality, as well as trust for that service. Correa et al. (2018) found that the traffic conditions have a mild relation with early deliveries and customer comments. Suhartanto, Ali, Tan, Sjahroeddin, & Kusdibyo (2019) found that food quality affects customer's loyalty toward that service. In the context of food delivery apps., researchers found that factors like information (user or firms), system quality, design quality, usefulness, ease-of-use (Lee, Lee, & Jeon, 2017), user trustworthiness, food choices, price, design, convenience (Cho, Bonn, & Li, 2018), online review, rating, online tracking, expected performance, hedonic factors, and monetary values (Alalwan, 2020) affect FDA usage.

Let us now have a look at all the important research works on the delivery of fresh food items. Mao, Wu, Li, and Li (2019) have utilized a customer expectation function to find out the optimal distance between the home delivery options. The main reason to find out the best route is that on the way for delivery, sometimes the fresh food items may get spoiled. Other researchers have also worked on developing various algorithms to find out the best routes and improve delivery operations, like, clustering techniques (Prajapati, Harish, Daultani, Singh, & Pratap, 2020), Bayesian network (Zhang, Liu, Zhang, Cui, & Xu, 2020), using a decision support system (Dellino, Laudadio, Mari, Mastronardi, & Meloni, 2017), using an ant colony model (Chen, Gui, Ding, Na, & Zhou, 2019), using analytical hierarchical process (Chaiyaphan, & Ransikarbum, 2020), etc. Nakandala, Lau, and Zhang (2016) also proposed a cost-optimization technique for the transportation of fresh food items. Bortolini, Faccio, Ferrari, Gamberi, and Pilati (2016) proposed another optimization technique for distributions of fresh food, taking into consideration three important aspects of operating costs, carbon footprints, and the delivery time. Wang and Yin (2018) proposed a fuzzy genetic algorithm for the delivery of fresh food items. In recent years, service providers have tried to use various innovative delivery mediums, like the use of drones, etc. for improving the delivery process. Kim and Hwang (2020) utilized a combination of norm activation model and the theory of planned behavior to understand the effect of product knowledge on attitudes and behavioral intentions in the context of drone food delivery services. These new innovative techniques can help in the context of the delivery of perishable food items. Hence, routing for vehicles to deliver the perishable food items within a limited period is an area of concern for service providers (Hsu, Hung, & Li, 2007).

6.3 Discussion

In an organizational hierarchy, the decisions are made at different levels. The various decision levels are discussed below:

- Strategic decisions are those, which have a long-term impact and are usually taken by senior managers. These decisions impact and shape the direction of the business. For example, the managers of a travel agency firm need to make decisions based on long-term forecasts of business turnover set against likely market conditions as to whether they can survive in the long run or they need to diversify their operations.

- Tactical decisions are those that help in implementing the strategy and are usually taken by middle management. For example, for the travel agency firm, a tactical decision would be whether to give the customers an option to have early check-in or late check-out to increase the business turnover.
- Operational decisions mainly deal with the day-to-day running of the business. The operational decisions are mainly taken by middle or junior managers. For example, a simple operational decision for the travel agency firm will be whether to keep more hotel rooms ready for next week, as based on previous records, more customers are likely to visit in the upcoming weeks.

For taking various managerial decisions and finding out if any service gaps exist, organizations need to understand the customer perspective better. To understand what the customers feel about the food delivery services, we have performed a qualitative based study using a thematic based analysis of semistructured interviews. The purposive sampling ensured proper representation of various relevant elements of the study population, like, participants from different educational and occupational backgrounds and the two different genders (male and female) (Uwizeyimana & Modiba, 2016). Purposive sampling helps in targeting those respondents who possess the research relevant information (Uwizeyimana & Modiba, 2016). Twelve online food delivery services users (refer Table 6.2) were interviewed based on their experience with online food delivery services in India to gain an understanding of the factors that users value when using the online food delivery services. The qualitative data were collected through semistructured face-to-face or in-depth telephonic interviews. The interviews took place between June–July 2019 with Indian participants. A semistructured interview guide was designed to explore how the users feel about online food delivery and what they feel can improve the services. The questions were designed to extract more information from the participants' personal experiences. The average duration of each interview was 15 min, and the interviews were recorded. Later the recorded

TABLE 6.2 Demographic characteristics of online food delivery users (qualitative study).

Sample characteristics (n = 12)		Responses	Percentage
Gender	Male	9	75%
	Female	3	25%
Age (in years)	20–25	4	33.33%
	26–35	4	33.33%
	>35 years	4	33.33%

interviews were transcribed on paper. After each interview, the points noted were discussed again with the respondents to make sure that the interviewer was able to capture the information that the respondents wanted to convey. Respondents were asked mainly about their experiences with online food delivery services, the factors that they feel affect their behavioral intentions, and what they feel can help in improving the online food delivery services.

A thematic based analysis for analyzing the qualitative data has been used since thematic-based analysis is not only helpful in capturing the most important themes and gives a more realist background (Vaismoradi, Turunen, & Bondas, 2013) but also have the capability to capture important patterns (Braun, Clarke, Hayfield, & Terry, 2019). Around 5−10 relevant statements from each interview were used for deriving the important themes. Statements conveying similar meanings were tagged to a single subdimension. Dimensions with higher frequencies were given more priority for avoiding research bias (Creswell, 2009). The steps followed for extracting the themes were proposed by Ravi (2013) are as follows. Initially, for each sentence spoken by the participants, a label/word was selected to convey the relevant meanings. For identifying common and unique labels from the full set of codes about each interview, a discussion was done between a team of researchers. A list that contains all the codes was formulated. Based on the second round of discussion and comparison of codes among a certain number of researchers and faculty members, a list of indispensable codes was selected, which will be useful for understanding the main objectives of the study. These codes are referred to as the focused codes. The team of researchers and faculties also decide the priority of the codes related to the study objective. This new list of codes is now referred to as the axial codes. The list of codes was rechecked to find out if the team has missed out on any codes and to see if any new relationship can exist. If a relationship exists among two codes, it is discussed as to whether code can be a subcode of the other. The list was checked to make sure that all the codes related to the study objective have been used, that is, whether theoretical saturation has been obtained. Then for each code, a definition was prepared based on the narratives on which they are based. These steps are also used by other researchers (Ray, Bala, Dasgupta, & Sivasankaran, 2019; Ray, Bala, Dasgupta, & Srivastava, 2020) for understanding user perspectives in the context of various e-Services.

The findings of the thematic analysis are shown in Table 6.3. Results of the thematic analysis revealed that people valued discounts and offer more while using online food delivery services. The other important themes/subthemes are the choice of products, delivery experience, experience, the packaging of food, food quality, convenience, health concerns, monetary benefits, and customer service.

TABLE 6.3 Themes and subthemes from qualitative analysis.

Sr. No.	Code	Definition	Exemplar	Percentage of respondents[a]
1.	Discounts and offers	Discounts offers, and cash-backs the OFD services provide.	*"Coming to the cost I think recently there are quiet discounts and cashbacks over there and which attracts the customer more toward itself. As compared to offline services they are not much cheaper but online can be sometimes."[Customer 2, Male, 29 years]*	[12 out of 12] i.e., 100%
2.	Choice of products	Variety of products, variety of categories the OFDs provide.	"... we also have lots of choices and range over there and these are most satisfying part" *[Customer 3, Male, 21years old]*	[11 out of 12] i.e., 91.67%
A.	Variety of products	Refers to the variety and number of products to choose from.	"We get variety of choices on food, as well as restaurant and we can get any food of any restaurant at our doorstep and in any quantity." *[Customer 12, Male, 20 years old]*	[11 out of 12] i.e., 91.67%
B.	Many platforms	There are many OFD providers.	"The first food delivery service was dominos who delivered only pizza but now a days there is swiggy you can eat variety of food at your home within a half an hour duration." *[Customer 6, Male, 31 years old]*	[4 out of 12] i.e., 33.33%
3.	Delivery Experience	Refers to the experience with the OFD services (includes packaging, delivery staff behavior, etc.)	" .... if the parcel is delivered at very less time from your expectation, so much time has been saved then feedback is must for the delivery boy also he is doing his job properly." *[Customer 4, Male, 20 years old]*	[11 out of 12] i.e., 91.67%

A.	Home delivery	Refers to doorstep delivery facilities.	"Yes online food service is good thing, we get food while sitting at home, home delivery service is there, therefore time saving, it is very easy to use, cost is quiet high and the discounts and offers are just to attract customer they provide these discounts and apart from that they do not have to do much advertising of that." *[Customer 1, Male, 42 years old]*	[11 out of 12] i.e., 91.67%
B.	Quick delivery	Fast and efficient food delivery.	"I prefer online service because we need not go outside or sometimes due to the situation, if its emergency I order food online and it gets delivered at a very less time. It is very easy process of ordering food." *[Customer 8, Female, 30 years old]*	[11 out of 12] i.e., 91.67%
C.	Food packaging	Refers to the packaging done for the food to be delivered.	" …. .food packaging is good and it is hygienic, also the food quality is too good same as that of the restaurant." *[Customer 10, Male, 36 years]*	[7 out of 12] i.e., 58.33%
4.	Food quality	Refers to acceptable quality of food provided through OFDs.	"… …. .the food quality is too good same as that of the restaurant." *[Customer 10, Male, 36 years]*	[10 out of 12] i.e., 83.33%
5	Convenience	State of using OFD without much difficulty. It can be in the form of making life easier, easy payment options, ease of use, etc.	"The convenience of ordering food online is it gets delivered right to our doorstep. Within 40–50 min. The service depends upon food ordering apps. I really liked the online ordering of food they provide cancellation of order and refund policy." *[Customer 11, Female, 24 years old]*	[8 out of 12] i.e., 66.67%

Continued

TABLE 6.3 Themes and subthemes from qualitative analysis.—cont'd

Sr. No.	Code	Definition	Exemplar	Percentage of respondents[a]
6	Health issues	Health concerns from consuming food from OFDs repeatedly.	"It has a negative impact on my health, its spicy and we intake more calories than it is required by our body as it is easily available, so fooding habit has also changed. We are more dependent on these services because of easy availability. It depends on mood how frequent I consume these food also sometimes because of friend circle. Cannot say anything about hygiene because we are not able to see how they are preparing food. In restaurant we can see how they are preparing food but not in online. I have gained weight after this service. amount of food consumption has also increased and no such health issue." *[Customer 7, Female, 34 years old]*	[8 out of 12] i.e., 66.67%
7	Monetary benefits	Various monetary benefits like cost benefit, value for money, etc.	"Most of the time I get offers like 40%–50% discount and during IPL I got very good offers and cashbacks.I get varieties of restaurants and food but mostly prefer good and branded restaurants." *[Customer 9, Male, 39 years old]*	[6 out of 12] i.e., 50%
8	Customer service	Customer service refers to the services the OFD providers provide in case of issues.	"Customer service should get better day by day because they are not frequent ..." *[Customer 2, Male, 29 years old]*	[6 out of 12] i.e., 50%
9	Past experiences	Experience in the past people had with OFD services.	"Yes once I ordered food from small hotel so there was a negative impact on my health but after that I started ordering from good restaurant then there is no health related issue." *[Customer 9, Male, 39 years old]*	[5 out of 12] i.e., 41.67%

[a]*Percentage of respondents = (number of respondents who considered the factor important/total number of respondents who chose to respond on the factor)*100%.*

Boyatzis (1998) formula, as given below, was used to calculate the percentage agreement of the presence of the code or subcode.

$$\frac{[2 * (\text{no. of times both coders A and B saw it present})]}{[(\text{no. of times coder A saw it present}) + (\text{no. of times coder B saw it present})]}$$

The rigor employed in this study has been examined using the criteria based on the suggestions of Charmaz (2006), and Guba and Lincoln (2005, pp. 191–215), namely, (1) credibility, (2) reflexivity, (3) transferability, and (4) originality. The validity of the instrument (semistructured interview) was ensured by the researchers using a heterogeneous sample (gender factors, economic status, various occupations, etc.), reviewing relevant literature, and discussing the research problem with other academicians. The reliability of the instrument was ensured by discussing the questions with various researchers to "fine-tune" the questions. Ethical considerations were ensured by assuring the respondents that their identities would not be revealed.

Since the data from a qualitative based semistructured interview can be limited, a natural language processing (NLP) based approach has been adopted for analyzing the textual data in the form of online reviews and social media posts. This is a relatively new technique, but this helps to capture the important themes that users value easily from a larger dataset. Additionally, user-generated content serves as a good information source and affect the views of potential customers (Chatterjee, 2019). This shows the importance of user-generated content, and hence, performing a thematic-based analysis will also help us gain an overall understanding of what customers feel about online food delivery services. The process of extracting the textual data from various online-based platforms is a web-mining process (Kosala, & Blockeel, 2000; Sakaki, Okazaki, & Matsuo, 2010; Crooks, Croitoru, Stefanidis & Radzikowski, 2013; Ray & Bala, 2020). Text mining is performed to extract important information from the user-generated data. This information can be useful for various managerial decisions. The text mining technique followed to extract important themes from the textual data is the topic modeling technique. Topic modeling is preferred because it builds a structure of topics from a set of documents by using the principle that the document set that represents the corpus belongs to the specific topic (Kunimoto & Saga, 2014). Although all topic models use the principle that each document is a collection of several topics and each topic is a collection of several words (Xu, 2018), there are different techniques for performing topic modeling. They are latent semantic indexing (LSI) or latent semantic analysis (LSA) or probabilistic latent semantic indexing or analysis (pLSI or pLSA) (Deerwester, Dumais, Landauer, Furnas, & Harshman, 1990; Sridharan & Sivakumar, 2018), latent Dirichlet allocation (LDA) (Blei, Ng, & Jordan, 2003), hierarchical LDA (hLDA) (Blei, Griffiths, & Jordan, 2010). The process of analyzing online textual data using various techniques can also be referred to as content analysis. The popularity of content analysis of online textual data rose with the accessibility of huge volumes of data (Neuendorf, 2017, p. 3) and the fact that content analysis helps in providing new insights from the captured data. In this study, the latent Dirichlet allocation technique to perform topic modeling (Blei, 2012; Blei et al., 2003) has been utilized since the latent Dirichlet allocation model helps in generating an estimated probability of a document being represented by a topic and the probability of a word being used to represent a topic (Hagen, 2018). Topic modeling is used because topic modeling helps to identify new themes from textual data, without chances of potentially biased perspectives (Hopkins & King, 2010; Jelveh, Kogut, & Naidu, 2015). The reviews given by people in Twitter and merchandise websites of online food delivery services from India, namely, Zomato and Swiggy, have been extracted to find out the important themes or topics, which affect people's intentions. The important themes extracted from the user-generated content are summarized in Table 6.4.

We found that customers look for better quality, safety measures for the food they consumer, ease of use of app., customer service, and differentiating features of apps. Let us now have a look at the main issues that affect online food delivery services and discuss them concerning various managerial decision levels.

6.3.1 Food quality and food safety

In layman terms, food quality refers to the quality characteristics of food. Researchers Cardello (1995) has stated that food quality is about assessing the "degree of excellence" of the food usually measured using various indices like nutrition, etc. Peri (2006) feels that food quality much more than just the quality of food alone. It also deals with the product, the production, and the packaging as well. The author Grunert (2005), using the total food quality model, found that food safety and food quality are vital issues in modern-day food economics, and these are dependent on consumer food choice and food demand. Researchers Wandel and Bugge (1997) explored the environmental role in the examination of food quality. The results of the study showed that women are more likely to consider environment-related aspects while evaluating food quality. Baron, Brulé, and Gautier (2016), chap. 9, argued that evaluating the food quality involves evaluating the microbiological quality of the food, which includes both hygiène and commercial quality. Additionally, the authors also stated that the color of the food also affects consumer's purchase intentions. Yaseen, Sun, and Cheng (2017) described the usefulness of utilizing advanced technologies that Raman imaging can help in the easy evaluation of food quality and food safety. Food quality and safety is not only important for governments and industries but also for customers. Gowen, O'Donnell, Cullen, and Frias (2007) have also demonstrated another technique for finding the image of food objects to find out food quality and safety through the detection of contaminations and defects. Liu, Pu, and Sun (2017) have utilized the hyperspectral imaging technique for evaluating food quality

TABLE 6.4 Themes generated from user-generated data.

Sr. No.	Theme	Description	Exemplar
1.	User-friendly app.	This refers to the design of the app., which makes the use of the software easy.	"App is superb. Ordering and tracking are easy. But needs to be more soothing to the eye." *[Male, rated 4 out of 5]*
2.	Good services	Refers to the customer service of the provider, which helps to solve various issues.	"Yes I ordered non veg thali. It was all fresh n tasty specially Gajar halwa was amazing. Quantity for rice could be more." *[Male, rated 5 out of 5]*
3.	Past experiences	Refers to the past experiences of a user while using the particular app.	"I got food poisoning after eating from here. Please avoid." *[Male, rated 1 out of 5]*
4.	Offers/ Discounts	Refers to the various offers and discounts provided by the service provider to attract customers.	"Great deals on food. Lots of offers from so many restaurants." *[Male, Rated 3 out of 5]*
5.	Information provided	Refers to the content available describing the product and the places, which provide the product that helps the user choose properly.	"It says in the menu that the sandwich and the roll will have hummus but dont expect any. I completely sauced on their harissa sauce. Im gonna ask for the recipe." *[Male, Rated 3 out of 5]*
6.	Order tracking	Refers to the ability of the app. to track the status of the order.	"Food was amazing and service was Fast!.. keep up your Good work.had order for Mexican chilli was really delicious … online process is very satisfactory, through that we come to know all status of our order and dispatch time ….cheers!!" *[Female, rated 5 out of 5]*

and safety. In the review paper on food quality and safety, Khan, Tango, Miskeen, Lee, and Oh (2017) described various food processing techniques, like, microwave heating, etc. and their shortcomings. The authors also discussed the effect of food processing on food nutrients. Kotsanopoulos, and Arvanitoyannis (2017) felt that due to the various disease outbreaks related to food, it is important to maintain adequate food quality and safety measures.

One way to achieve this is through audits in the food industry not only toevaluate the management systems but also to understand the food standards followed and the hygiene of the place where the food is prepared or stored. My, Rutsaert, Van Loo, and Verbeke (2017) also states that food quality certifications are necessary to understand the food quality and safety measures taken by various shop owners, but the problem lies in the fact that customers are not familiar with these certifications. One interesting thing to note in this case is that the packaging of the food affects customer's perceptions about the quality of food (Magnier, Schoormans, & Mugge, 2016). Sohail, Sun, & Zhu (2018) and Fuertes, Soto, Carrasco, Vargas, Sabattin, and Lagos (2016) described the use of intelligent packaging of food items for improving food quality and food safety. Various other researchers (Kiani, Minaei, & Ghasemi-Varnamkhasti, 2016; Rossi et al., 2017; Yousefi, Su, Imani, Alkhaldi, Filipe, & Didar, 2019) have also mentioned the use of modern intelligent techniques to examine the food quality and safety measures. Röhr, Lüddecke, Drusch, Müller, and Alvensleben (2005) based on a survey in Kiel, however, feels that over the years the quality of food has improved, but at present, there are two types of customers, one who is price sensitive and the other who is safety-sensitive. Whatever be the case, it is important in the service industry to focus on the quality and safety aspects because customers value them a lot (Vukasović, 2016). Petrescu, Vermeir, and Petrescu-Mag (2019) feel that values customer views are very important for service providers. They found that customers use the freshness, taste, and appearance of the food item to examine the food quality. The crucial hints for them are the components, mainly the nutrients and the additives. Additionally, the packaging and the origin of production also matters.

6.4 Sustainability

In layman terms, sustainability refers to the ability to be maintained at a certain level over some time. But scholars Schaltegger and Burritt (2006) lay importance on corporate sustainability because these initiatives can help the service-providers provide differentiated services and a means to take measures to reduce costs and improving operations. Costanza and Patten (1995) argued that assessing sustainability is more of a prediction problem. The authors stated that assessing sustainability involves assessing the systems or subsystems that need to be sustained, assessing how long these needs to be sustained, and after a period assessing if the system or subsystem has been sustained. Pope, Annandale, and Morrison-Saunders (2004) feel that assessing sustainability is an important activity for organizations and hence should be carefully managed. Researchers van Marrewijk, and Werre (2003) have discussed various levels of corporate sustainability in their paper. The researchers have also argued that corporate sustainability can be defined in many ways. The researchers also stated that scholars often describe corporate sustainability as a corporate social responsibility that takes into account both the social and environmental concerns that affect business operations and stakeholders. The authors have also described various techniques to measure corporate sustainability in their work. Researchers Baumgartner and Ebner (2010) stated that organizations need to have proper identification of the various dimensions of sustainability, namely economic, ecological, and social. Additionally, the researchers also state that organizations need to focus on internal, as well as external strategies. Singh, Murty, Gupta, and Dikshit (2009) have discussed various indices for measuring sustainability in their paper.

Based on the customer views and the studies by earlier researchers, the important points that can be noted is that:

- Customers look for more offers, discounts, and good deals.
- Customers prefer good services and efficient delivery and customer care staff.
- Customers look for safety and hygiene.
- Customers want the platform to be easy to use.

6.4.1 Food delivery process

A simple network design is shown in Fig. 6.1 for the food delivery system operations. As the customer places an order on the application or website of the service provider, the restaurant either accepts the order or rejects it based on the availability of items. Once the restaurant accepts the order, the customer can view the estimated time of food delivery. To able to display this estimated

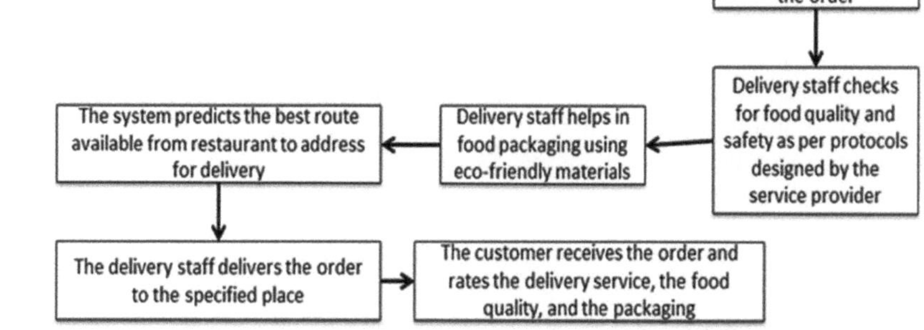

FIGURE 6.1 Network design for food delivery system delivery.

time, the food delivery service provider need to train the algorithms by taking into considerations the approximate time taken for preparation of the items ordered, the demand in the restaurant, the approximate time that takes for a vehicle to cover the distance between the restaurant and the place of delivery at a certain speed, the traffic conditions during that time, etc. The delivery staff reaches the restaurant and verifies the cooking conditions, the quality of food items used, and the hygiene of the restaurant kitchen. Once the food is prepared, the delivery staff helps in the packaging of the food item using mainly eco-friendly packaging materials. Then the delivery staff delivers the items ordered to the address mentioned. Once the order is received, the delivery staff verifies the payment and completes the order. The customer then can rate the delivery services, the attitude of the delivery staff and also the quality of the food items and the packaging. These data can be utilized by the customer service team to improve their services.

6.4.2 Strategic planning

Now for the top management for taking the strategic decision, it is important to envision how the business will shape up shortly. The management team needs to focus on three main aspects: food quality and safety, efficient delivery, and addressing customer issues efficiently. In this era of dynamic competition, it will be a tough decision for the top management to decide between attracting more customers through offers and discounts, or try to create a stronger customer base and focus on better services. For improving the delivery network the top-level managers need to bring in innovative options like delivery through drones, developing a fuzzy genetic algorithm for better route allocations, using intelligent packaging for food safety and hygiene, plans for regular auditing of storehouses or production houses to ensure quality and safety standards are maintained, etc. Additionally, to stay ahead of the competition, the service providers need to bring in differentiated schemes for their customers and also focus on addressing service gaps that exist. The strategic decision-making process usually consists of five main steps:

- To define the problems that exist and evaluate the current performance.
- To gather information from all the stakeholders involved (directors, consumers, employees, etc.).
- To assess the internal (structure, culture, resources) and external environment (natural, societal, task).
- To analyze the internal factors (strengths and weaknesses) and external factors (opportunities and threats).
- To develop and evaluate options and select the best option and monitor the decision.

Now in case of making strategic decisions for better food quality and food safety, the service providers have to keep various intelligent systems that can make sure that the food items involved are of good quality. There needs to be regular surprise audits to make sure that the food items involved are of good quality and the safety standards are maintained. This not only involves taking note of the nutrition content present but also taking note of the microbiological content. In the case of service providers who are involved in delivering prepared food to customers, the service providers need to train the delivery staff to keep regular checks on the restaurants they visit to find out if the restaurants use hygienic ingredients while cooking the food. Since there are cases of customers complaining about the health issues they face while consuming food from certain restaurants, there is a need for service providers to take proper note of the restaurants' hygiene. This is important for service providers because complaints related to health-related issues can affect the brand image of the service provider. One easy way to achieve this is to train the delivery staff properly to take care of this added responsibility. In the future, however, the installation of smart cameras and surprise checks of those recorded clips can improve the safety and hygiene-related issues. Now to deliver the food, researchers have suggested the use of intelligent packaging. The restaurants and delivery staff should be trained properly to pack the food properly because, in the earlier paragraphs, we have seen researchers expressing their views that packaging and color of food affect user's behavioral intention. Now to solve the issue related to the delivery of food, it is important to keep a note of choosing the best optimal path, which takes note of note only the traffic, the operational costs, but also the perishable time of the food item. The algorithm that most suits to solve this issue is the fuzzy genetic algorithm (Wang & Yin, 2018). In the modern-day scenario, although Google maps are the best bet to understand traffic conditions and the shortest route to a destination, the delivery staff needs to be trained properly to understand the routes properly and also work on their behavior. Companies need to keep proper background checks of the delivery staff and also make sure that they get adequate breaks between deliveries. Additionally, while calculating the optimal route, a check needs to be kept that the vehicle speed does not exceed a predefined speed since it is important for service providers to also focus on employee and staff safety. Finally, to stay in the long run, service providers need to understand and address the issues that the customers face at the earliest possible. For achieving this, service providers need to have a trained customer service team.

6.4.3 Tactical Planning

Tactical planning generally refers to the short term decisions, which are usually taken by the middle-level managers to achieve the long term goals. The main point that needs to be kept in mind while making tactical decisions is that the plan needs to be flexible, and there should be good communication among the various functional managers involved. Since the strategic plan is to have more focus on food safety, food quality, and sustainability, the middle-level managers need to make decisions that can help them achieve these goals. Now to make sure that the food is of high quality, the middle-level managers need to keep monthly plans to have quality checks in various restaurants or stores in various localities. The plans must be properly made since it will be difficult to check food quality and food safety in most restaurants unless proper plans are made. Additionally, the plan needs to be made, keeping in mind the operational costs involved. One way to achieve this is to recruit the right delivery staff and train them properly so that they can also report quality issues, if any. Doing this will not only reduce costs but also help organizations keep a proper check on the restaurants that provide quality food. But organizations also need to take note of any fraudulent behavior of the delivery staff. For ensuring the safety of the delivery staff, organizations can attach sensors to the vehicles so that the delivery staff cannot cross the predefined speed limits. Improving delivery operations is another headache. Organizations can divide the area of operations into various zones and try to recruit staff who are well accustomed to those areas so that they can provide faster deliveries. Proper background checks also help to choose the best staff. To ensure that the plan is sustainable, it is important to understand the customer needs and solve the issues that the customers face. In a locality where there is more number of family members present, the service providers can have a tie-up with restaurants who provide more combo options, and in localities with more number of bachelors, the service providers can have several fast-food options. Thus the plans need to be flexible and need to be aligned with the main strategic plans.

6.4.4 Operational planning

Operational planning refers to the plan that is mostly concerned about day-to-day activities that will help the organization achieve its overall strategic goals. The operational plan for improving food quality and food safety is to have a surprise check by lower-level managers on various restaurants or food stores in various localities. These surprise checks need to be planned as to which restaurants or stores in each locality to be targeted daily. One more thing that needs to be closely monitored is the customer views about the services. If any service gaps exist, the lower level managers need to work in coordination with the customer service team to make sure that the issues that the customers face are solved as soon as possible. Managers also need to keep a note of the various festivals in various localities and provide offers accordingly to increase sales on festivals. But to make sure that the deliveries are made properly and within a time assigned, intelligent fuzzy-based routing needs to be made considering the traffic conditions and the demand of the restaurants. All these plans need to be predefined in an algorithm to make sure that the machine learning algorithms can help in better operations. Daily checks on the packaging need to be done so that the customers receive food that is packaged well.

The corporate sustainability using the 4P-matrix of corporate sustainability has also been analyzed (van Marrewijk & Werre, 2003). The organizations face various situations and operate from different value systems, and they can develop six different ambition levels that reflect their motivation involved in implementing sustainability in their various operations. They are as follows:

1. Precorporate sustainability: At this level, there is no ambition for corporate sustainability. However, corporate sustainability can be initiated when it is instigated by the external environment. For this, the activities need to be closely monitored.
2. Compliance driven corporate sustainability: At this level, it is more about looking into societal welfare and charity. It is more of an obligation and duty for the organizations.
3. Profit-driven corporate sustainability: This level focuses on integrating the three main aspects, namely, social, ecological, and ethical aspects with business operations and decisions, as long as it helps in contributing to the organization financially.
4. Caring corporate sustainability: At this level, it is more about the economic, social, and ecological concerns. It is more about the human potential, societal responsibilities, and caring attitude for the environment.
5. Synergistic corporate sustainability: It is more about creating a well-balanced synergy and a win-together attitude among various stakeholders involved.
6. Holistic corporate sustainability: It involves contributing to the quality of life of every being and entity in the present and the future. Hence it needs to be integrated and embedded in every aspect of the organization.

Table 6.5 discusses the various internal and external drivers of corporate sustainability based on the 4P-matrix of corporate sustainability as driven by the six levels of corporate sustainability ambition levels. Again, for achieving sustainability, different decisions need to be taken at different levels and under mainly three categories, namely, economic, environmental, and social, as proposed by Garza (2013). The sustainability indicators and related decisions at different levels for food delivery services are shown in Fig. 6.2.

It is to be noted that for food delivery service providers to be sustainable, from an economic perspective the firm needs to look into policies that can reduce costs and improve financial performance. However, it is also important for organizations to look into ways of improving delivery performance. Second, food delivery providers also need to provide offers and discounts to not only increase sales but also attract prospective customers. The third headache for the service providers, while they try to improve financial performance, will be the costs involved in training employees and delivery staff. Because of the increasing concern on environmental issues, researchers (Chouinard, D'Amours, & Ait-Kadi, 2005) have stated that environmental factors can drive the various strategic sustainability initiatives. Apart from obliging with the

TABLE 6.5 Internal and external drivers of corporate sustainability based on the different ambition levels for online food delivery services.

Ambition levels	Internal drivers	External drivers
Precorporate sustainability	Taking decisions related to increasing a good hold of the market scenario and increasing personal power. Service providers can provide deals better than their competitors and bring differentiated offerings to create a strong customer base.	Outside forces like government policies regarding food delivery, fake news about food quality, etc. can lead to distrust, and hence, proper control of organizational, operational activities is needed.
Compliance driven corporate sustainability	It involves making decisions by mainly the board of directors or higher authorities in the organization to decide on the impact of various operations on society. Hence for food delivery businesses, it is important to take good care of food quality and safety.	Various instructions from higher authorities like the government needs to comply. But the social welfare-related compliances is more of a responsibility for the state than the service provider. But for legal compliances, the organization needs to take appropriate steps.
Profit-driven corporate sustainability	It mainly involves taking care of the issues related to profits, like, fake news, scandals, etc. Food delivery providers need to take proper precautions to avoid such issues.	Proper coordination needs to be maintained among the service provider and higher authorities to avoid negative coverage or bad reputation in the market. Since profits are involved, shareholders are given first preference over other stakeholders.
Caring for corporate sustainability	Decisions need to be taken with a focus on people, planet, and profit, taking into consideration the views of all stakeholders.	It involves coordination among the staff/employees and the customers/ stakeholders involved in social and environmental care. It can involve using eco-friendly packaging, avoiding plastics, etc.
Synergistic corporate sustainability	The top level managers need to take properly balanced decisions taking into consideration not only sustainability but also the social and environmental aspects	It involves coordination with various authorities and stakeholders to help in implementing corporate sustainability better.

TABLE 6.5 **Internal and external drivers of corporate sustainability based on the different ambition levels for online food delivery services.—cont'd**

Ambition levels	Internal drivers	External drivers
	given long term perspectives.	
Holistic corporate sustainability	It is a responsibility of the food delivery services to have a holistic view of the surroundings and society in particular because it is more of a responsibility for everyone toward all other beings or entities.	It involves mainly the governmental role in creating an environment for the sustainability of various businesses, as well as taking into consideration the other aspects, namely, the social, ethical, and ecological components.

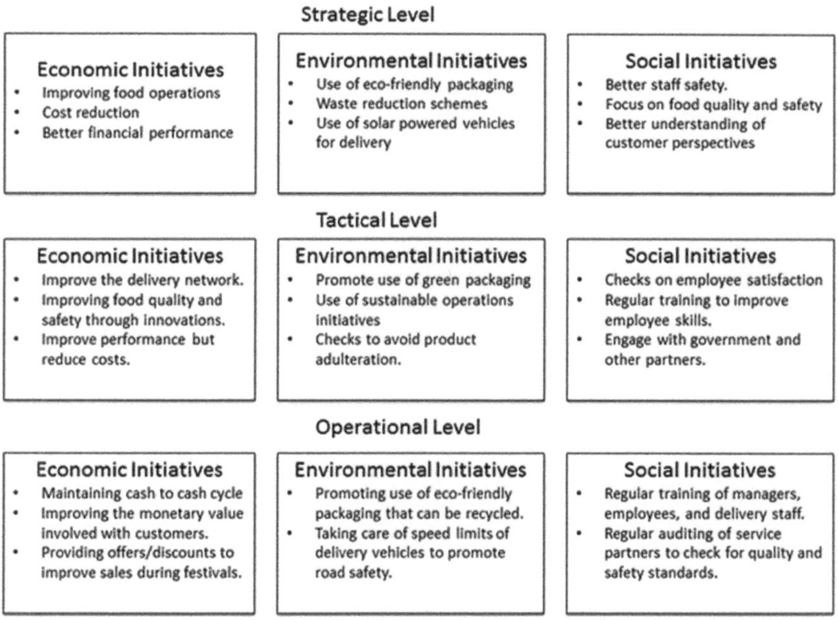

FIGURE 6.2 The sustainability indicators and related decisions at different levels for food delivery services.

various regulatory issues and ethical considerations, food delivery operators should have environmental considerations in their operations, like using eco-friendly packaging, reducing wastage of food, etc. Apart from economic and environmental initiatives, food delivery operators also need to take care of the social aspect, which involves taking care of employees, looking at delivery staff safety, having regular training programs, etc. Thus it is to be noted that for better operations, the service providers should not only understand customer perspectives better but also look at ways to improve sustainability.

But what happens when a pandemic issue, like the coronavirus, hampers the operations? Two main questions arise during such situations, which are as follows: First, how to handle the rise in demand since people prefer to stay at home and make online orders, and second, is it ethical to hire a person to perform the delivery duties for someone else when people are advised to stay at home (Pardes, 2020; Purdy, 2020b). The author has also highlighted more striking issues. Since most delivery providers have contracted staff, so when there is not much demand for food, they do not get much pay, and it affects their livelihood. Pardes (2020), based on a survey of the six-hundred delivery staff, revealed that although 43% of the delivery staff was worried about getting infected while performing delivery duties, 53% of the workers were worried about their cut of pay, and the ratings they receive. Although "contactless" delivery (Statt, 2020; Thompson, 2020a) can reduce the risks, still the delivery staff is at risk of getting the disease while in contact with anything during his/her duty hours. Additionally, the thresholds set by various providers like DoorDash (getting a rating of less than 4.2 will dismiss the worker's association with DoorDash), etc. (Pardes, 2020) adds to the pressure of performing well under such risks. Since most service-providers also do not provide much-paid leaves, these delivery staff is forced to perform their duties in such risky conditions since most staff are the only bread earners for their family (Pardes, 2020). The health officials of various countries are now forcing providers to change their paid leave policies (Marcellus, 2020). This will make the service-provider allot separate fund for employee health issues, which will be an added pressure on financial goals. Apart from the issues faced by delivery staff, service-providers also need to manage the increasing demand among customers since customers now prefer to stay at home and place online orders (Keshner, 2020; Purdy, 2020a). Although some service-providers notify that the delay in delivery of orders, the question arises as to how long can the delay be made in case of an increasing number of orders and the shortage of delivery staff. The third issue that organizations are likely to face is the transmission of diseases from restaurant staff, staff who packs the order, the delivery staff, etc. to customers (Cheng, 2020; Pardes, 2020). The less number of options for people also adds to the headache of service-providers (Hang, 2020). So in such situations, the whole dynamics change. The service-providers now need to formulate their strategies as to how to take care of their employees, and at the same time, how to manage the delivery operations. Fourth, the disruption in the supply of items is also a threat to the food delivery industry (Hahn, 2020; Marcellus, 2020; Parsons, 2020).

In such situations, service providers need to keep their Plan B ready related to Strategic Planning, Tactical Planning, and Operational Planning. The board members need to make decisions as to how to cope with the recent crisis and make sure no employee is getting infected. Buying suits for the employees, giving them paid leaves, taking care of hospital bills in case they get infected, checking the health of restaurant staff, etc. will add to the financial statements for the company. The costs need to be adjusted in the bills of the customers ordering food online. But the board members need to rethink their strategy as to whether customers will be willing to pay hefty charges to get the food delivered. The economy slowdown adds to the pressure on the service-providers since they have to bear the pain, in case they opt for giving more discounts to customers. The strategic planning in such situations will be to have regular checks on the delivery staff, provide them suitable equipment to reduce risks of getting infected, have a proper check on hygiene and food safety, have proper checks on restaurant staff and food store staffs, and have plans to manage supply and demand properly in such situations. Additionally, the top-level management can also decide on whether they can provide an option to customers to pay online and have their food left at the doorstep by delivery staff for avoiding "contact" in such situations (Marcellus, 2020). This will reduce the risk of providers getting infected to some extent. Additionally, service providers can also devise various methods, like, providing information regarding the cook and the delivery staff health (Marcellus, 2020; Purdy, 2020a), providing information related to the disinfection of the packaging and delivery boxes (Marcellus, 2020), etc. Although researchers have stated that the delivery of food items is completely safe, they have expressed concerns over the transmission of the virus during the handling of the packaged item (Weiss, 2020). These will increase the consumer's trust for the company. Tactical planning will be to have short term goals and try to evaluate the situation every week. Service providers need to have weekly training sessions involving all the workers to give them proper sessions on how to effectively deliver items without coming in contact, how to make the least number of contacts while on delivery, how the virus spreads, the procedures of how to stay safe through the use of sanitizers, and the symptoms of the viral infection. Similarly, the middle-level managers need to provide proper instructions to the restaurant owners or store owners to take appropriate actions to minimize the risks. The middle-level managers also need to keep details of the locations where the risk of getting infected is high, the restaurants where the staff has got infected, the number of delivery staff available, and whether any delivery staff is facing health issues. Additionally, the middle-level managers also need to take note of the various procedures that need to be followed to clean the restaurants where the staff got infections before they can resume their operations. If a delivery staff gets infected, the service-provider needs to make sure that the doctors are informed, and proper financial care is taken for the employee until he/she recovers. If a delivery staff gets infected, the service-provider needs to make sure that all the persons with whom the delivery staff was in contact also get proper health checkups. This shows how a natural or man-made situation or issue can complicate the whole process and increase the operating costs of providers. To have effective operational planning, the lower-level managers will need to have a daily update on the staff health, the locations where the impact is more, the operational restaurants, and the delivery staff available for delivery. Additionally, the managers also need to make sure that for managing the demand and supply gaps, additional delivery time needs to be notified to users as and when required. So we also present the case where service providers need to manage their operations and decisions based on situations. The sustainability of the company lies in how well the providers will be able to manage the operations and financial health of the company in various situations. However, the question of whether it is ethical for service providers to let their employees deliver the food despite knowing the risks of infection varies from organization to organization. While IT firm employees have the benefit of working from home, the medical staff needs to perform their duties by being among the patients despite knowing the fact they have a high chance of getting infected. Since, in the corporate world, employees need to perform their duties, the ethical question related to whether delivery staff should be allowed to deliver food for others is not relevant. We feel that the ethical issue gets solved if the service-provider can take appropriate steps to ensure the safety of their employees and minimize their risks of getting affected. Some organizations are even looking at avenues like the use of robots to avoid human to human contact (Purdy, 2020a; O'Brien, 2020; Osaki, 2020). But the fact that the use of robots as a replacement for human delivery staff is another ethical debate. Even in such emergencies depending on the policies of the different providers, they are likely to take different actions of ensuring the safety of all the stakeholders (Thompson, 2020b).

Thus, the food delivery industry is a fast-growing industry, and although the overall delivery model will be the same for all the providers, the underlying policies and strategies will be different for every provider. Different providers need to have proper strategies and make situation based managerial decisions for not only gaining an upper-hand in this highly competitive world but also adapt to various unavoidable emergency scenarios. The ability to adapt and provide services in any situation will decide the sustainability of the services.

6.5 Summary

Due to the advancement of technological innovations, there has been a shift from traditional ways of providing services to online-based services in various sectors, like, education, food, healthcare, etc. Online food delivery services have revolutionized the food industry. However, the main questions that the service providers face are related to not only optimization of the distribution network but also related to food quality, safety, and also sustainability. But for examining these issues, it is important for organizations to under the customer perspectives better. Customers also have concerns related to food quality, customer service, and food safety. The issues related to food quality, food safety, and sustainability has been based on three decision levels: strategic network design, tactical network planning, and operational transportation planning. This study has also discussed an emergency scenario that disrupts all the supply chains and affects not only the service-providers but also the lives of employees and customers. This study will help various academicians and practitioners working on the food delivery domain. Additionally, this study will also help various food scientists, technologists, agriculturists, and industrialists to think about the aspects discussed in this chapter, they can formulate various strategies.

References

Alalwan, A. A. (2020). Mobile food ordering apps: An empirical study of the factors affecting customer e-satisfaction and continued intention to reuse. *International Journal of Information Management, 50*, 28–44.

Baroková, A., Kryvinska, N., & Strauss, C. (2016). from traditional services to eServices – a business value analysis. In *International conference information systems 2016 special interest group on big data proceedings* [online]. https://aisel.aisnet.org/sigbd2016/3.

Baron, F., Brulé, G., & Gautier, M. (2016). Evaluation of the physical-chemical properties and quality of food. *Handbook of Food Science and Technology*, 205–234. https://doi.org/10.1002/9781119268659.ch9.

Baumgartner, R. J., & Ebner, D. (2010). Corporate sustainability strategies: Sustainability profiles and maturity levels. *Sustainable Development, 18*(2), 76–89. https://doi.org/10.1002/sd.447.

Bhattacharya, A. (2018). *India's food-delivery startups are back in business, but there's still reason to worry.* QuartzIndia Accessed on: 10 September 2019. Published on: 6 August 2018. Available at: https://qz.com/india/1347865/what-indian-food-delivery-startups-like-swiggy-and-zomato-need-to-do/.

Bhushan, R. (2019). *Zomato, Swiggy & others eating into our business: Restaurants.* ETtech Accessed on: 10 February 2018. Published on: 3 January 2020. Available at: https://tech.economictimes.indiatimes.com/news/internet/zomato-swiggy-others-eating-into-our-business-restaurants/67358478.

Blei, D. M. (2012). Probabilistic topic models. *Communications of the ACM, 55*(4), 77–84.

Blei, D. M., Griffiths, T. L., & Jordan, M. I. (2010). The nested Chinese restaurant process and bayesian nonparametric inference of topic hierarchies. *Journal of the ACM, 57*(2), 1–30.

Blei, D. M., Ng, A. Y., & Jordan, M. I. (2003). Latent dirichlet allocation. *Journal of Machine Learning Research, 3*, 993–1022.

Bortolini, M., Faccio, M., Ferrari, E., Gamberi, M., & Pilati, F. (2016). Fresh food sustainable distribution: Cost, delivery time and carbon footprint three-objective optimization. *Journal of Food Engineering, 174*, 56–67. https://doi.org/10.1016/j.jfoodeng.2015.11.014.

Boyatzis, R. E. (1998). *Transforming qualitative information: Thematic analysis and code development*. New York, NY: Sage.

Braun, V., Clarke, V., Hayfield, N., & Terry, G. (2019). Thematic analysis. In P. Liamputtong (Ed.), *Handbook of research methods in health social sciences*. Singapore: Springer.

Cardello, A. V. (1995). Food quality: Relativity, context and consumer expectations. *Food Quality and Preference, 6*(3), 163–170. https://doi.org/10.1016/0950-3293(94)00039-X.

Çavuşoğlu, M. (2007). Electronic commerce and Turkish patterns of online food delivery system. In *Conference international tourism biennial 2007*. Available at: https://www.journalagent.com/iuyd/pdfs/IUYD_3_1_45_62.pdf.

Chaiyaphan, C., & Ransikarbum, K. (2020). Criteria analysis of food safety using the analytic hierarchy process (AHP) - a case study of Thailand's fresh markets. *E3S Web Conference, 141*, 1–7. https://doi.org/10.1051/e3sconf/202014102001.

Chai, L. T., & Yat, D. N. C. (2019). Online food delivery services: Making food delivery the new normal. *Journal of Marketing Advances and Practices, 1*(1), 62–77.

Charmaz, K. (2006). *Constructing grounded theory* (1st ed.). UK: Sage.

Chatterjee, S. (2019). Explaining customer ratings and recommendations by combining qualitative and quantitative user generated contents. *Decision Support Systems, 119*, 14–22. https://doi.org/10.1016/j.dss.2019.02.008.

Cheng, E. (2020). *Coronavirus batters China restaurants and eateries as millions of workers struggle to return to work*. CNBC. Published on: 13 March 2020. Accessed on: 14 march 2020. Available at: https://www.cnbc.com/2020/03/13/coronavirus-batters-china-restaurants-as-millions-struggle-to-return-to-work.html.

Chen, J., Gui, P., Ding, T., Na, S., & Zhou, Y. (2019). Optimization of transportation routing problem for fresh food by improved ant colony algorithm based on tabu search. *Sustainability, 11*(23), 1–22. https://doi.org/10.3390/su11236584.

Cho, M., Bonn, M. A., & Li, J. (2018). Differences in perceptions about food delivery apps between single-person and multi-person households. *International Journal of Hospitality Management, 77*, 108–116.

Chouinard, M., D'Amours, S., & Ait-Kadi, D. (2005). Integration of reverse logistics activities within a supply chain information system. *Computers in Industry, 56*, 105–124.

Correa, J. C., Garzón, W., Brooker, P., Sakarkar, G., Carranza, S. A., Yunado, L., et al. (2018). Evaluation of collaborative consumption of food delivery services through web mining techniques. *Journal of Retailing and Consumer Services, 46*, 45–50. https://doi.org/10.1016/j.jretconser.2018.05.002.

Costanza, R., & Patten, B. C. (1995). Defining and predicting sustainability. *Ecological Economics, 15*(3), 193–196. https://doi.org/10.1016/0921-8009(95)00048-8.

Creswell, J. W. (2009). *Research design: Qualitative, quantitative, and mixed methods approaches*. Thousand Oaks, CA: Sage.

Crooks, A., Croitoru, A., & Radzikowski, J. (2013). #Earthquake: Twitter as a distributed sensor system. *Transactions in GIS, 17*(1). https://doi.org/10.1111/j.1467-9671.2012.01359.x.

Deerwester, S., Dumais, S. T., Landauer, T. K., Furnas, G. W., & Harshman, R. (1990). Indexing by latent semantic analysis. *Journal of the American Society for Information Science, 41*, 391–407.

Dellino, G., Laudadio, T., Mari, R., Mastronardi, N., & Meloni, C. (2017). A reliable decision support system for fresh food supply chain management. *International Journal of Production Research, 56*(4), 1458–1485. https://doi.org/10.1080/00207543.2017.1367106.

Dua, A. (2017). *Challenges that online food delivery restaurants and services face*. YourStory. Published on: 26 October 2017. Available at: https://yourstory.com/2017/10/challenges-that-online-food-delivery-restaurants-and-services-face . (Accessed 10 February 2019).

Elvandari, C. D. R., Sukartiko, A. C., & Nugrahini, A. D. (2017). Identification of technical requirement for improving quality of local online food delivery service in yogyakarta. *Journal of Industrial and Information Technology in Agriculture (JIITA), 1*(2), 1–7.

Fuertes, G., Soto, I., Carrasco, R., Vargas, M., Sabattin, J., & Lagos, C. (2016). Intelligent packaging systems: Sensors and nanosensors to monitor food quality and safety. *Journal of Sensors, 2016*, 1–8. https://doi.org/10.1155/2016/4046061.

Garza, F. A. (2013). A framework for strategic sustainability in organizations: A three pronged approach. *Journal of Comparative International Management, 16*(1), 23–36.

Gowen, A., O'Donnell, C., Cullen, P., Downey, G., & Frias, J. (2007). Hyperspectral imaging – an emerging process analytical tool for food quality and safety control. *Trends in Food Science & Technology, 18*(12), 590–598. https://doi.org/10.1016/j.tifs.2007.06.001.

Grunert, K. G. (2005). Food quality and safety: Consumer perception and demand. *European Review of Agricultural Economics, 32*(3), 369–391. https://doi.org/10.1093/eurrag/jbi011.

Guba, E. G., & Lincoln, Y. S. (2005). *Paradigmatic controversies, contradictions, and emerging confluences, The sage handbook of qualitative research.* CA: Sage Publications Ltd.

Hagen, L. (2018). Content analysis of e-petitions with topic modeling: How to train and evaluate LDA models? *Information Processing & Management, 6*, 1292–1307. https://doi.org/10.1016/j.ipm.2018.05.006.

Hahn, S. M. (2020). *Coronavirus (COVID-19) supply chain update*. U.S. Food & Drug. Published on: 27 February 2020. https://www.fda.gov/news-events/press-announcements/coronavirus-covid-19-supply-chain-update. (Accessed 14 March 2020).

Hang, K. (2020). *Coronavirus concerns lead to reduced hours, limited supplies for SGV restaurants.* Eater Los Angeles. Published on: 13 March 2020. Accessed on: 14 March 2020. Available at: https://la.eater.com/2020/3/13/21177299/coronavirus-impact-san-gabriel-valley-restaurants-la-employees-reduced-hours.

He, Z., Han, G., Cheng, T. C. E., Fan, B., & Dong, J. (2018). Evolutionary food quality and location strategies for restaurants in competitive online-to-offline food ordering and delivery markets: An agent-based approach. *International Journal of Production Economics, 215*, 61–72. https://doi.org/10.1016/j.ijpe.2018.05.008.

Hector, D. T. (2017). *Will Alibaba's planned investment in Zomato disrupt the food-tech space?* NewsCorp VCCiRCLE. Published on: 5 September 2017. Available at: https://www.vccircle.com/will-alibabas-planned-investment-in-zomato-disrupt-the-food-tech-space/. (Accessed 10 February 2019).

Hopkins, D. J., & King, G. (2010). A method of automated nonparametric content analysis for social science. *American Journal of Political Science, 54*(1), 229–247.

Hsu, C.-I., Hung, S.-F., & Li, H.-C. (2007). Vehicle routing problem with time-windows for perishable food delivery. *Journal of Food Engineering, 80*(2), 465–475. https://doi.org/10.1016/j.jfoodeng.2006.05.029.

Jelveh, Z., Kogut, B., & Naidu, S. (2015). *Political language in economics. (SSRN scholarly paper No. ID 2535453).* Rochester, NY: Social Science Research Network [online]. http://papers.ssrn.com/abstract=2535453.

Johari, S. (2019). *NRAI for indirect regulation of online food delivery cos, open to making a pact.* MediaNama. Published on: 6 February 2019. Available at: https://www.medianama.com/2019/02/223-nrai-food-delivery-regulation/. (Accessed 10 February 2020).

Keshner, A. (2020). *If the coronavirus spreads in America, food delivery companies could see a surge in demand — are they ready?* MarketWatch. Published on: 2 March 2020. https://www.marketwatch.com/story/if-the-coronavirus-spreads-in-america-food-delivery-companies-could-see-a-surge-in-demand-are-they-ready-2020-02-28. (Accessed 14 March 2020).

Khan, I., Tango, C. N., Miskeen, S., Lee, B. H., & Oh, D.-H. (2017). Hurdle technology: A novel approach for enhanced food quality and safety – a review. *Food Control, 73*, 1426–1444. https://doi.org/10.1016/j.foodcont.2016.11.010.

Kiani, S., Minaei, S., & Ghasemi-Varnamkhasti, M. (2016). Fusion of artificial senses as a robust approach to food quality assessment. *Journal of Food Engineering, 171*, 230–239. https://doi.org/10.1016/j.jfoodeng.2015.10.007.

Kim, J. J., & Hwang, J. (2020). Merging the norm activation model and the theory of planned behavior in the context of drone food delivery services: Does the level of product knowledge really matter? *Journal of Hospitality and Tourism Management, 42*, 1–11. https://doi.org/10.1016/j.jhtm.2019.11.002.

Klein, D. (2019). *Fighting for Share in the $16.6 billion delivery app market*. QSR. Published on: June 2019. Available at: https://www.qsrmagazine.com/technology/fighting-share-166-billion-delivery-app-market . (Accessed 24 February 2020).

Kosala, R., & Blockeel, H. (2000). *Web mining research: A survey*. SIGKDD Explorations.

Kotsanopoulos, K. V., & Arvanitoyannis, I. S. (2017). The role of auditing, food safety, and food quality standards in the food industry: A review. *Comprehensive Reviews in Food Science and Food Safety, 16*(5), 760–775. https://doi.org/10.1111/1541-4337.12293.

Kundu, S., & Chatterjee, S. (2018). Role of digitalisation in hospitality industry with special reference to online food delivery services. *BIMS Journal of Management, 3*(2), 58–66.

Kunimoto, R., & Saga, R. (2014). Purchase factor expression for game software using structural equation modeling with topic model in user's review texts. *International Journal of Innovation, Management and Technology, 5*(6), 417–421.

Lee, E.-Y., Lee, S.-B., & Jeon, Y. J. J. (2017). Factors influencing the behavioral intention to use food delivery apps. *Social Behaviour and Personality, 45*(9), 1461–1473.

Liu, Y., Pu, H., & Sun, D.-W. (2017). Hyperspectral imaging technique for evaluating food quality and safety during various processes: A review of recent applications. *Trends in Food Science & Technology, 69*, 25–35. https://doi.org/10.1016/j.tifs.2017.08.013.

Magnier, L., Schoormans, J., & Mugge, R. (2016). Judging a product by its cover: Packaging sustainability and perceptions of quality in food products. *Food Quality and Preference, 53*, 132–142. https://doi.org/10.1016/j.foodqual.2016.06.006.

Maimaiti, M., Zhao, X., Jia, M., Ru, Y., & Zhu, S. (2018). How we eat determines what we become: Opportunities and challenges brought by food delivery industry in a changing world in China. *European Journal of Clinical Nutrition, 72*(9), 1282–1286. https://doi.org/10.1038/s41430-018-0191-1.

Mao, M., Wu, Z., Li, L., & Li, Z. (2019). Home delivery distance and free shipping policy for fresh food considering the loss. In *Sixth International Conference on Transportation Engineering*.

Marcellus, S. (2020). *How the coronavirus outbreak is affecting your online food delivery*. Yahoo Finance. Published on: 13 March 2020 https://finance.yahoo.com/news/how-coronavirus-affects-your-online-food-delivery-134401016.html. (Accessed 13 March 2020).

Marrewijk, van M., & Werre, M. (2003). Multiple levels of corporate sustainability. *Journal of Business Ethics, 44*, 107–119. https://doi.org/10.1023/A:1023383229086.

Mundy, S. (2018). *Competition heats up in India's online food delivery market*. Financial Times. https://www.ft.com/content/365617e6-f239-11e8-ae55-df4bf40f9d0d. (Accessed 14 October 2019).

My, N. H. D., Rutsaert, P., Van Loo, E. J., & Verbeke, W. (2017). Consumers' familiarity with and attitudes towards food quality certifications for rice and vegetables in Vietnam. *Food Control, 82*, 74–82. https://doi.org/10.1016/j.foodcont.2017.06.011.

Nakandala, D., Lau, H., & Zhang, J. (2016). Cost-optimization modelling for fresh food quality and transportation. *Industrial Management & Data Systems, 116*(3), 564–583. https://doi.org/10.1108/IMDS-04-2015-0151.

Nandita, V. (2018). *For food ordering & delivery, India market with tremendous potential.* FnBnews.com. Published on: 14 October 2018. Available at: http://www.fnbnews.com/Interview/for-food-ordering–delivery-india-market-with-tremendous-potential-43798. (Accessed 10 February 2019).

Neuendorf, K. A. (2017). *The content analysis guidebook* (2nd ed.). Los Angeles: SAGE.

O'Brien, S. A. (2020). *Now you can get your food delivered without any human contact.* CNN Business. Published on: 6 March 2020. https://edition.cnn.com/2020/03/05/tech/instacart-leave-at-my-door/index.html. (Accessed 13 March 2020).

Osaki, T. (2020). *Japan delivery services go 'zero contact' in age of new coronavirus.* TheJapanTimes. Published on: 10 March 2020. https://www.japantimes.co.jp/news/2020/03/10/business/corporate-business/japan-delivery-zero-contact-covid-19/#.Xmy5u3IzbIU. (Accessed 14 March 2020).

Pardes, A. (2020). *Is it Ethical to order delivery During a pandemic? Wired.* Published on: 11 March 2020. https://www.wired.com/story/coronavirus-food-delivery-gig-economy/. (Accessed 13 March 2020).

Parsons, T. (2020). *How CoronaVirus will affect the global supply chain.* HUB. Published on: 6 March 2020. https://hub.jhu.edu/2020/03/06/covid-19-coronavirus-impacts-global-supply-chain/. (Accessed 13 March 2020).

Peri, C. (2006). The universe of food quality. *Food Quality and Preference, 17*(1–2), 3–8. https://doi.org/10.1016/j.foodqual.2005.03.002.

Petrescu, D. C., Vermeir, I., & Petrescu-Mag, R. M. (2019). Consumer understanding of food quality, healthiness, and environmental impact: A cross-national perspective. *International Journal Of Environmental Research and Public Health, 17*(1), 169–189. https://doi.org/10.3390/ijerph17010169.

Pigatto, G., Machado, J. G.d-C. F., Negreti, A. d-S., & Machado, L. M. (2017). Have you chosen your request? Analysis of online food delivery companies in Brazil. *British Food Journal, 119*(3), 639–657. https://doi.org/10.1108/BFJ-05-2016-0207.

Pope, J., Annandale, D., & Morrison-Saunders, A. (2004). Conceptualising sustainability assessment. *Environmental Impact Assessment Review, 24*(6), 595–616. https://doi.org/10.1016/j.eiar.2004.03.001.

Prajapati, D., Harish, A. R., Daultani, Y., Singh, H., & Pratap, S. (2020). A clustering based routing heuristic for last-mile logistics in fresh food E-commerce. *Global Business Review.* https://doi.org/10.1177/0972150919889797. Ahead of print.

Purdy, C. (2020a). *Coronavirus forced Chinese food delivery to evolve. Will it do the same in the US?* Quartz. Published on: 28 February 2020. https://qz.com/1809381/will-grubhub-and-uber-eats-learn-from-meituans-coronavirus-plan/. (Accessed 13 March 2020).

Purdy, C. (2020b). *Food delivery during a pandemic exposes the gig economy's biggest rift.* Quartz. Published on: 28 February 2020. https://qz.com/1809629/the-coronavirus-exposes-the-biggest-rift-in-the-gig-economy/. (Accessed 13 March 2020).

Ravi, P. K. (2013). *The encyclopedia of cross-cultural psychology* (1st ed.). US: John Wiley & Sons, Inc.

Ray, A., & Bala, P. K. (2020). Social media for improved process management in organizations during disasters. *Knowledge and Process Management, 27*(1), 63–74. https://doi.org/10.1002/kpm.1623.

Ray, A., Bala, P. K., & Dasgupta, S. A. (2019). Role of authenticity and perceived benefits of online courses on technology based career choice in India: A modified technology adoption model based on career theory. *International Journal of Information Management, 47*, 140–151. https://doi.org/10.1016/j.ijinfomgt.2019.01.015.

Ray, A., Bala, P. K., & Dasgupta, S. A. (2020). Psychological analytics based technology adoption model for effective educational marketing. In N. Rana, et al. (Eds.), *Digital and social media marketing. Advances in theory and practice of emerging markets*. Cham: Springer.

Ray, A., Bala, P., Dasgupta, S., & Sivasankaran, N. (2019). Factors influencing adoption of e-services in rural India – perspectives of consumers and service providers. *Journal of Indian Business Research*. https://doi.org/10.1108/JIBR-11-2018-0295. ahead-of-print No. ahead-of-print.

Ray, A., Bala, P. K., Dasgupta, S. A., & Srivastava, A. (2020). Understanding the factors influencing career choices in India: From the students' perspectives. *International Journal of Indian Culture and Business Management*. https://doi.org/10.1504/IJICBM.2020.10023811. Ahead of print.

Ray, A., Dhir, A., Bala, P. K., & Kaur, P. (2019). Why do people use food delivery apps (FDA)? A uses and gratification theory perspective. *Journal of Retailing and Consumer Services, 51*, 221–230.

Roh, M., & Park, K. (2018). Adoption of O2O food delivery services in South Korea: The moderating role of moral obligation in meal preparation. *International Journal of Information Management, 47*, 262–273. https://doi.org/10.1016/j.ijinfomgt.2018.09.017.

Röhr, A., Lüddecke, K., Drusch, S., Müller, M. J., & Alvensleben, R.v. (2005). Food quality and safety—consumer perception and public health concern. *Food Control, 16*(8), 649–655. https://doi.org/10.1016/j.foodcont.2004.06.001.

Rossi, M., Passeri, D., Sinibaldi, A., Angjellari, M., Tamburri, E., Sorbo, A., et al. (2017). Nanotechnology for food packaging and food quality assessment. *Advances in Food & Nutrition Research, 82*, 149–204. https://doi.org/10.1016/bs.afnr.2017.01.002.

Sakaki, T., Okazaki, M., & Matsuo, Y. (2010). Earthquake shakes twitter users: real-time event detection by social sensors. In *Proceedings of the international conference on the World Wide Web (WWW '10)* (pp. 851–860). Raleigh, North Carolina.

Schaltegger, S., & Burritt, R. (2006). Corporate sustainability. In H. Folmer, & T. H. Tietenberg (Eds.), *The international yearbook of environmental and resource economics*. Edward Elgar Publishing.

See-Kwong, G., Soo-Ryue, N. G., Shiun-Yi, W., & Lily, C. (2017). Outsourcing to online food delivery services: Perspective of F&B business owners. *Journal of Internet Banking and Commerce, 22*(2), 1–13.

Singh, R. K., Murty, H. R., Gupta, S. K., & Dikshit, A. K. (2009). An overview of sustainability assessment methodologies. *Ecological Indicators, 9*(2), 189–212. https://doi.org/10.1016/j.ecolind.2008.05.011.

Sjahroeddin, F. (2018). The Role of E-S-Qual and food Quality on customer Satisfaction in online food delivery service. In *9th international research workshop and national seminer*. Available at: https://jurnal.polban.ac.id/index.php/proceeding/article/viewFile/1097/898.

Sohail, M., Sun, D.-W., & Zhu, Z. (2018). Recent developments in intelligent packaging for enhancing food quality and safety. *Critical Reviews in Food Science and Nutrition, 58*(15), 1–13. https://doi.org/10.1080/10408398.2018.1449731.

Sridharan, K., & Sivakumar, P. (2018). A systematic review on techniques of feature selection and classification for text mining. *International Journal of Business Information Systems, 28*(4), 504–518.

Statista Report. (2018). *eServices Report 2019 - Online Food Delivery*. Published in: November 2018. Available at: https://www.statista.com/study/40457/food-delivery/.

Statt, N. (2020). *Postmates says it will start 'non-contact' meal deliveries during the coronavirus outbreak.* The Verge. Published on: 6 March 2020. https://www.theverge.com/2020/3/6/21168699/postmates-coronavirus-no-contact-deliveries. (Accessed 13 March 2020).

Suhartanto, D., Ali, M. H., Tan, K. H., Sjahroeddin, F., & Kusdibyo, L. (2019). Loyalty toward online food delivery service: The role of e-service quality and food quality. *Journal of Foodservice Business Research, 22*(1), 81–97. https://doi.org/10.1080/15378020.2018.1546076.

Tandon, S. (2018). *India's food delivery firms are under fire for ignoring safety and hygiene.* Published on: 3 August 2018. Available at: https://qz.com/india/1346827/zomato-swiggy-and-the-rest-can-no-longer-ignore-the-quality-of-food-they-deliver/ . (Accessed 10 September 2018).

Thompson, L. (2020a). *Amidst coronavirus fears, U.S. Food delivery companies launch contact-free services.* San Antonio Current. Published on: 11 March 2020. https://www.sacurrent.com/Flavor/archives/2020/03/11/amidst-of-coronavirus-fears-us-food-delivery-companies-launch-contact-free-services. (Accessed 13 March 2020).

Thompson, K. (2020b). *Here's how popular food delivery services are tackling coronavirus concerns.* Thrillist. Published on: 13 March 2020. https://www.thrillist.com/news/nation/food-delivery-coronavirus. (Accessed 14 March 2020).

Uwizeyimana, D. E., & Modiba, N. S. (2016). Managing institutions differently in the 21st century for efficacious grounds: Making schooling a societal exercise. *Journal of Social Sciences, 46*(2), 123–129.

Vaismoradi, M., Turunen, H., & Bondas, T. (2013). Content analysis and thematic analysis: Implications for conducting a qualitative descriptive study. *Nursing and Health Sciences, 15*(3), 398–405.

Varma, V. (2018). *In Kerala, restaurant owners cross swords with food delivery apps.* The Indian Express. Published on: 29 November 2018. Available at: https://indianexpress.com/article/india/kerala-restaurant-owners-cross-swords-with-food-delivery-apps-5471387/. (Accessed 10 February 2018).

Verma, A. (2018). *Zomato shocker: Restaurateurs, delivery boys, customers weigh in on food tampering scare.* HindustanTimes. Published on: 12 December 2018. Available at: https://www.hindustantimes.com/delhi-news/food-app-in-a-pickle-netizens-grossed-out/story-fFUyeDshGoTLNbWk9ljS0M.html. (Accessed 10 February 2018).

Vukasović, T. (2016). Food quality and safety: Added value in a customer-oriented concept. *International Journal of Value Chain Management, 7*(3), 241–254. https://doi.org/10.1504/ijvcm.2016.079208.

Wandel, M., & Bugge, A. (1997). Environmental concern in consumer evaluation of food quality. *Food Quality and Preference, 8*(1), 19–26. https://doi.org/10.1016/s0950-3293(96)00004-3.

Wang, Y. M., & Yin, H. L. (2018). Cost-optimization problem with a soft time window based on an improved fuzzy genetic algorithm for fresh food distribution. *Mathematical Problems in Engineering, 2018*, 1–16. https://doi.org/10.1155/2018/5743287.

Weiss, S. (2020). *Is food delivery safe during the coronavirus pandemic?* New York Post. Published on: 13 Mrach 2020. https://nypost.com/2020/03/13/is-food-delivery-safe-during-the-coronavirus-pandemic/. (Accessed 14 March 2020).

Xu, J. (2018). *Topic modeling with LSA, PLSA, LDA & lda2Vec.* NanoNets. Published on: 25 May 2018. [online]. https://medium.com/nanonets/topic-modeling-with-lsa-psla-lda-and-lda2vec-555ff65b0b05.

Yaseen, T., Sun, D.-W., & Cheng, J.-H. (2017). Raman imaging for food quality and safety evaluation: Fundamentals and applications. *Trends in Food Science & Technology, 62*, 177–189. https://doi.org/10.1016/j.tifs.2017.01.012.

Yeo, V. C. S., Goh, S.-K., & Rezaei, S. (2017). Consumer experiences, attitude and behavioral intention toward online food delivery (OFD) service. *Journal of Retailing and Consumer Services, 35*, 150–162.

Yousefi, H., Su, H.-M., Imani, S. M., Alkhaldi, K., Filipe, C. D. M., & Didar, T. F. (2019). Intelligent food packaging; a review of smart sensing technologies for monitoring food quality. *ACS Sensors*. https://doi.org/10.1021/acssensors.9b00440.

Yusra, Y., & Agus, A. (2018). Online food delivery service quality: Does personal innovativeness matter? *Asia Proceedings of Social Sciences, 2*(3), 251–255. Retrieved from: http://readersinsight.net/APSS/article/view/453.

Zhang, H., Liu, Y., Zhang, Q., Cui, Y., & Xu, S. (2020). A Bayesian network model for the reliability control of fresh food e-commerce logistics systems. *Soft Computing*. https://doi.org/10.1007/s00500-020-04666-5.

Chapter 7

Blockchain in agriculture

Andreas Kamilaris[1,2], Ian R. Cole[1,4], Francesc X. Prenafeta-Boldú[3]
[1]*Research Centre on Interactive Media, Smart Systems, and Emerging Technologies (RISE), Nicosia, Cyprus;* [2]*Pervasive Systems Group, University of Twente, Enschede, The Netherlands;* [3]*Institute of Agri-Food Food Research and Technology (IRTA), Barcelona, Spain;* [4]*University of Cyprus, Nicosia, Cyprus*

7.1 Introduction

The seminal whitepaper "Bitcoin: A Peer-to-Peer Electronic Cash System," published by a pseudonymous author (Nakamoto, 2008), seems to have changed our reality about reconsidering our capacity to perform real-life transactions of goods, products, and services. This whitepaper inspired the creation of Bitcoin, the first cryptocurrency that gained popularity based on the promise that it can enable financial transactions without the need of trusted central authorities, such as bank institutions and other financial organizations (Tschorsch & Scheuermann, 2016). Bitcoin solved the double-spending problem, which relates to the flaw of digital tokens, essentially that digital information can be easily duplicated or falsified. A blockchain is a ledger of digital transactions, which can be maintained by a group of computing devices that do not rely on a trusted third party. Blocks of information related to financial transactions are managed through various software protocols and algorithms that enable the transmission, processing, storage, and presentation of the data in a user-friendly way. In its original bitcoin configuration, each block contains a header with a timestamp, transactional information, and a link to the previous block. Based on its contents, a unique code (called "hash") is generated for each block. This hash is then used to link the blocks together in a linked list structure, where each block has a referral to the hash of the previous block, as Fig. 7.1 shows. Thus, any manipulation of a given block would result in a hash mismatch with all successive blocks.

All transactions are disseminated through the network of computational machines running the blockchain protocol and validated by all computational nodes. The principal feature of a blockchain is its ability to maintain consistency and agreement among the participants (i.e., *consensus*) (Bano, 2017), even if some of them are dishonest (Castro & Liskov, 1999). The scientific community has extensively studied consensus; however, its use in the domain

Food Technology Disruptions. https://doi.org/10.1016/B978-0-12-821470-1.00003-3

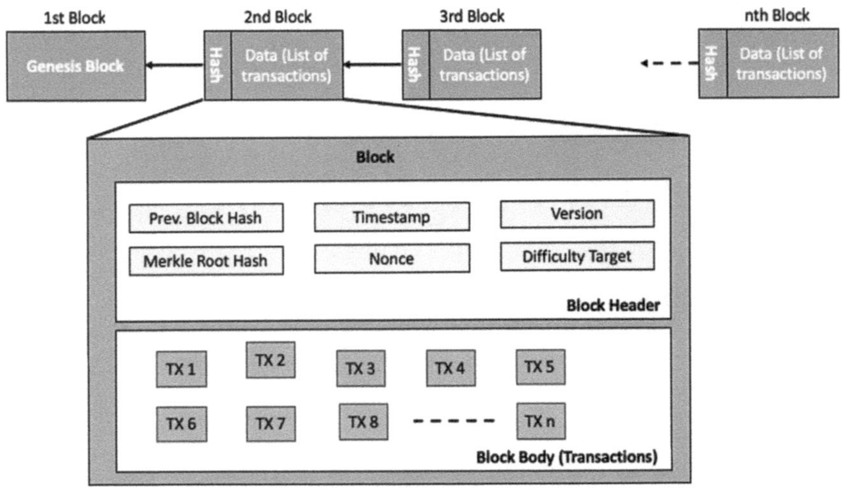

FIGURE 7.1 Example of a blockchain comprising n blocks.

of blockchain led to a resurgence in its research, leading to novel proposals for the design of blockchain systems. Given its use in Bitcoin, the most well-known system is called "Proof of Work" (PoW). PoW requires that computer nodes, *miners*, solve complex computational tasks before transactions are validated and added to the blockchain (Bentov, Gabizon, & Mizrahi, 2016). The first miner to solve the task bundles the block to the chain and is rewarded newly minted coins plus a small transaction fee after validation of the blockchain by the other miners.

Common criticisms of the PoW include the regular competition of miners and the excessive use of computing power required in its operation, which is itself associated with high hardware and energy costs, leading to a significant environmental footprint (Becker et al., 2013) (Krause & Tolaymat, 2018). Another approach mitigating this issue and growing in popularity is *Proof of Stake* (PoS). The PoS approach delegates decision-making to entities who have power and resources (e.g., coins) within the system, asking them to put these resources "at stake" during the process of transaction validation (Bentov et al., 2016). In PoS, the nodes are termed "validators," and as opposed to mining the blockchain, they validate the transactions in return for a transaction fee. Mining is not required because coins exist from day one. In short, nodes are selected randomly in order to approve blocks, while this random selection depends on the number of stakeholders. Thus, PoS applies distributed consensus without mining and the associated expenditure of large amounts of computing power and energy (BitFury Group, 2015). Other consensus mechanisms proposed include Proof of Elapsed Time (PoET), Simplified Byzantine Fault Tolerance (SBFT), and Proof of Authority (PoA).

In the wake of this development, hundreds of alternative digital tokens have appeared, aiming either to address specific disadvantages of the aforementioned popular cryptocurrencies or to target a specific application area, including insurance, health, and nutrition, gambling, and agriculture industries, to name a few (Coinmarketcap, 2017). Blockchain has been adopted by some banking institutes as well. Many banking organizations across the globe are investing resources in exploring how blockchains can impact their businesses (IBM, 2017).

Since 2014, it has increasingly been realized that the applications of blockchain reach far beyond cryptocurrency and financial transactions. Several new applications have emerged (Tayeb & Lago, 2018), including management and administration of assets; digital signatures, authentication, and voting systems; verification and tracking of IP and patent rights; health records in hospitals and the healthcare sector; charity-driven purposes. The application of blockchain is revolutionizing many aspects of business, government, and society, but such significant change poses new challenges and threats that need to be anticipated. In order to mitigate this, the combination of distributed ledger technologies (DLTs) with smart contracts is commonplace among these new applications (Buterin, 2015). The food supply chain is an example of such a complex network of "transactions" between numerous untrusted stakeholders, which can potentially benefit from the full potential of blockchain technology. The present chapter is based on a previous review (Kamilaris, Fonts, & Prenafeta-Boldú, 2019) and presents an updated overview of recent developments for the implementation of blockchain in food supply chains.

7.2 food supply chain

The global food supply chain is highly multiactor based and distributed in its nature, while numerous different actors are involved, such as farmers, shipping companies, wholesalers, retailers, distributors, and grocery stores, until food products reach the final consumer at the end of the chain. A generic agri-food supply chain is described in five primary phases below (Caro, Ali, Vecchio, & Giaffreda, 2018):

1. ***Production***: All agricultural activities implemented within the farm are included in this phase. Organic material (fertilizers, seeds, animal breeds, and feeds) are used by farmers to grow crops and livestock. Depending on the cultivations and animal production cycle, there may be one or more harvest/yield per annum.
2. ***Processing***: The transformation of primary products into secondary ones. A subsequent packaging phase is expected, allowing for the unique identification of products through production batch codes, containing various relevant information such as day of production, raw materials used.

3. ***Distribution*:** After the product is packaged and labeled, it is released for the distribution. Depending on the product, there may be temporal requirements for storage/delivery.
4. ***Retailing*:** Following the distribution, the products arrive for retail by traders. Traders sell the products to customers, the end-users in the chain.
5. ***Consumption*:** The consumer purchases the product. The consumer may request information such as quality standards, country of origin, methods of production that will inform their decision to buy.

Fig. 7.2 (top section, physical flow) shows a simple food supply chain system. Nowadays, the system in operation is not considered efficient or reliable (Tripoli & Schmidhuber, 2018). The documented exchange of goods involves complex processes and much paperwork. The established processes are of questionable transparency, often correlated with high risks for buyers during the exchange of goods. The involvement of intermediaries in order to reduce these risks usually results in increased costs (Lierow, Herzog & Oest, 2017). It is estimated that two-thirds of the final cost of goods can be attributed to the operational costs of supply chains. Thus, there is much space for optimization of the supply chains, by effective reduction of operational costs. Moreover, consumers are rarely aware of the origins of their purchased goods, or the overall environmental footprint associated with its production and distribution.

7.3 The blockchain in agri-food systems

Blockchain technology seems suitable for application in scenarios where various untrusted actors are involved during the distribution of a resource along a supply chain (Manski, 2017; Sharma, 2017). Two crucial and highly

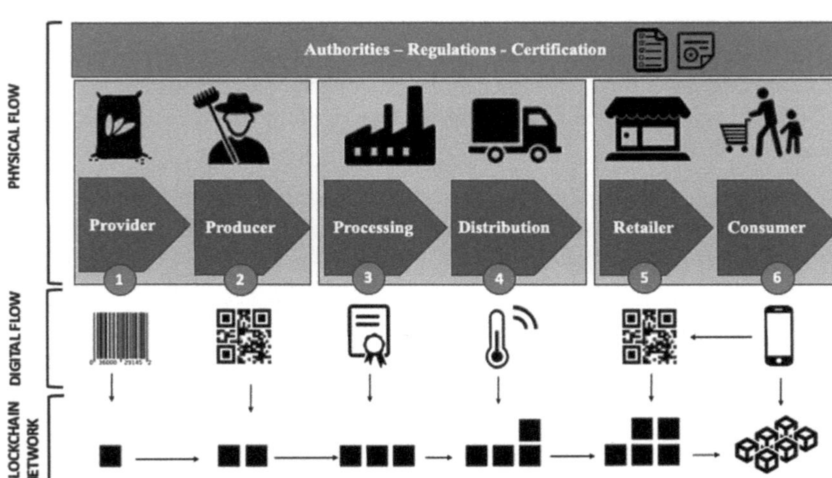

FIGURE 7.2 A simplified food supply chain system.

relevant areas for the application of the technology are *agriculture* and *food supply chain* (Dujak & Sajter, 2019; Tripoli & Schmidhuber, 2018). The is codependency between agriculture and the food supply chain as agricultural products are typically distributed along various supply chains before reaching the consumer (Maslova, 2017).

There is evidence that blockchain applications were adopted for use in supply chain management soon after the technology appeared (Tribis, El Bouchti, & Bouayad, 2018). Today, significant growth in the use of blockchain is being observed. In financial terms, this growth has an annual rate of 87%. The business value of $45 million in 2018 is expected to be $3315 million by 2023 (Chang, Iakovou, & Shi, 2019). An overview of the growth of blockchain use in this sector can be seen in survey papers published in scientific journals (Kamilaris et al., 2019; Antonucci et al., 2019; Bermeo-Almeida et al., 2018).

A successful example is a company AgriDigital, which executed the world's first settlement of the sale of grain on a blockchain, i.e., 23.46 tons in December 2016 (ICT4Ag, 2017). Since then, transactions of more than 1.6 million tons have been performed, with payments of $360 million globally and more than 1000 entities involved. AgriDigital uses blockchain for building trust in agricultural supply chains (AgriDigital, 2017). Two years later, Louis Dreyfus Co (LDC), together with Dutch and French banks, performed the first blockchain-enabled commodity trade, which was about a cargo of soybeans from the US to China (Hoffman & Munsterman, 2018). According to LDC, the time taken for document processing in such transactions was reduced by 80% by automatically matching data in real-time, avoiding duplication, and manual checks.

Fig. 7.2 depicts how blockchain technology could be used to digitize the food supply chain. The digital layer is below the physical layer, where the physical flow of goods occurs. The digital layer integrates various digital technologies, (QR codes, RFID, NFC, trusted online certification, digital signatures, sensors, and actuators, mobile phones) these technologies communicate either directly or via the Internet (Kamilaris & Pitsillides, 2016). Every action, process, measurement, or transaction performed, empowered by the use of the aforementioned digital technologies, is added to a shared blockchain (bottom layer of Fig. 7.2). The blockchain contains information, which is accepted by all participating parties as to the ground truth. Various participating business partners validate the transactions forming a global consensus inside the chain network. After validation by each business partner (physical flow in Fig. 7.2), using a number of different technologies for recognition and tracking of goods (digital flow in Fig. 7.2), each block is added to the chain of transactions (blockchain network in Fig. 7.2), stored as a permanent record. At each stage (numbers 1–6 in Fig. 7.2), a range of different tracking technologies are involved, where different types of information are added to the shared distributed ledger. Example information to be recorded for the six stages is described below:

1. ***Provider*:** Data about crops, fertilizers, pesticides, people involved and labor, machinery, methods, processes, and transactions are recorded.
2. ***Producer*:** Data about the farm and the practices employed therein, such as locality and certified production methods. Additional information about crop cultivation processes, weather conditions, animals, and their welfare may also be recorded.
3. ***Processing*:** Data about the factory and its equipment, the processing methods, unique identifiers, as well as the financial transactions occurring between the producers and distributors are recorded.
4. ***Distribution*:** Data about shipping, item itineraries, storage conditions (e.g., temperature, humidity), distance, and time in transit for all transportation methods, Transactions between the distributors and retailers.
5. ***Retailer*:** Data about food products; historical and current quality and quantity; conditions of storage; shelf time.
6. ***Consumer*:** Access to all information associated with the product. Full transparency for conditions, processes, and methods from the producer and provider to the retail store.

In order to locate relevant initiatives where blockchain was employed in real-life agri-food systems, a keyword-based bibliographic search was performed using *Web of Science* and *Google Scholar*, using keywords such as *Blockchain, Agriculture, Food, Food Supply Chain, as well as combinations.*

The focus was on *on-going* initiatives, pilots, and studies; aiming to capture the applicability and potential of blockchain in the agri-food industry. Fifty-nine documents were identified with the above search, and 47 of them deemed relevant in terms of using blockchain technology in the food supply chain. In order to widen the bibliographic search, the bibliography of the original 59 papers was examined, followed by a keyword-based search in the Google search engine. This allowed increasing the identified efforts and initiatives to 80. These 80 projects/initiatives were divided into six main categories based on their purpose and overall target/goal. These are detailed alongside popularity below:

(a) food security (3 papers, 4%),
(b) food safety (9 papers, 11%),
(c) food integrity (31 papers, 39%),
(d) support of small farmers (12 papers, 15%),
(e) waste reduction, circular economy and environmental awareness (12 papers, 15%),
(f) better supervision and management of the supply chain (13 papers, 16%).

7.3.1 Food security

FAO defines food security as *sufficient access to nutritious food for all people, for an active and healthy life*. This objective can be jeopardized in scenarios of pandemics, environmental disasters, wars, political conflicts, and other unpredictable events and hazards. Blockchain presents an alternative for transparent management and distribution of international aid; facilitating the delivery process, and making records and assets verifiable and accessible. Blockchain can enable rapid and efficient response during humanitarian emergencies (AID Tech, 2017). Examples of its use to this end include digital food coupons distributed to Palestinian refugees in Jordan's Azraq camp (Blockchain for Zero Hunger, 2017). This was done with an Ethereum-based blockchain (Ethereum, 2015), where the coupons could be redeemed via biometric data (Built to Adapt, 2018), providing help and support to more than 100 × 1000 refugees.

7.3.2 Food safety

Food safety comprises the processing, management, and storage of food in hygienic ways, in order to prevent illness in the human population. Food safety and quality assurance have become increasingly difficult as the exchange of goods has globalized and increased in complexity (Creydt & Fischer, 2019). CDC claimed that food-borne contamination causes 48 million Americans to become ill with three ×1000 deaths each year (Tripoli & Schmidhuber, 2018; CDC, 2018). Oceana's research into seafood fraud during 2016, revealed that 20% of seafood products were mislabeled (Oceana, 2013). It is a general observation that food supply chains have long shipment distances, high complexity, and long processing times, which sometimes affects the quality of food products (Lee, Mendelson, Rammohan, & Srivastava, 2017). Blockchain can be a reliable solution, satisfying the need for improved safety and reliability. As shown in Fig. 7.2, when data about food products is recorded at each stage of the supply chain to a distributer ledger whose contents are accepted by all participating entities, better hygienic conditions, easier identification of contaminated products, and early identification of fraud and related issues are ensured.

Walmart and Kroger include blockchain technology in their food chains (Insights, 2017), initially working on case studies focused on Chinese pork and Mexican mangoes (Kamath, 2018). Initial results from these studies showed that the tracking of a package of mangoes from the supermarket to the farm of origin took 6.5 s with traditional methods, yet was available in just a few seconds with blockchain (Wass, 2017).

In an effort to tackle food fraud in the dairy supply chain, the company CyberSecurity developed the Milk Verification Project prototype (CyberSecurity, 2019), implemented with blockchain technology. In this project, IT tools were developed to automate the acquisition of the registration of information in the supply chain processes. Another example of the application of blockchain in the dairy sector can be found in (Kasten, 2019).

Recently, the use of the Internet of Things (IoT), in combination with blockchain, has been proposed for real-time data acquisition for products along the supply chain (Tian, 2017). IoT technology seems to be advantageous for effective management and operation of the cold-chain in the distribution logistics of spoilable food products. This has been demonstrated by ZetoChain, who has performed environmental condition monitoring at each phase of the cold chain utilizing IoT devices (Zeto, 2018). This enables the real-time identification of problems, and notifications can be sent immediately to the parties involved to ensure fast action taking and corrective measures. Smart contracts are harnessed in this project to increase the safety of and confidence in sales and deliveries of goods. Consumers can use mobile applications to scan *Zeto labels* on products to learn about the product's history.

Mohan's model used blockchain to track chicken products (Mohan, 2018). The model utilized existing food quality systems and technology already present in the various stages of the supply chain. Blockchain technology was integrated to enhance the operations and methods in place. In this model, all stakeholders connected their production systems to the shared blockchain-enabled network. This model revealed various advantages over the traditional tracking systems in terms of food traceability and enhanced food safety.

Finally, George, Harsh, Ray and Babu (2019) focused on the implementation of a blockchain model in restaurants, where storage time was identified as a key variable for various fresh products, such as pork meat. A prototype system was developed for the collection of data from various actors across the supply chain, generating a Food Quality Index (FQI) value. This FQI value was used to aid the determination of whether the food was fit for consumption and was generated based on the standard storage and handling regulations specified by food safety authorities. The system automated checking whether the derived value was within the permissible range.

7.3.3 Food integrity

Food integrity is about the exchange of food in a reliable way. Throughout the food supply chain, each member is supposed to provide elaborate information about the origin and management of goods. Examples of this information are illustrated in Fig. 7.2. This particular challenge of food integrity is of great concern in China, where the fast-paced economic growth has led to serious transparency issues (Tian, 2017; Tse, Zhang, Yang, Cheng, & Mu, 2017).

Food safety and integrity can be effectively enhanced with better traceability via the use of blockchain technology (Galvez, Mejuto, & Simal-Gandara, 2018; Creydt & Fischer, 2019). Food companies can avoid or at least mitigate frauds on goods and products by identifying and tracking issues early enough, tracing back to the original sources and actors who produced the fraud (Levitt, 2016). The food traceability market was estimated to be worth $14 billion in 2019 (Markets and Markets Research, 2016). The following paragraphs list various examples of companies (including a number of start-ups), which have deployed blockchain technology in an aim to increase the integrity of the food chains in which they operate.

Cargill Inc. used blockchain for enabling shoppers to trace turkeys from the store back to their farm of origin, aiming to bring greater transparency to the transaction (Bunge, 2017). Turkeys and associated issues of animal welfare have been considered in a recent pilot involving blockchain (Genetics, 2018). Here, welfare was associated with the animals having more room and some covered outdoor space. Carrefour has announced the improved verification of standards and tracing of various food origins in different categories, such as meat, fish, fruits, vegetables, and dairy products, via the use of blockchain (Carrefour, 2018).

Downstream beer (Ireland Craft Beers, 2017) is the first company in the beer industry to make use of blockchain technology. It is used to make transparent all information related to its beers, including the ingredients used and the brewing processes employed. Each step in the processing of their craft beer products was recorded to a blockchain in order to provide enhanced transparency and ensure product authenticity. Customers can scan the QR code on the front of the bottle, using their mobile phones, and then visit a website, which lists all relevant information, from raw ingredients to bottling.

San Domenico roastery (Foodchain, 2019) adopted blockchain technology to ensure that its coffee products are accompanied by reliable, untampered documentation and to guarantee full transparency to their customers. By using blockchain, each processing step until the point of sale is recorded and finalized before being moving to the next step. Product quality is thus ensured, and the entire production chain and associated information concerning a given coffee product are available by association with a QR code. This allows for traceable access to information, news, videos, images, and certifications from the supply chain to the final consumer. Fig. 7.3 shows a snapshot of the mobile application after scanning a QR code of an existing coffee product. Information about the origin of the coffee is presented. After tailoring the system to the specific needs of the company, the cost savings of certification ranged between 70% and 90%, according to Foodchain (Foodchain, 2019). Foodchain is an Italian company offering traceability services for food supply chains using blockchain technology.

The Italian company Aldo Cozzi claimed that blockchain technology-enabled consumers to find out about the entire supply chain of the pasta they were buying (Cozzi, 2019). By scanning the QR code on the product label (similar to the coffee product demonstrated in Fig. 7.3), information regarding the whole supply chain was presented, including info about the manufacturer, products, and flours used, drying method, and transportation information).

"Paddock to a plate," a research project tracking beef throughout the production-consumption chain, is focused on helping to increase the reputation of Australian meat produce and its association with high quality (Campbell, 2017). The project uses BeefLedger as its technology platform (BeefLedger Limited, 2017). As another example, the beef produced in inner Mongolia and distributed to different provinces of China is monitored by the e-commerce

FIGURE 7.3 A snapshot of the San Domenico roastery mobile application. After the QR code of the product, an existing coffee, has been scanned, information about the origin of the coffee is presented. *Foodchain, "San Domenico roastery: The first case of a coffee supply chain fully traced with blockchain," 2019. [Online]. Available https://food-chain.it/public/case/san-domenico/.*

platform JD.com (JD.com Blog, 2018). Consumers can find out about the animals involved in the production, their feed, the slaughtering and meat packaging dates, and the results of food safety tests. This is implemented by QR code scanning. The Gogochicken company provides evidence to support their statement that their chickens are free-range by making available GPS tracking data acquired by ankle bracelets to monitor the chickens' movements and behavior (Peter, 2017). This information is then available through the web. The company intends to build trust with its customers by documenting and making available information on the origins of its food produce. To date, 100 × 1000 chickens have been fitted with GPS tracking anklets, and there are plans to extend the practice to include 23 million birds in the project over the next 3 years.

The Grass Roots Farmers Cooperative (2017) sells a meat subscription box, which incorporates blockchain technology to ensure that their consumers are reliably informed about the welfare of their animals. A pilot was performed in San Francisco, where cases of chicken were labeled with QR codes that linked to information about the meat inside.

Intel Hyperledger is an essential technology for facilitating traceability in the seafood supply chain. Hyperledger Sawtooth is a platform for creating and managing blockchains (Hyperledger, 2018). The study used sensors to record information about current and historical locations of fish, as well as storage conditions. Hyperledger is considered among the most promising efforts in this domain as it provides a complete range of high-quality software-based services and tools while being supported by a large user base that includes significant global commercial entities, including IBM. Hyperledger provides solutions for professional use of blockchain by enterprises, and at the same time, has also been harnessed in academic efforts such as AgriBlockIoT (Caro, Ali, Vecchio, & Giaffreda, 2018). Hyperledger supports the development of open-source software frameworks, based on DLT, suitable for business solutions. Fabric (for permission blockchain networks) and Sawtooth (for both permissioned and permissionless blockchain networks) are two of the most mature Hyperledger frameworks, supporting, among others, smart contracts languages. Although Hyperledger Fabric is the better established, Hyperledger Sawtooth is the more advanced framework, facilitating integration with other blockchain frameworks (Suprunov, 2018). A demonstrator application based on the Hyperledger Fabric framework implemented in (Bechtsis, Tsolakis, Bizakis, & Vlachos, 2019) indicated that blockchain technology has matured to the point of commercial application, further compounding the point that its adoption in various operations of the food supply chain could add significant value by authenticating critical parameters and providing enhanced traceability.

The AgriOpenData blockchain (Galvez, Mejuto, & Simal-Gandara, 2018) constitutes an innovative digital technology for traceability in the food supply chain, focused primarily on organic products, allowing the transparent and secure processing of various agri-food products. AgriOpenData certifies the digital identity products, enhancing trust and quality in the agri-food business. Transparency in the organic food supply chain was also addressed in (Basnayake & Rajapakse, 2019), where smart contract instances were created for each physical product and deployed to a blockchain network. Each transaction and event associated with a product was validated by peers in the blockchain system using tokens as a mechanism to assess the reputation of farmers and product traceability. Certification requests can be placed by farmers, which can be approved by the peers participating in the blockchain network, adding to their reputation.

An interesting approach has been presented in (Salah, Nizamuddin, Jayaraman, & Omar, 2019), which leverages smart contracts in combination with the Ethereum blockchain, in order to efficiently perform business transactions for soybean tracking and traceability across the agricultural supply chain.

The Blockchain Supply Chain Traceability Project (WWF, 2018), supported by WWF, targets the elimination of illegal tuna fishing via the incorporation of blockchain in industry transactions. RFID e-tagging is used by fishermen to declare their catch on the shared blockchain. Another project focusing on the traceability of tuna was that of the company Balfegó (Balfegó Group, 2017).

A food quality network created by ripe.io, the Blockchain of Food (Ripe.io, 2017), maps the journey of food from its production to the consumer's plate. $2.4 million in seed funding has recently been raised by ripe.io; in a funding round led by the venture capital arm of Maersk (AgFunder News, 2018).

OriginTrail is a company providing services that enable consumers to find the origins of their ingredients, including the growing conditions of poultry (OriginTrail, 2018). A project, "blockchain for agri-food," developed a proof-of-concept blockchain-based application for the management and tracking of table grapes from South Africa, with a focus on consumer transparency (Ge et al., 2017). Blockchain technology has been used to develop a framework for greenhouse farming with enhanced security features (Patil et al., 2017). Nestlé has recently joined the IBM Food Trust partnership, which aims to become established as an *ecosystem of stakeholders toward a more sustainable food system for all* (ITUNews, 2018), having a first large pilot based on canned pumpkin and mango.

The authors in (Kamilaris & Pitsillides, 2016) proposed the combination of blockchain with other digital technologies (such as IoT, RFID, and NFC), toward higher food traceability. Combining RFID and blockchain technologies were discussed in (Tian, 2016), while a system based on IoT devices and smart contracts was proposed in (Kim, Hilton, Burks, & Reyes, 2018). The combination of blockchain technology with Near Field Communication (NFC) and verified users is proposed (Boehm, Kim, & Hong, 2017) toward an updated traceability system.

A permissioned blockchain network for tracking of the cannabis supply chain has been developed recently in Canada (Abelseth, 2018). Health Canada aims to use this technology to enforce regulations more conveniently.

Finally, blockchain technology has been assessed for use in tracing the production of non-edible crops, which are associated with regulation and legal issues. The digital traceability of wood via a blockchain, together with RFID sensors, from the standing tree to the final user, was discussed in (Figorilli et al., 2018), using only open source technologies for a prototype implementation. Multiple aspects of wood cutting were recorded to the blockchain, including tree felling, harvesting, and sawmill processes.

7.3.4 Small farmers support

Competition in the agriculture industries of developing countries can be raised with small cooperatives of farmers (Chinaka, 2016). Operating with a cooperative, individual farmers can access a larger share of their crop value (FarmShare, 2017). FarmShare proposes new structures of property ownership, the cooperation of communities, and the creation of resilient self-sustaining local economies. FarmShare envisions an evolution of the existing model of community-supported agriculture, exploiting blockchain to achieve and maintain distributed transparency, consensus, and equity shares, as well as automatic digital governance, reducing managerial burdens while at the same time fostering greater community engagement (FarmShare, 2017).

AgriLedger has demonstrated how using a distributed crypto-ledger can increase trust among small cooperatives in Africa (AgriLedger, 2017). A new approach for trusted applications and services among farmers and other entities in the agro-food chain has also been presented (Davcev et al., 2018). OlivaCoin, a B2B platform for the trade of olive oil, supports the olive oil market by offering the reduced overall cost, increased transparency, and easier access to global markets (OlivaCoin, 2016). Various case studies using blockchain technology to secure better financial resilience for Kenyan smallholders affected by climate change have been presented (Bolt, 2019).

Furthermore, the industry has seen several startups, which offer various software solutions targeting the increased traceability of food products. Solutions suitable for supporting small farmers and communities, include Provenance, Arc-Net, Bart. Digital and Bext360. Recently, the Soil Association Certification decided to utilize the Provenance solution in a pilot initiative for tracking the story of organic food products along the chain (Soil Association Certification, 2018).

Advantages for Brazilian agriculture exports obtained with the implementation of a blockchain are highlighted in (Lucena, Binotto, Momo, & Kim, 2018). Such platforms could help producers track stored grains to help optimize trading with global exporters allowing for enhanced collaboration between members of the business network and removing requirement for intermediaries in some business processes. In (Chong, Perez, Castilla, & Rosario, 2020), a similar concept was applied to the cocoa export supply chain of Peru, demonstrating how to trust in international buyers can be generated via the use of the blockchain.

Medium-size farmers could also benefit from blockchain technology and the initiatives outlined above, although they form a differentiable category large corporations (FarmShare, 2017). Also, cooperatives may be formed from either small- or medium-size farmers, growing to become large entities representing tens or hundreds of farmers. The transparency of information brought about by blockchain has useful applications for such organizations as the associated levels of transparency can mitigate conflict among the partners in a fair way (Chinaka, 2016; AgriDigital, 2017). Fig. 7.4 shows how blockchain can be applied for an automatic transaction between a farmers' cooperative and a retailer by using smart contracts. In the hypothetical scenario provided as an example, a cooperative of farmers based in Africa uses a smart contract to achieve the sale of its cereal products. Automatic access of the buyer to a storage room containing the crops is included in this particular execution.

Blockchain can also be used to facilitate the rollout of insurance programs, securing farmers (as cooperative members) against unpredictable and unavoidable events that negatively affect their harvests, such as extreme weather or natural disasters (Jha, Andre, & J. O, 2018). The ARBOL project puts forward the idea that, via customized agreements, farmers can receive payments in times of unavoidable reduced harvest (ArbolMarket, 2019).

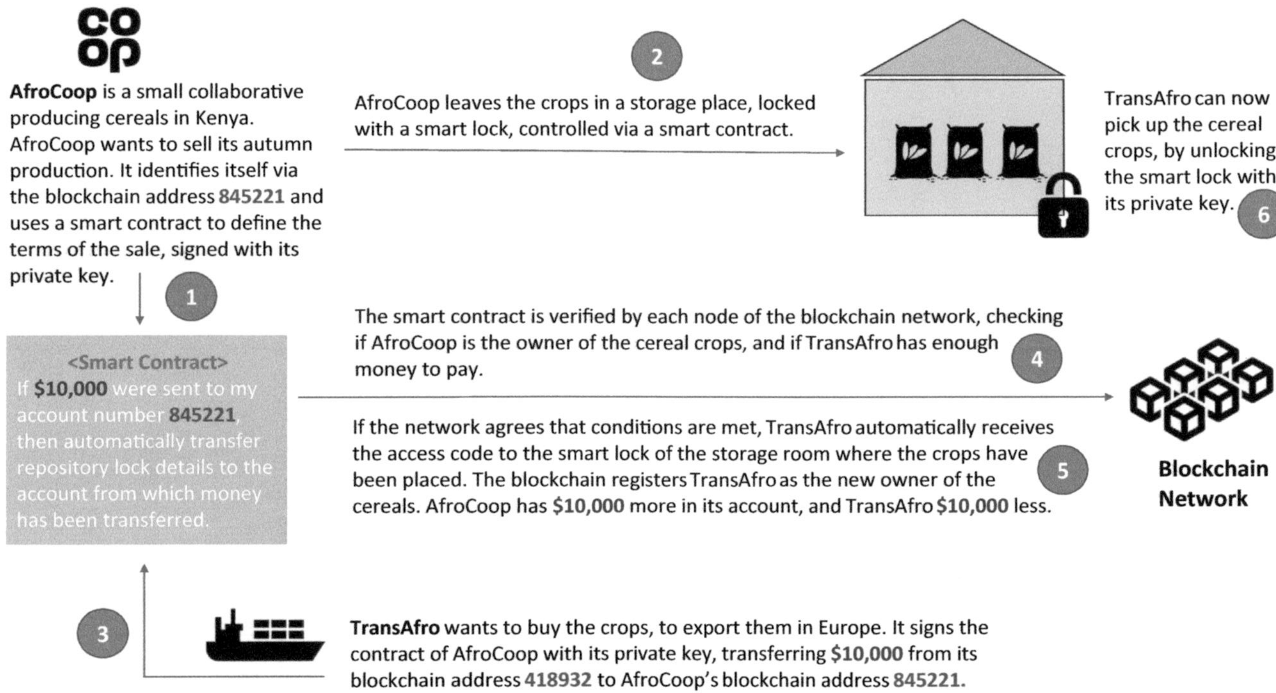

FIGURE 7.4 An example of a smart contract execution.

Finally, in (Kim & Laskowski, 2018a, 2018b), blockchain applications across the agricultural sector beyond the typical finance use cases are explored with a strong focus on developing farmers, sustainable agriculture, and local economy cooperatives. Pilot programs in Kenya, Myanmar, and Papua New Guinea are presented.

7.3.5 Waste reduction, environmental awareness, and circular economy

Various waste management initiatives incorporating blockchain technology have been reported. An initiative of some significance is the Plastic Bank (Bank, 2019), a global recycling company from Canada, whose operations focus on reducing plastic waste in developing countries, with current activity in Haiti, Peru, and Colombia, and plans for the extension of operations into Indonesia and Philippines. Plastic Bank rewards people who bring plastic waste to bank recycling centers using blockchain-secured digital tokens. These tokens can be exchanged for goods or services such as food or phone-charging services, in any shop, using the Plastic Bank mobile application (Steenmans & Taylor, 2018). The Plastic Bank initiative has been successful to date, with over one million participants, over two 1000 collector units, and approximately three million kilograms of plastic collected for recycling in Haiti since 2014. A similar mission is found with Agora Tech Lab (Agora Tech Lab, 2018), which aims to promote circular economy initiatives by rewarding responsible behavior.

Another example of blockchain technology being used to improve system operations is found in waste management at railway stations. The traditional waste management practices have been chaotic, producing hundreds of tonnes of waste each year. In order to tackle this issue, a system was developed by SNCF subsidiary Arep that used blockchain in the collection of detailed information over Bluetooth and the continual updating of categorized waste quantities (SNCF, 2017).

There exist several other commercial solutions that use blockchain to improve systems of recycling and waste sorting throughout the food chain, including Recereum, (2017) and Swachhcoin, (2018).

An application of blockchain technology for incentivizing the efficient use of agricultural waste has been proposed (Zhang, 2019), whereby trading of biomass energy and agricultural products across the waste-to-energy ecosystem is promoted. A case study was performed in Changzhi City, Shanxi Province, China, with waste such as crop straw and animal residues.

In (Lin, Shen, Zhang, & Chai, 2018), an ecological food system based on blockchain and IoT technologies is proposed, involving various untrusted parties of a smart self-organized agri-food system, which is open and trustful. IoT devices replaced manual recording and verification methods in order to minimize human intervention.

Moreover, blockchain can help to raise awareness about the environmental consequences and characteristics of the food produced. The degradation of land, soil, and water associated with food production is an issue of growing concern. Soil quality has been highlighted as an important issue toward the realization of the United Nations Sustainable Development Goals (SDG) (Keesstra et al., 2016). In attaining these goals, sustainable development, effective management, and responsible use of agricultural fields, water resources, and soils are of great importance (Keesstra et al., 2018). The functionality of tracing such information throughout the supply chain and making it transparent and visible to the public is essential for adding pressure to incentivize the environmentally responsible behavior of producers and policymakers.

Finally, focusing on the circular economy, a new model of the supply chain via blockchain has been proposed (Casado-Vara, Prieto, De la Prieta, & Corchado, 2018), which enables the implementation of a circular economy and eliminates many existing disadvantages of the food chain. A multiagent system has been proposed to organize in a more efficient way the transactions happening in the supply chain. The review in (Kouhizadeh, Zhu, & Sarkis, 2019) presents different studies on the interactions of blockchain and circular economy in various industrial sectors, including agricultural and food systems.

7.3.6 Supervision and management

Harnessing blockchain for credit evaluation purposes can add robustness to supervision and management practices in the food supply chain and improve the monitoring of international agreements, such as the Paris Agreement on Climate Change (Tripoli & Schmidhuber, 2018) and World Trade Organization. A system that gathers credit evaluation text from traders by smart contracts, based on the Hyperledger blockchain, has been developed (Mao, Wang, Hao, & Li, 2018b). Trader credit in this system can be used as a reference for regulators to evidence credibility. In this way, traders are held accountable for their actions in the processes of a transaction; and credit evaluation by regulators is improved and streamlined. AgriBlockIoT is another example of a fully decentralized, blockchain-based solution used in agriculture and food supply chain management (Caro, Ali, Vecchio, & Giaffreda, 2018). The solution offers the seamless integration of IoT devices generating and analyzing data along the chain. A system has been applied to the management of a grape farm near the City of Skopje, North Macedonia, combining IoT sensors and cloud technologies (Davcev, Kocarev, Carbone, Stankovski, & Mitresk, 2018).

Labor exploitation in agriculture is also tackled by blockchain products, protecting sector workers with temporary agreements and employment relationships (Pinna & Ibba, 2018). With labor agreements as part of a blockchain, authorities can better ensure fairness in payment and taxation. As an example, Coca-Cola has attempted to identify forced labor in the sugarcane industry using blockchain (Reuters, 2018).

Other areas where blockchain can be used is in the trusted management of water in irrigation communities (Bordel, Kocarev, Carbone, Stankovski, & Mitresk, 2019). A project integrating a fuzzy logic algorithm for a smart irrigation system, based on blockchain, is presented in (Munir, Bajwa, & Cheema, 2019).

Quality assurance embodies the avoidance of system failures such as delays of deliveries to final destinations, poor or insufficient monitoring, fraud, and the maintenance of product quality along the food chain. This fact requires good practice throughout the chain, such as suitable storage conditions and avoidance of contamination. For example, several properties defining a good quality of grain production are listed in (Brooker, Bakker-Arkema, & Hall, 1992). The preliminary results in (Lucena, Binotto, Momo, & Kim, 2018) support a potential demand for a blockchain-based certification. It is reported that this would lead to an added value in the retail of approximately 15% for genetically modified- (GM-)free soy in an example business network for grain exports in Brazil. This added value arises from more reliable and efficient quality assurance processes, facilitated by blockchain. The rice value chain has also adopted blockchain to record events taking place in the transportation of produce, strengthening the security and quality of rice (Kumar & Iyengar, 2017).

The integration of IoT and blockchain is found in a conceptual extension to a mushroom farm distributed process control system, enabling the collection of distributed data on environmental parameters affecting mushroom production (Branco, Moreira, Martins, Au-Yong-Oliveira, & Gonçalves, 2019). The approach is focused on complementing the existing production control system with added functionality.

Finally, in (Shaji, Shaji, Rony, Kuriakose, & Rawther, 2019), blockchain is used in a system that helps farmers in India let agricultural land from landlords easily and securely. The system acted as a bridge between landlords and farmers, using blockchain technology to achieve transparency and security of transactions. This demonstrates that the management of common resources such as energy, land, and water can benefit from the use of blockchain technology, adding security and preventing speculation in the trading of these resources (Poberezhna, 2018).

7.4 Analysis of the findings

Table 7.1 lists various projects, case studies, and pilots, which incorporate blockchain technology in the agri-food sector, based on the initiatives described in the previous sections of this chapter. These studies tend to have a focus on automation of production, productivity and transparency (CyberSecurity, 2019; Peter, 2017; Basnayake & Rajapakse, 2019; WWF, 2018; Tian, 2016; Kim et al., 2018; Boehm, Kim, & Hong, 2017; Lin et al., 2018; Chen, Shi, Ren, Yan, Shi, & Zhang, 2017).

TABLE 7.1 Goods and products concerning projects using blockchain technology and their overall objectives.

Goods, products, resources	Initiative/project/ company involved	Objectives
Soybeans	LDC (Hoffman and Munsterman, 2018; Salah et al., 2019)	Financial, Faster Operations, Traceability
Grains	(AgriDigital, 2017); GEBN study (Lucena, Binotto, Momo, & Kim, 2018	Financial, supervision, and management
Olive oil	OlivaCoin (2016)	Financial, small farmers support
Dairy milk	Milk verification project prototype (CyberSecurity, 2019), Kasten (Kasten, 2019)	Food safety
Turkeys	Cargill Inc. (Bunge, 2017); (Genetics, 2018)	Traceability, animal welfare
Mangoes	Walmart, Kroger, IBM (Insights, 2017), (Kamath, 2018), Nestle (ITUNews, 2018)	Traceability
Canned pumpkin	Nestle (ITUNews, 2018)	Traceability
Pork	Walmart, Kroger, IBM (Insights, 2017; Kamath, 2018; George et al., 2019)	Traceability
Sugar cane	Coca-Cola (Reuters, 2018)	Supervision and management
Beer	Downstream (Ireland Craft Beers, 2017)	Traceability
Coffee	San Domenico roastery (Foodchain, 2019)	Traceability
Pasta	(Cozzi, 2019)	Traceability
Beef	"Paddock to plate" project (Campbell, 2017; BeefLedger Limited, 2017; JD.com; Peter, 2017)	Traceability

TABLE 7.1 Goods and products concerning projects using blockchain technology and their overall objectives.—cont'd

Goods, products, resources	Initiative/project/ company involved	Objectives
Cannabis	Medical cannabis Tracking (MCT) system (Abelseth, 2018)	Traceability
Chicken	Gogochicken (Peter, 2017; Grass Roots Farmers Cooperative, 2017; OriginTrail, 2018; Mohan, 2018)	Traceability
Wood (Chestnut trees)	(Figorilli et al., 2018)	Traceability
Sea-food	Intel (Hyperledger, 2018; WWF, 2018; Balfegó Group, 2017)	Environmental impact, Traceability
Table grapes	"Blockchain for agri-food" project (Ge et al., 2017), a grape farm near the City of Skopje (Davcev et al., 2018)	Experimental feasibility study, supervision and management
Organic food	AgriOpenData blockchain (Galvez, Mejuto, & Simal-Gandara, 2018; Basnayake & Rajapakse, 2019; Soil Association Certification, 2018)	Financial, Traceability, small farmers support
Cocoa	(Chong et al., 2020)	Financial, Traceability, small farmers support
Food waste	(Bank, 2019; Agora Tech Lab, 2018; SNCF, 2017; Recereum, 2017; Swachhcoin, 2018)	Waste reduction
Agricultural by-products, residues, and wastes	Crop straw and animal residues (Zhang, 2019)	Waste reduction
Water	Global water assets (Poberezhna, 2018), management of	Supervision and management

Continued

TABLE 7.1 Goods and products concerning projects using blockchain technology and their overall objectives.—cont'd

Goods, products, resources	Initiative/project/ company involved	Objectives
	irrigation communities (Bordel et al., 2019), smart irrigation system (Munir et al., 2019)	
Rice	Quality of rice in transportation (Kumar & Iyengar, 2017)	Supervision and management
Mushrooms	Mushroom farm process control system (Branco et al., 2019)	Supervision and management
Agricultural land	The lending of land in India (Shaji et al., 2019)	Supervision and management
The food chain in general	(AgriLedger, 2017; FarmShare, 2017; Carrefour, 2018; Ripe.io, 2017; OriginTrail, 2018), AgriBlockIoT (Caro, Ali, Vecchio, & Giaffreda, 2018), food supply chain prototypes enhanced with other technologies (Tian, 2017; Bechtsis et al., 2019; Kim et al., 2018; Boehm, Kim, & Hong, 2017), ecological food system (Lin et al., 2018), local economy cooperatives (Kim & Laskowski, 2018a, 2018b), financial resilience of smallholders affected by climate change (Bolt, 2019)	Financial, Traceability, food safety, small farmers support, waste reduction, supervision, and management
Supply chain and circular economy	Casado-Vara et al., 2018; Kouhizadeh et al., 2019)	Waste reduction, environmental impact, circular economy

7.4.1 Technology

The technology underlying the 80 different projects, initiatives, and papers reviewed herein is summarized here. The most popular technology is Ethereum (15%, 19%), followed by Hyperledger Fabric (8%, 10%). Eight were committed to developing their own blockchain solution (AgriDigital, 2017; Zeto, 2018; Carrefour, 2018; JD.com Blog, 2018; Ripe.io, 2017; OriginTrail, 2018; OlivaCoin, 2016; Munir et al., 2019). Other technologies referred to include: BigchainDB in (Tian, 2017), the Bitcoin protocol in (Bunge, 2017), BeefLedger in (Campbell, 2017; Foodchain, 2019), the ZhongAn blockchain open platform in (Peter, 2017), Provenance in (Grass Roots Farmers Cooperative, 2017; Soil Association Certification, 2018; Kim & Laskowski, 2018a, 2018b; Hyperledger, 2018), the Azure Blockchain Workbench together with Ethereum in (Figorilli et al., 2018) and a combination of Ethereum and Hyperledger Sawtooth in (Caro, Ali, Vecchio, & Giaffreda, 2018). Thirty-eight projects (47%) did not explicitly define the underlying structure of their blockchain-based solutions, possibly as they are at the conceptual stage, as detailed further below.

7.4.2 Maturity and sustainability

Fig. 7.5 depicts the maturity level of the related work as identified through this survey, from the conceptual stage (17%, 21%) to fully integrated operations (5%, 6%). The majority of the projects are in the implementation (23%, 29%) or proof-of-concept stage (18%, 22%), where small pilot studies are under the demonstration. Projects stemming from the research domain tend to sit at the pilot implementation phase or below, as most are either conceptual or innovation-based studies. The 8 (10%) demonstrating large-scale case studies

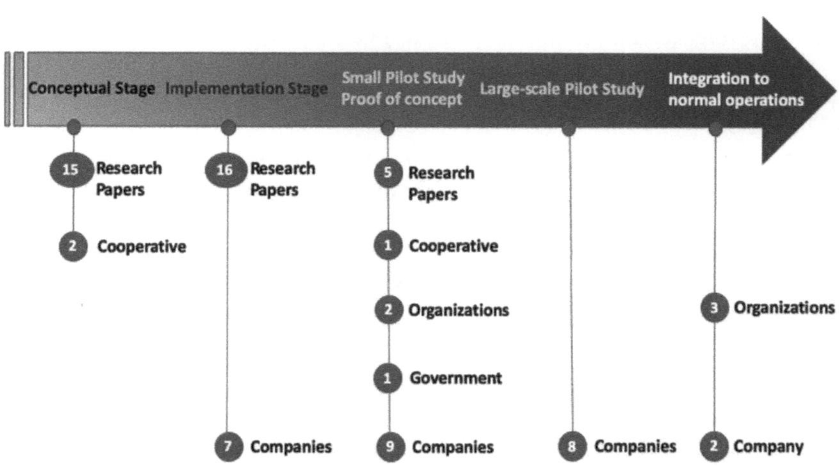

FIGURE 7.5 Maturity level and a number of projects, initiatives, and research papers, as identified in this study.

(hundreds of thousands of goods/products involved, interaction with thousands of consumers, and involvement of tens to hundreds of intermediate actors in the supply chain) are published or supported by big companies. From the limited number of initiatives (5%, 6%) demonstrating complete integration with normal operations (AID Tech, 2017; Blockchain, 2017; Ireland Craft Beers, 2017; Foodchain, 2019; Bank, 2019), it can be concluded that blockchain technology is still, for the most part, perceived as an emerging technology with its potential applications under consideration. No maturity level was recorded for 11 (14%) of the research papers included in the study since no implementation or deployment details were provided.

It is worth reviewing whether the aforementioned initiatives are ongoing or have stopped and failed. This data would give a useful indication of the economic viability of blockchain-empowered projects. However, it is difficult to address at this particular moment this question, because some initiatives are recent and still ongoing. From the 80 efforts identified in this chapter, seven of them (8.7%) started in 2016, 12 (15%) in 2017, and 22 (27.5%) in 2018. Due to their short lifetimes, most projects are ongoing. Thus a fair assessment is difficult. By reviewing project update, news articles, and other recorded activity, it is suspected that 7 (24%) of the 29 commercial initiatives (governmental and NGO-based) are now inactive; namely (AID Tech, 2017; Jha et al., 2018; Grass Roots Farmers Cooperative, 2017; WWF, 2018; Balfegó Group, 2017; FarmShare, 2017; Soil Association Certification, 2018; SNCF, 2017; Recereum, 2017). Such a large proportion of fallouts might be an indication of the overall complexity of blockchain technology, or of the immaturity of the market for complete integration of these solutions with companies' everyday operations. It should be noted that some companies/organizations may simply have finished their pilots and are still considering the possibility of massive adoption, time will show if this is the case. The preliminary findings reported here are in line with others (Behnke & Janssen, 2019), where it is noted that the *use of blockchain remains limited despite its promises*.

7.5 Potential benefits

By providing a secure, distributed way to perform transactions among different untrusted parties, blockchain technology offers many benefits (Yuan, Qiu, Bi, Chang, & Lam, 2019; Pearson et al., 2019; Creydt & Fischer, 2019). This feature is key in agriculture and food supply chains, as many actors are involved from the stages of raw product to the supermarket shelves (Lin et al., 2017; Tripoli & Schmidhuber, 2018). A decentralized ledger helps to improve traceability in value chains by connecting inputs, suppliers, producers, buyers, and regulators, which are otherwise distant and operating under different programs and rules (policies) while using distinct and separate applications (Lee et al., 2017). Smart contracts can improve the overall effectiveness of manufacturing services and empower manufacturers to develop scalable and flexible businesses at a lower cost (Li, Liu, Liu, He, & Huang, 2018).

Blockchain has the potential to improve monitoring of social and environmental responsibility; and provenance of information. Furthermore, it can facilitate mobile payments, credits, and financing; decrease of transaction fees; and facilitate real-time management of supply chain transactions in a secure and trustworthy way (Lee et al., 2017; Zhang, 2019) (Branco et al., 2019). In the case of outbreaks of an animal- or plant-based diseases, contaminated products could be traced quickly (Tripoli & Schmidhuber, 2018) in systems using blockchain. Blockchain can even be used to secure agricultural robotic swarm operations, making them more autonomous and flexible (Ferrer, 2018). The global distribution and trade of botanical material used as medicines, health foods, cosmetics, and other applications can be augmented by blockchain, empowering high-value agricultural value chains (Heinrich et al., 2019). The use of blockchain could also improve the management of nitrogen in crops (Tao & Bullock, 2019). In the future, animal and crop farmers could trade organic fertilizers through a blockchain platform, adding security and assurance to the transactions by tamper-free traceability provision.

Specifically, for the developing world, blockchain seems very suitable for use, particularly regarding supporting small farmers. Other potential scenarios include the finance and insurance of rural farmers (Chinaka, 2016), financial exploitation of agricultural waste (Zhang, 2019), and the facilitation of reliable, trustworthy transactions in developing countries. Cash-based transactions lack traceability, ultimately hindering the ability of small- and medium-sized enterprises to access credit and new markets and to grow. Blockchain introduces a new method of accounting for transfers of value in ways that minimize uncertainty, and by operating as a decentralized and shared ledger, it functions as a digital institution of trust with reduced transaction costs, if any (Tripoli & Schmidhuber, 2018). Although over 80% of goods in developing countries are produced by small farmers produce, they do not typically have access to services such as finance and insurance (Chinaka, 2016). Blockchain could constitute a strong tool in order to fight corruption and the typically insufficient environmental, social, and economic regulatory frameworks of developing countries (Rejeb, 2018). Further examples of how blockchain could empower the poor in developing countries are listed in (Thomason et al., 2018), with a focus on tracking climate-related issues.

Concerning the developed world, various problems have historically limited the environmental/economic sustainability of smaller farms, such as unfair pricing and the influence of large commercial entities. Blockchain could help in facilitating fairer pricing through the whole value chain. An example of the use of blockchain in monitoring the water quality in a catchment area is discussed in (IWA, 2018).

Moreover, the transparency provided by blockchain technology could facilitate the development of trading systems based on reputation, rewarding fair practice. Reputation in trading systems has been shown to improve the behavior of participating parties and increases their reliability, responsibility, and commitment (Khaqqi, Sikorski, Hadinoto, & Kraft, 2018; Sharma, 2017), as demonstrated by its use in various other trading systems (e.g., eBay, Alibaba).

Further, there is the potential benefit of increasing consumer awareness and empowerment, with the consumer as the market driving force. Increased consumer awareness would lead to increased pressure for more transparent, sustainable, safe, and fair practices in food production. Given that consumers are typically overwhelmed by the amount and complexity of certification labels, blockchain technology can have positive influences on consumers' purchasing decisions (Sander, Semeijn, & Mahr, 2018). Finally, the case study performed in (Perboli, Musso, & Rosano, 2018) shows that the cost of implementing a blockchain is highly sustainable, given the resulting benefits.

7.6 Challenges and open issues

There are various challenges for the wider adoption of blockchain technology in related work (Chang et al., 2019; Galvez, Mejuto, & Simal-Gandara, 2018; Hald & Kinra, 2019; Tribis et al., 2018; Zhao et al., 2019; Pearson et al., 2019; Linsner, Kuntke, Schmidbauer-Wolf, & Reuter, 2019). Table 7.2 lists potential benefits and existing barriers for the use of blockchain in agriculture and the food supply chain, as identified in the sections below, as well as in (Chang et al., 2019) (Pearson et al., 2019). These aspects should be further researched in the food sector so as to generate more resilient blockchain architectures and ameliorate the themes identified in this review (Antonucci et al., 2019). A case study in the Netherlands revealed that SMEs lack the necessary size, scale, or know-how needed securely to invest in blockchain (Ge et al., 2017). Eighteen boundary conditions categorized under business, regulation, quality, and traceability; have been identified (Behnke & Janssen, 2019). Boundary conditions should be met before blockchain can be used. Some boundary inherent in all supply chains, whereas others were domain-specific. Blockchain technology requires standardization and data governance; hence, it entails organizational transformations.

7.6.1 Accessibility

Blockchain needs to become more accessible, which is a big challenge considering the complexities underlying its digital technology, especially considering that more and more components are being integrated into the blockchain (IoT, RFID, sensors and actuators, robots, biometric data, big data, 5G, edge computing) (Tian, 2016; Figorilli et al., 2018; Kim et al., 2018; Lin et al., 2018; Bordel et al., 2019; Li et al., 2018; Rabah, 2018; Mistry, Tanwar, Tyagi, & Kumar, 2020; Chen et al., 2017). The functionality of blockchains relies on external systems to obtain accurate information from the real-world. These are the so-called *oracles* that connect the physical and digital worlds, usually associated with automated sensor readings (i.e., hardware oracles), datasets from web applications (i.e., software oracles), and manual records (i.e., human oracles). However, the dependency on these third-party

TABLE 7.2 Potential benefits and existing barriers to the use of blockchain in agriculture.

Opportunities and potential benefits	Challenges and barriers
Traceability in value chains	SME have difficulties in adopting the technology
Support for small farmers	Information infrastructure might prevent access to markets for new users
Finance and insurance of rural farmers	Lack of expertise by small SME
Facilitation of financial transactions in developing countries	High uncertainties and market volatility
Fairer pricing through the whole value chain	Limited education and training platforms
A useful platform in emission reduction efforts	No regulations in place
Consumer awareness and empowerment	Lack of understanding among policymakers and technical experts
More informed consumer purchasing decisions	Open technical questions and scalability issues (e.g., the latency of transactions)
Increased sustainability and reduction of waste	Digital divide among developed and developing world
Reduced transaction fees and less dependence on intermediaries	Decline of cryptocurrencies in market share and high volatility (reputation issues)
More transparent transactions and less frauds	Cost of computing/IoT equipment required
Better quality of products, lower probability for foodborne diseases	Design decisions might reduce overall flexibility
	Privacy issues
	Some quality parameters of food products cannot be monitored by objective analytical methods, especially environmental indicators
	Ownership of infrastructure and maintenance
	Distribution of profits and advantages
	Certification of the inserted data

intermediaries might compromise the blockchain concept of building decentralized trust. This observation is known as the *oracle problem* and focuses on substantial research, particularly for finance and smart contract-related applications. The proposed solutions generally rely on developing decentralized and consensus-based oracle solutions; and novel methods of authenticating oracle data.

The information infrastructure required to operate and maintain the systems of complex global supply chains can be seen as a barrier to access to markets for new users or food suppliers. This fact could lead to a reduction in market competition and access (Pearson et al., 2019).

Moreover, there is a general lack of awareness and knowledge of blockchain technology (Zhao et al., 2019), coupled with a currently limited provision of training platforms (ICT4Ag, 2017). Besides policymakers, the capacitation of blockchain technology is also fundamental for the food value chain stakeholders. Conceptual metaphors for understanding and accepting blockchain are discussed (Swan & De Filippi, 2017). Various startups are developing software to make blockchain technology easier for farmers to use, such as 1000 EcoFarms (1000EcoFarms, 2017), which has aggregated all the important blockchain processes relevant to food, farming, and agriculture, using FoodCoin as the proposed ecosystem (FoodCoin, 2017). OriginChain is a software system that aspires to restructure the current central database systems with blockchain, offering decentralized solutions that offer transparency and high performance (Xu et al., 2019).

7.6.2 Governance and sustainability

Despite the rather long list of initiatives presented in this review, convincing business cases for blockchain are still scarce. This can be attributed to a large number of uncertainties involved and the early stages of the technology. This observation was also made in a relevant survey (Tribis et al., 2018). The long-term impact of blockchain on governance, economic sustainability, and social aspects still needs to be assessed. It has been noted that an excess of information transparency and the immutability of the data stored in blockchains might bring new challenges for the performance of supply chains (Hald & Kinra, 2019). Permanent data visibility could be seen to compromise privacy issues and strengthen the surveillance power of centralized entities. On the other hand, large corporations could conceivably implement private and permissioned blockchains, supporting oligopolistic practices (Pearson et al., 2019). Economic models, which could be applied in order to self-start the supply-chain and the relative blockchain infrastructure, need to be carefully defined (Antonucci et al., 2019).

Perhaps paradoxically, blockchain has also been described as a potentially *deskilling* technology for workers and organizations (Hald & Kinra, 2019). The increased automation of tasks and procedures throughout supply chains and the elimination of transaction intermediaries might significantly reduce human intervention, leading to a loss of skilled jobs. The justification of human intervention in blockchain-managed supply chains could be reduced significantly. However, we must consider that such phenomena have occurred in all previous technological revolutions, which have typically demanded new skills and capacities in the labor market.

Advantages generated by cryptocurrency (e.g., profits) is an important aspect that needs to be carefully considered, especially when untrusted parties are involved (Antonucci et al., 2019). Potential strategies and techniques that could be applied in order to take advantage of the blockchain in the certification of data is also an important consideration (Antonucci et al., 2019).

Finally, it is worth adding that the strong association of quality parameters of food products (made more transparent to the consumer employing the blockchain) can themselves justify higher prices. Such data is a known target of food fraudsters (Creydt & Fischer, 2019). Thus, governance is also important.

7.6.3 Regulation

Policy development and regulation with blockchain practices are both a necessity for and a significant barrier to its wider adoption (Zhao et al., 2019; Pearson et al., 2019; Linsner et al., 2019). Cryptocurrencies form the complete global blockchain case study to date (Yli-Huumo, Ko, Choi, Park, & Smolander, 2016). Current analysis of these cryptocurrencies indicates that they are vulnerable to speculation and that their price has large fluctuations almost daily. The recent decline in market share, coupled with the high volatility of the financial value of the most popular cryptocurrencies, reduces the overall trust of the public in the underlying blockchain technology, negatively affecting its reputation (Gaurav, 2019). Hence, without regulation, cryptocurrencies are not generally trusted enough to be used in food supply chains as a complete solution in the current environment. A lack of (common) understanding among policymakers and technical experts still exists on how blockchain technology and transactions based on some currency should be used (ICT4Ag, 2017). The first realization of the infrastructure with the different smart contracts and the responsible entities (governmental or certified third party), respecting regulations in place, and action constitutes an important challenge (Antonucci et al., 2019).

7.6.4 Technical challenges and design decisions

There are many design decisions affecting blockchain technology, both existing and under development (e.g (AgriDigital, 2017; AgriLedger, 2017; FarmShare, 2017; Ripe.io, 2017; OriginTrail, 2018)). For example, permission (i.e., participants are trusted) versus permissionless, open (i.e., everyone can join) versus closed (Jayachandran, 2017), and ownership (Pearson et al., 2019). In existing permissionless blockchains, the latency of transactions can be from several minutes to some hours, awaiting all participants to update their ledgers before the smart contracts become publicly accessible. Such design decisions affect the operation of the blockchain system and limit the flexibility, introducing inefficiencies under certain forms of operation when compared to the equivalent conventional centralized approaches. The ownership and maintenance duties of the infrastructure are other important considerations (Antonucci et al., 2019).

Moreover, although objective analytical methods can monitor some of the quality parameters of food products, it is not possible to do so for all of them (Creydt & Fischer, 2019). Some parameters, especially environmental ones (Keesstra, Mol, de Leeuw, Okx, & de Cleen, 2018), are difficult to include, assess, and audit in an automated way. Although Life cycle assessment (LCA) has reached a standard methodology for the assessment of environmental impacts associated with all the stages of a product system (raw material, production, distribution, consumption, and final disposal), collecting reliable data through supply chains remains a major challenge. A system architecture that integrates blockchain; IoT; and big data analytics and visualization has been proposed (Zhang et al., 2020). A preliminary system approach using blockchain in supply chain management for quality improvement was introduced (Chen et al., 2017). The system, constructed on four layers, tackled both the technological and supply chain complexity by adopting blockchain in combination with IoT devices for the supply chain management.

Existing blockchain protocols, as applied in popular cryptocurrencies, face serious scalability obstacles (Eyal et al., 2016; Pearson et al., 2019; Linsner et al., 2019). Scalable operation is hindered by some protocols' constraints in terms of parallel processing and acceptance of transactions (e.g., in the BitCoin case), which are limited by the size and interval of the transaction block (Tribis et al., 2018). This problem is tackled in (Mao, Hao, Wang, & Li, 2018a) by helping users find suitable transactions and improve transaction efficiency. Most of the proposed blockchain-based frameworks have seen limited scale implementations, being tested under 'laboratory conditions or in small scale pilots. Although blockchain offers advanced security, there are high associated risks related to loss of funds. For example, the account owner might have lost the private keys needed to access and manage the account.

Privacy issues are also an important consideration (Zhao et al., 2019; Linsner et al., 2019). Since every transaction is recorded on a common ledger, users can be identified by their public keys. Although this ensures transparency and builds trust, it does not protect users' privacy. This privacy is considered particularly important in the food supply ecosystem since many actors are in competition with each other. Maintaining a reasonable level of privacy is an existing challenge of blockchain technologies. A review of various methods for privacy protection in blockchain systems can be found in (Feng, He, Zeadally, Khan, & Kumar, 2018).

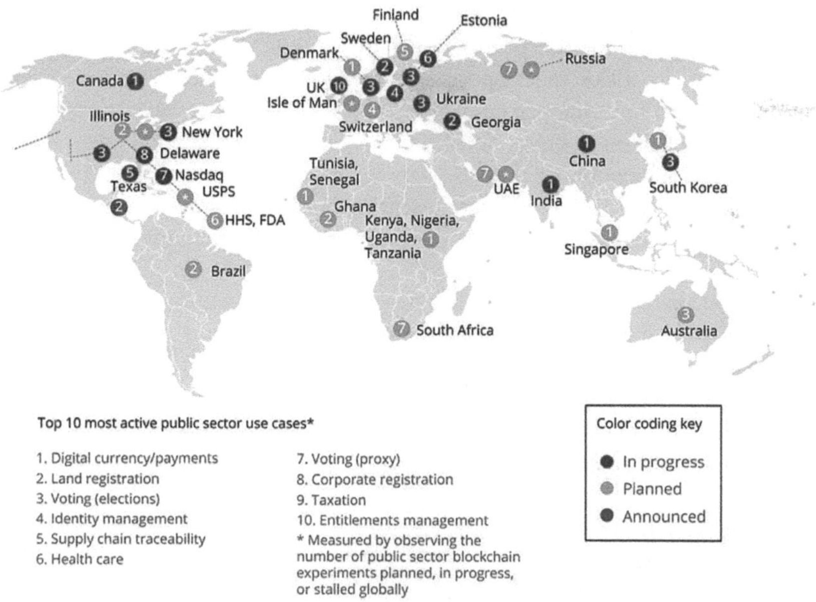

FIGURE 7.6 Blockchain in the public sector in 2017. *J. Killmeyer, M. White and B. Chew, "Will blockchain transform the public sector?," Deloitte center for government Insights, Deloitte University Press, 2017.*

Finally, issues associated with different data standards among different stages in the supply chain were addressed in (Kim & Laskowski, 2018a, 2018b) in the context of a decentralized blockchain network. An ontology-based blockchain modeling approach was introduced, integrating IoT devices for data capture and data sharing for supply chain provenance. This blockchain technology was built upon Internet technologies, with a web browser as a natural interface. This fact could be considered an early sign that blockchain-driven initiatives in food supply systems would likely also embrace the IoT and the relevant concepts of "Web of Things" (W. E., 2007; Kamilaris, Pitsillides, & Trifa, 2011).

7.6.5 Digital gap between developed and developing countries

As mentioned above, farmers need to establish an effective understanding of blockchain before adopting it (Tribis et al., 2018). However, in many parts of the world, farmers' efforts are dedicated to subsistence and expectations of expertise in cutting edge technologies are unreasonable. Blockchain technologies require significant computing resources (Zhao et al., 2019), and these are not typically available in developing countries. There is a gap between the developed and developing world regarding digital competence and access to blockchain technology (Maru et al., 2018). Many of the bibliographic sources reviewed herein come from developed countries with a well-organized and wealthy primary sector (such as the USA, Australia, Europe). This digital divide was also observed in the use of big data in agriculture (Kamilaris et al., 2017). Some authors do make the important observation that most of the current projects are found in developed countries, but no real significance is attributed to this. Since blockchain is being heralded as a technology that will solve many challenges in the developing world— *why the gap?*— is an important question to ask and a legitimate area for future research. Fig. 7.6 illustrates the number of blockchain experiences in the public sector in various countries around the world (Killmeyer, White, & Chew, 2017).

It can be that indeed most of the on-going experiments with blockchain happen in developed regions. Considering that blockchain may well bring significant opportunity for small farmers, developmental aid should focus on technology transfer, deployment, and training to farmers in developing areas with the view of bringing real solutions to the specific conditions that restrain their socio-economic progression.

7.7 Conclusion

This book chapter demonstrates that blockchain technology is entering a mature phase in its development and is making steady progress. Blockchain is already being used by many projects and initiatives aiming to establish a proven and trusted environment to build a transparent and sustainable food production and distribution system, integrating key stakeholders into the supply chain. However, there are still many issues and challenges beyond those at the technical level, that remain open need to be addressed moving forward.

To reduce barriers of use, governments must *lead by example* and foster the digitalization of public administration processes. More effort and money should be invested in research and innovation, as well as in education and training, in order to produce and demonstrate evidence, based on real-world deployment and large-scale pilots, for the social and environmental benefits of this technology. The possible transition of governments toward the use of the blockchain is discussed in (Gupta, 2017), where it is noted that governments and their relevant departments should observe and understand the particular "pain points" and address them accordingly.

From a policy perspective, various actions can be taken to aid in the development of blockchain and help achieve the benefits promised by the technology. Such actions include: encouraging the growth of blockchain-minded ecosystems in agri-food chains, supporting the technology as part of the general goals of optimizing competition and ensuring sustainability in the agri-food supply chain, and designing a clear regulatory framework for blockchain implementations.

Given their infancy, the economic sustainability of the existing initiatives still needs to be assessed. The outcomes of these economic studies are expected to influence the popularity of blockchain technologies in the food supply chain domain shortly.

In summary, blockchain is a promising technology toward a transparent food supply chain. However, many barriers and challenges still exist, which hinders its current popularity among farmers and food supply systems. The near future will show if and how these challenges can be addressed by governmental and private efforts in order to establish blockchain technology as a secure, reliable, and transparent way to ensure food safety and integrity throughout the supply chain. It interesting to see research innovation combining blockchain with other emerging technologies (big data, robotics, IoT, RFID, NFC, hyperspectral imaging, 5G, edge computing) toward higher automation of the food supply processes, enhanced with full transparency and traceability; and promoting the position of blockchain as a technology of the future.

Acknowledgments

Andreas Kamilaris has received funding from the European Union's Horizon 2020 research and innovation program under the grant agreement No. 739578, complemented by the Government of the Republic of Cyprus through the Directorate-General for European Programs, Coordination, and Development.

References

1000EcoFarms. (2017) [Online]. Available https://www.1000ecofarms.com/.

Abelseth, B. (2018). Blockchain tracking and cannabis regulation: Developing a permissioned blockchain network to track Canada's cannabis supply chain. *Dalhousie Journal of Interdisciplinary Management, 14*.

AgFunder News. (2018). *Maersk leads blockchain of food startup Ripe.io $2.4m seed round* [Online]. Available https://agfundernews.com/maersk-leads-blockchain-of-food-startup-ripeio-2-4m-seed-round.html.

Agora Tech Lab. (2018). *Creating circular economies by rewarding responsible behavior* [Online]. Available https://www.agoratechlab.com/about.

AgriDigital. (2017) [Online]. Available https://www.agridigital.io/blockchain.

AgriLedger. (2017) [Online]. Available http://www.agriledger.com/.

AID Tech. (2017). *How blockchain technology is enabling international aid to be delivered transparently* [Online]. Available https://v3.aid.technology/aid-in-lebanon/.

Antonucci, F., Figorilli, S., Costa, C., Pallottino, F., Raso, L., & Menesatti, P. (2019). A Review on blockchain applications in the agri-food sector. *Journal of the Science of Food and Agriculture, 99*(14), 6129–6138. https://doi.org/10.1002/jsfa.9912.

ArbolMarket. (2019). *ARBOL* [Online]. Available https://www.arbolmarket.com.

Balfegó Group. (2017) [Online]. Available https://balfego.com/ca/trasabilitat/.

Bank, P. (2019) [Online]. Available https://www.plasticbank.com.

Bano, S. (2017). Consensus in the age of blockchains. *arXiv preprint arXiv:1711.03936, 1711*(03936).

Basnayake, B., & Rajapakse, C. (2019). A Blockchain-based decentralized system to ensure the transparency of organic food supply chain. In *In international research conference on smart computing and systems engineering (SCSE), colombo, Sri Lanka*.

Bechtsis, D., Tsolakis, N., Bizakis, A., & Vlachos, D. (2019). A blockchain framework for containerized food supply chains. *Computer Aided Chemical Engineering, 46*(1), 1369–1374.

Becker, J., Breuker, D., Heide, T., Holler, J., Rauer, H. P., & Böhme, R. (2013). Can we afford integrity by proof-of-work? scenarios inspired by the bitcoin currency. In *Economics of information security and privacy, berlin, heidelberg*.

BeefLedger Limited. (2017) [Online]. Available http://beefledger.io.

Behnke, K., & Janssen, M. (2019). Oundary conditions for traceability in food supply chains using blockchain technology. *International Journal of Information Management, 52*(101969). https://doi.org/10.1016/j.ijinfomgt.2019.05.025.

Bentov, I., Gabizon, A., & Mizrahi, A. (2016). Cryptocurrencies without proof of work. In *International conference on financial cryptography and data security, berlin, heidelberg*.

Bermeo-Almeida, O., Cardenas-Rodriguez, M., Samaniego-Cobo, T., Ferruzola-Gómez, E., Cabezas-Cabezas, R., & Bazán-Vera, W. (2018). Blockchain in agriculture: A systematic literature review. In *International conference on technologies and innovation*.

BitFury Group. (2015). Proof of stake versus proof of work. *White Paper*.

Blockchain. (2017). Blockchain for Zero hunger [Online]. Available https://innovation.wfp.org/project/building-blocks.

Boehm, V., Kim, J., & Hong, J. (2017). Holistic tracking of products on the blockchain using NFC and verified users. *International Workshop on Information Security Applications, 10763*(1), 184–195.

Bolt, J. (2019). *Financial resilience of Kenyan smallholders affected by climate change, and the potential for blockchain technology*. Netherlands: Wageningen Environmental Research.

Bordel, B., Martín, D., Alcarria, R., & Robles, T. (2019). A blockchain-based water control system for the automatic management of irrigation communities. In *Proc. of international conference on consumer electronics (ICCE), taiwan*.

Branco, F., Moreira, F., Martins, J., Au-Yong-Oliveira, M., & Gonçalves, R. (2019). Conceptual approach for an extension to a mushroom farm distributed process control system: IoT and blockchain. In *Proc. Of the world conference on information systems and technologies (WorldCist), La toja island, galicia, Spain*.

Brooker, D., Bakker-Arkema, F., & Hall, C. (1992). *Drying and storage of grains and oilseeds, springer science & business media*.

Built to Adapt. (2018). *Can blockchain help feed the hungry?* [Online]. Available https://builttoadapt.io/can-blockchain-help-feed-the-hungry-161b8fef7e24.

Bunge, J. (2017). Latest use for a bitcoin technology: Tracing turkeys from farm to table. *The Wall Street Journal*.

Buterin, V. (2015). *The problem of censorship* [Online]. Available https://blog.ethereum.org/2015/06/06/the-problem-of-censorship/.

Campbell, A. (2017). *Sustainability from paddock to plate* [Online]. Available https://www.sciencealert.com/sustainability-from-paddock-to-plate.

Caro, M., Ali, M., Vecchio, M., & Giaffreda, R. (2018). Blockchain-based traceability in agri-food supply chain management: A practical implementation. In *In IoT vertical and topical summit on agriculture-tuscany (IOT tuscany)* (pp. 1–4). IEEE.

Carrefour. (2018). *The food blockchain*. https://actforfood.carrefour.com/Why-take-action/the-food-blockchain.

Casado-Vara, R., Prieto, J., De la Prieta, F., & Corchado, J. (2018). How blockchain improves the supply chain: Case study alimentary supply chain. *Procedia computer science, 134*(1), 393–398.

Castro, M., & Liskov, B. (1999). Practical Byzantine Fault tolerance. *OSDI, 99*, 173–186.

CDC. (2018). *Centers for disease control and prevention* [Online]. Available https://www.cdc.gov/foodborneburden/2011-foodborne-estimates.html.

Chang, Y., Iakovou, E., & Shi, W. (2019). Blockchain in global supply chains and cross border trade: A critical synthesis of the state-of-the-art, challenges and opportunities. In *arXiv:1901.02715.*

Chen, S., Shi, R., Ren, Z., Yan, J., Shi, Y., & Zhang, J. (2017). A blockchain-based supply chain quality management framework. In *IEEE 14th international conference on e-business engineering (ICEBE), shanghai, China.*

Chinaka, M. (2016). Blockchain technology - applications in improving financial inclusion in developing economies: Case study for small scale agriculture in Africa. Doctoral dissertation *Massachusetts Institute of Technology.*

Chong, M., Perez, E., Castilla, J., & Rosario, H. (2020). Blockchain technology applied to the cocoa export supply chain: A Latin America case. In *Handbook of research on emerging technologies for effective project management* (pp. 323–339). IGI Global.

Coinmarketcap. (2017) [Online]. Available http://www.coinmarketcap.com.

Cozzi, A. (2019). *BlockChain arrives in the food sector: The packaging of pasta carries all its history with it* [Online]. Available https://www.aldocozzi.it/news/la-blockchain-sbarca-nel-settore-alimentare-la-confezione-di-pasta-porta-con-se-tutta-la-sua-storia/.

Creydt, M., & Fischer, M. (2019). Blockchain and more-algorithm driven food traceability. *Food Control, 105*(1), 45–51. https://doi.org/10.1016/j.foodcont.2019.05.019.

CyberSecurity. (2019). *Milk verification project* [Online]. Available https://www.greatitalianfoodtrade.it/consum-attori/blockchain-nella-filiera-alimentare-il-prototipo-di-bari.

Davcev, D., Kocarev, L., Carbone, A., Stankovski, V., & Mitresk, K. (2018). Blockchain-based distributed cloud/fog platform for IoT supply chain management. In *Eighth international conference on advances in computing, electronics and electrical technology (CEET).*

Dujak, D., & Sajter, D. (2019). Blockchain applications in supply chain. *SMART Supply Network,* 21–46.

Ethereum. (2015). *The Ethereum project* [Online]. Available https://www.ethereum.org.

Eyal, I., Gencer, A., Sirer, E., & Van Renesse, R. (2016). Bitcoin-ng: A scalable blockchain protocol. In *Proceedings of the 13th USENIX symposium on networked systems design and implementation (NSDI), santa clara, CA, USA.*

FarmShare. (2017) [Online]. Available http://farmshare.org.

Feng, Q., He, D., Zeadally, S., Khan, M., & Kumar, N. (2018). A survey on privacy protection in blockchain system. *Journal of Network and Computer Applications, 126*(1), 45–58. https://doi.org/10.1016/j.jnca.2018.10.020.

Ferrer, E. C. (2018). The blockchain: A new framework for robotic swarm systems. In *Proceedings of the future technologies conference.*

Figorilli, S., Antonucci, F., Costa, C., Pallottino, F., Raso, L., Castiglione, M., et al. (2018). A blockchain implementation prototype for the electronic open source traceability of wood along the whole supply chain. *Sensors, 18*(9), 3133.

Foodchain. (2019). *San Domenico roastery: The first case of a coffee supply chain fully traced with blockchain* [Online]. Available https://food-chain.it/public/case/san-domenico/.

FoodCoin. (2017). *The FoodCoin ecosystem* [Online]. Available https://www.foodcoin.io/.

Galvez, J., Mejuto, J., & Simal-Gandara, J. (2018). Future challenges on the use of blockchain for food traceability analysis. *TRAC Trends in Analytical Chemistry, 107*, 222–232. https://doi.org/10.1016/j.trac.2018.08.011.

Gaurav, S. (2019). The market for cryptocurrencies. *Economic and Political Weekly, 54*(2), 13.

Ge, L., Brewster, C., Spek, J., Smeenk, A., Top, J., van Diepen, F., et al. (2017). *"Blockchain for agriculture and food," wageningen economic research, No. 2017-112, wageningen.*

Genetics, H. (2018). *New blockchain project involving turkeys and animal welfare* [Online]. Available https://www.hendrix-genetics.com/en/news/new-blockchain-project-involving-turkeys-and-animal-welfare/.

George, R., Harsh, H., Ray, P., & Babu, A. (2019). Food quality traceability prototype for restaurants using blockchain and food quality data index. *Journal of Cleaner Production, 240*(1), 118021.

Grass Roots Farmers' Cooperative. (2017). *How we 're using blockchain tech for total transparency* [Online]. Available https://grassrootscoop.com/blog/how-we-use-blockchain-technology-to-give-you-total-transparency.

Gupta, V. (2017). *Building the hyperconnected future on blockchains, world government summit.* http://internetofagreements.com/files/WorldGovernmentSummit- Dubai2017.pdf.

Hald, K., & Kinra, A. (2019). How the blockchain enables and constrains supply chain performance. *International Journal of Physical Distribution and Logistics Management, 49*(4), 376–397. https://doi.org/10.1108/IJPDLM-02-2019-0063

Heinrich, M., Scotti, F., Booker, A., Fitzgerald, M., Kum, K., & Löbel, K. (2019). Unblocking high-value botanical value chains: Is there a role for blockchain systems? *Frontiers in Pharmacology, 10*(1), 396.

Hoffman, A., & Munsterman, R. (2018). *Dreyfus teams with banks for first agriculture blockchain trade* [Online]. Available https://www.bloomberg.com/news/articles/2018-01-22/dreyfus-teams-with-banks-for-first-agriculture-blockchain-trade.

Hyperledger. (2018). *Bringing traceability and accountability to the supply chain through the power of hyperledger sawtooth's distributed ledger technology* [Online]. Available https://sawtooth.hyperledger.org/examples/seafood.html.

IBM. (2017). Leading the pack in blockchain banking: Trailblazers set the pace. *IBM Institute for Business Value, The Economist Intelligence Unit.*

ICT4Ag. (2017). *Perspectives for ICT and Agribusiness in ACP countries: Start-up financing, 3D printing and blockchain* [Online]. Available http://www.fao.org/e-agriculture/events/cta-workshop-perspectives-ict-and-agribusiness-acp-countries-start-financing-3d-printing-and.

Insights, C. B. (2017). *How blockchain could transform food safety* [Online]. Available https://www.cbinsights.com/research/blockchain-grocery-supply-chain/.

Ireland Craft Beers. (2017). *Downstream beer* [Online]. Available http://www.down-stream.io.

ITUNews. (2018). *Food Trust' partnership uses blockchain to increase food safety* [Online]. Available https://news.itu.int/food-trust-blockchain-food-safety/.

IWA. (2018). *Demystifying blockchain for water professionals* [Online]. Available http://www.iwa-network.org/demystifying-blockchain-for-water-professionals-part-1/.

Jayachandran, P. (2017). *The difference between public and private blockchain* [Online]. Available https://www.ibm.com/blogs/blockchain/2017/05/the-difference-between-public-and-private-blockchain/.

JD.com Blog. (2018). *JD blockchain open platform* [Online]. Available https://jdcorporateblog.com/jd-launches-blockchain-open-platform/.

Jha, S., Andre, B., & J. O. (2018). *Arbol: Smart contract weather risk protection for agriculture* [Online]. Available https://www.arbolmarket.com/wp-content/uploads/2019/02/ARBOL_WP-3.pdf.

Kamath, R. (2018). Food traceability on blockchain: Walmart's pork and mango pilots with IBM. *The JBBA, 1*(1), 3712.

Kamilaris, A., Fonts, A., & Prenafeta-Boldú, F. X. (2019). The rise of blockchain technology in agriculture and food supply chains. *Trends in Food Science & Technology Journal, 91*(1), 640–652.

Kamilaris, A., Kartakoullis, A., & Prenafeta-Boldu, F. X. (2017). A review on the practice of big data analysis in agriculture. *Computers and Electronics in Agriculture, 143*, 23–37.

Kamilaris, A., & Pitsillides, A. (2016). Mobile phone computing and the internet of things: A survey. *IEEE Internet of Things (IoT) Journal, 3*(6), 885–898.

Kamilaris, A., Pitsillides, A., & Trifa, V. (2011). The smart home meets the web of things. *International Journal of Ad Hoc and Ubiquitous Computing, 7*(3), 145–154.

Kasten, J. (2019). Blockchain application: The dairy supply chain. *Journal of Supply Chain Management Systems, 8*(1).

Keesstra, S., Bouma, J., Wallinga, J., Tittonell, P., Smith, P., Cerdà, A., et al. (2016). The significance of soils and soil science towards realization of the United Nations Sustainable Development Goals. *Soils, 2*, 111–128.

Keesstra, S., Mol, G., de Leeuw, J., Okx, J., de Cleen, M., & Visser, S. (2018). Soil-related sustainable development goals: Four concepts to make land degradation neutrality and restoration work. *Land, 7*(4), 133.

Khaqqi, K. N., Sikorski, J. J., Hadinoto, K., & Kraft, M. (2018). Ncorporating seller/buyer reputation-based system in blockchain-enabled emission trading application. *Applied Energy, 209*, 8–19.

Killmeyer, J., White, M., & Chew, B. (2017). *"Will blockchain transform the public sector?," deloitte center for government insights*. Deloitte University Press.

Kim, M., Hilton, B., Burks, Z., & Reyes, J. (2018). Integrating blockchain, smart contract-tokens, and IoT to design a food traceability solution. In *9th annual information technology, electronics and mobile communication conference (IEMCON)*.

Kim, H., & Laskowski, M. (2018a). Agriculture on the blockchain: Sustainable solutions for food, farmers, and financing. In *Supply chain revolution, barrow books*.

Kim, H., & Laskowski, M. (2018b). Toward an ontology-driven blockchain design for supply-chain provenance. *Intelligent Systems in Accounting, Finance and Management, 25*(1), 18–27.

Kouhizadeh, M., Zhu, Q., & Sarkis, J. (2019). Blockchain and the circular economy: Potential tensions and critical reflections from practice. *Production Planning and Control*, 1–7.

Krause, M., & Tolaymat, T. (2018). Quantification of energy and carbon costs for mining cryptocurrencies. *Nature Sustainability, 1*, 711–718.

Kumar, M., & Iyengar, N. (2017). A framework for Blockchain technology in rice supply chain management. *Advanced Science Technology Letters, 146*, 125–130.

Lee, H., Mendelson, H., Rammohan, S., & Srivastava, A. (2017). *Technology in agribusiness: Opportunities to drive value*. White paper, Stanford Graduate School of Business.

Levitt, T. (2016). *Blockchain technology trialled to tackle slavery in the fishing industry* [Online]. Available https://www.theguardian.com/sustainable-business/2016/sep/07/blockchain-fish-slavery-free-seafood-sustainable-technology.

Lierow, M., Herzog, C., & Oest, P. (2017). Blockchain: The backbone of digital supply chains. *Oliver Wyman*.

Li, Z. W. W., Liu, G., Liu, L., He, J., & Huang, G. (2018). Toward open manufacturing: A cross-enterprises knowledge and services exchange framework based on blockchain and edge computing. *Industrial Management & Data Systems, 118*(1), 303–320.

Lin, Y. P., Petway, J. R., Anthony, J., Mukhtar, H., Liao, S. W., Chou, C. F., et al. (2017). Blockchain: The evolutionary next step for ICT E-agriculture. *Environments, 4*(3), 50.

Lin, J., Shen, Z., Zhang, A., & Chai, Y. (2018). Blockchain and IoT based food traceability for smart agriculture. In *Proc. Of the 3rd international conference on crowd science and engineering (ICCSE), Singapore*.

Linsner, S., Kuntke, F., Schmidbauer-Wolf, G., & Reuter, C. (2019). Blockchain in agriculture 4.0-an empirical study on farmers expectations towards distributed services based on distributed ledger technology. In *Proc. of Mensch und Computer (MUC), Hamburg, Germany.*

Lucena, P., Binotto, A., Momo, F., & Kim, H. (2018). A case study for grain quality assurance tracking based on a Blockchain business network. *arXiv preprint arXiv:1803.07877, 1803*(07877).

Manski, S. (2017). Building the blockchain world: Technological commonwealth or just more of the same? *Strategic Change, 26*(5), 511–522.

Mao, D., Hao, Z., Wang, F., & Li, H. (2018a). Innovative blockchain-based approach for sustainable and credible environment in food trade: A case study in shandong province, China. *Sustainability, 10*(9), 3149.

Mao, D., Wang, F., Hao, Z., & Li, H. (2018b). Credit evaluation system based on blockchain for multiple stakeholders in the food supply chain. *International Journal of Environmental Research and Public Health, 15*(8), 1627.

Markets and Markets research. (2016). *Markets and Markets research, food traceability market worth $14 billion by 2019* [Online]. Available https://www.marketsandmarkets.com/PressReleases/food-traceability.asp.

Maru, A., Berne, D., Beer, J., Ballantyne, P., Pesce, V., Kalyesubula, S., et al. (2018). Digital and data-driven agriculture: Harnessing the power of data for smallholders. *Global Forum on Agricultural Research and Innovation.* https://doi.org/10.7490/f1000research.1115402.1.

Maslova, A. (2017). *Growing the garden: How to use blockchain in agriculture* [Online]. Available https://cointelegraph.com/news/growing-the-garden-how-to-use-blockchain-in-agriculture.

Mistry, I., Tanwar, S., Tyagi, S., & Kumar, N. (2020). Blockchain for 5G-enabled IoT for industrial automation: A systematic review, solutions, and challenges. *Mechanical Systems and Signal Processing, 135*(1), 106382.

Mohan, T. (2018). Improve food supply chain traceability using blockchain. Doctoral Dissertation. In *The Pennsylvania state university. State college, PA, Pennsylvania, USA.*

Munir, M., Bajwa, I., & Cheema, S. (2019). An intelligent and secure smart watering system using fuzzy logic and blockchain. *Computers & Electrical Engineering, 77*(1), 109–119.

Nakamoto, S. (2008). *Bitcoin: A peer-to- peer electronic cash system* [Online]. Available https://bitcoin.org/bitcoin.pdf.

Oceana. (2013). *Oceana study reveals seafood fraud nationwide* [Online]. Available http://oceana.org/reports/oceana-study-reveals-seafood-fraud-nationwide.

OlivaCoin. (2016) [Online]. Available http://olivacoin.com/.

OriginTrail. (2018) [Online]. Available https://origintrail.io/use-cases.

Patil, A. S., Tama, B. A., Park, Y., & Rhee, K. H. (2017). A framework for blockchain based secure smart green house farming. In *Advances in computer science and ubiquitous computing, Singapore*.

Pearson, S., May, D., Leontidis, G., Swainson, M., Brewer, S., Bidaut, L., et al. (2019). Are distributed ledger technologies the panacea for food traceability? *Global Food Security, 20*, 145–149.

Perboli, G., Musso, S., & Rosano, M. (2018). Blockchain in logistics and supply chain: A lean approach for designing real-world use cases. *IEEE Access, 6*, 62018–62028.

Peter, A. (2017). *Fast company, "in China, you can track your chicken on—you guessed it— the blockchain* [Online]. Available https://www.fastcompany.com/40515999/in-china-you-can-track-your- (chicken-on-you-guessed-it-the-blockchain).

Pinna, A., & Ibba, S. (2018). A blockchain-based Decentralized System for proper handling of temporary Employment contracts. In *Science and information conference*.

Poberezhna, A. (2018). Addressing water sustainability with blockchain technology and green finance. In *Transforming climate finance and green investment with blockchains*.

Rabah, K. (2018). Convergence of AI, IoT, big data and blockchain: A review. *The Lake Institute Journal, 1*(1), 1–18.

Recereum. (2017) [Online]. Available http://recereum.com.

Rejeb, A. (2018). Blockchain potential in Tilapia supply chain in Ghana. *Acta Technica Jaurinensis, 11*(2), 104–118.

Reuters, G. C.-D., (2018). *Coca-Cola, U.S. State Dept to use blockchain to combat forced labor* [Online]. Available https://www.reuters.com/article/us-blockchain-coca-cola-labor/coca-cola-u-s-state-dept-to-use-blockchain-to-combat-forced-labor-idUSKCN1GS2PY?mc_cid=d7c996d219&mc_eid=4c123074ea.

Ripeio. (2017) [Online]. Available http://www.ripe.io/.

Salah, K., Nizamuddin, N., Jayaraman, R., & Omar, M. (2019). Blockchain-based soybean traceability in agricultural supply chain. *IEEE Access, 7*(1), 73295–73305.

Sander, F., Semeijn, J., & Mahr, D. (2018). The acceptance of blockchain technology in meat traceability and transparency. *British Food Journal, 120*(9), 2066–2079.

Shaji, A., Rony, A., Kuriakose, A., & Rawther, F. (2019). Decentralized land lending system using blockchain. *International Journal of Information, 8*(no. 2).

Sharma, S. (2017). Climate change and blockchain. *ISO, 690*.

SNCF. (2017). *How Blockchain simplifies waste disposal* [Online]. Available https://www.digital.sncf.com/actualites/data-tritus-comment-la-blockchain-simplifie-le-tri-des-dechets.

Soil Association Certification. (2018) [Online]. Available https://www.soilassociation.org/certification/.

Steenmans, K., & Taylor, P. (2018). *A rubbish idea: How blockchains could tackle the world's waste problem* [Online]. Available https://theconversation.com/a-rubbish-idea-how-blockchains-could-tackle-the-worlds-waste-problem-94457.

Suprunov, P. (2018). *Medium.com* [Online]. Available https://medium.com/practical-blockchain/5-hyperledger-projects-in-depth-3d14c41f902b.

Swachhcoin. (2018). *Decentralized waste management system* [Online]. Available https://swachhcoin.com.

Swan, M., & De Filippi, P. (2017). Toward a philosophy of blockchain: A symposium: Introduction. *Metaphilosophy, 48*(5), 603–619.

Tao, H., & Bullock, D. (2019). Using digital agriculture technologies to improve nitrogen management and wheat yield. *Cereal Foods World, 64*(6).

Tayeb, S., & Lago, F. (2018). Blockchain technology: Between high hopes and challenging implications. *MENA Business Law Review First quarter*.

Thomason, J., Ahmad, M., Bronder, P., Hoyt, E., Pocock, S., Bouteloupe, J., et al. (2018). Blockchain—powering and empowering the poor in developing countries. *Transforming Climate Finance and Green Investment with Blockchains*, 137–152.

Tian, F. (2016). An agri-food supply chain traceability system for China based on RFID & blockchain technology. In *13th international conference on service systems and service management (ICSSSM)*.

Tian, F. (2017). A supply chain traceability system for food safety based on HACCP, blockchain & Internet of things. In *In international conference on service systems and service management (ICSSSM)*.

Tribis, Y., El Bouchti, A., & Bouayad, H. (2018). Supply chain management based on blockchain: A systematic mapping study. *MATEC Web of Conferences, 200*.

Tripoli, M., & Schmidhuber, J. (2018). Emerging opportunities for the application of blockchain in the agri-food industry. *FAO and ICTSD: Rome and Geneva, Vols. Licence*. CC BY-NC-SA 3.

Tschorsch, F., & Scheuermann, B. (2016). Bitcoin and beyond: A technical survey on decentralized digital currencies. *IEEE Communications Surveys and Tutorials, 18*(3), 2084–2123.

Tse, D., Zhang, B., Yang, Y., Cheng, C., & Mu, H. (2017). Blockchain application in food supply information security. In *International conference on industrial engineering and engineering management (IEEM)*.

Wass, S. (2017). Global trade review [Online]. Available https://www.gtreview.com/news/fintech/food-companies-unite-to-advance-blockchain-for-supply-chain-traceability/.

W. E. (2007). *"Putting Things to REST," UCB iSchool report 2007-015, Berkeley, USA*.

WWF. (2018). *New blockchain project has potential to revolutionise seafood industry* [Online]. Available https://www.wwf.org.nz/media_centre/?uNewsID=15541.

Xu, X., Lu, Q., Liu, Y., Zhu, L., Yao, H., & Vasilakos, A. V. (2019). Designing blockchain-based applications a case study for imported product traceability. *Future Generation Computer Systems, 92*, 399–406.

Yli-Huumo, J., Ko, D., Choi, S., Park, S., & Smolander, K. (2016). Where is current research on blockchain technology?—a systematic review. *PloS One, 11*(10).

Yuan, H., Qiu, H., Bi, Y., Chang, S., & Lam, A. (2019). Analysis of coordination mechanism of supply chain management information system from the perspective of block chain. *Information Systems and E-Business Management*, 1–23.

Zeto. (2018). ZetoChain [Online]. Available https://www.zeto.ie/.

Zhang, D. (2019). Application of blockchain technology in incentivizing efficient use of rural wastes: A case study on yitong system. *Energy Procedia, 158*(1), 6707–6714.

Zhang, Z. R., Farooque, A.,M., Kang, K., & Venkatesh, V. (2020). Blockchain-based life cycle assessment: An implementation framework and system architecture. *Resources, Conservation and Recycling, 152*(104512).

Zhao, G., Liu, S., Lopez, C., Lu, H., Elgueta, S., Chen, H., et al. (2019). Blockchain technology in agri-food value chain management: A synthesis of applications, challenges and future research directions. *Computers in Industry, 109*, 83–99.

Chapter 8

Digital extension service: quick way to deliver agricultural information to the farmers

Mahantesha B.N. Naika[1,2], Manjunath Kudari[1,2], Maguluri Sree Devi[1,2], Dhanush Swaroop Sadhu[1,2], Suma Sunagar[1,2]

[1]*Kittur Rani Channamma College of Horticulture, Arabhavi, Paramaddi, Karnataka, India;* [2]*University of Horticultural Sciences, Bagalkot, Karnataka, India*

8.1 Introduction

The world population is likely to rise by two billion from 7.6 billion in 2018 (UN DESA, 2019) in the next 30 years, which will be accompanied by increasing demand for food and nutritional security. However, the increase in urbanization, land degradation, quality water, and other natural resources are the main challenges to meet the demand (FAO, 2017). Agricultural growth is based on the use of improved technologies, and it is a continuous process that involves technology transformation and implementation. Therefore, to increase the productivity, research and extension service play important roles. At the dawn of the new century, it is not an overstatement to summarize that digitalization is an indispensable phenomenon that changed the economy as a whole. Digitalization in any industry is predominantly dependent on the intelligent use of information technology (IT) that sustainably maximizes productivity. It is the most happening event in India where the emphasis has been laid upon improving online infrastructure through Internet connectivity, thereby empowering the nation in the field of technology. Looking into such scope and opportunities, access to agricultural information for the benefit of smallholder farmers, recently, several digital services have been developed (Baumüller, 2018; Fabregas, Kremer, & Schilbach, 2019). The majority of the available solutions concentrate on supporting specific business sharing information via SMS (Batchelor, Scott, Manfre, & Edwards, 2014).

In India, since 2013, the data costs were seen to fall by 95 %. Thus, it was envisioned that there would be a rise in Internet users by 40 %, and smartphones' numbers would also double by 2023. In India, it was estimated that around 560 million people subscribe Internet globally; this is the biggest and

Food Technology Disruptions. https://doi.org/10.1016/B978-0-12-821470-1.00006-9

fastest-growing market for digital clients after China (McKinsey Global Institute Report, 2019). So far, digitalization in India has been characterized by four aspects viz., social media, mobility, analytics, and cloud, commonly called SMAC (https://www.digitalindia.gov.in/). (Table 8.1). The ICT development conducted Index rankings showed that among 157 countries in that India got 1389th position amongst 175 countries (MISR, 2016).

TABLE 8.1 Overview of digitalization in India (2018).

Internet subscribers	560 million
Average data consumption per month	8.3 GB
Mobile phone subscriptions	1.2 billion
App downloads	12 billion
People with biometric digital Identity	1.2 billion
Businesses with digital platforms	10 million
Number of smartphones per 100 persons	26.2
Number of cashless transactions per person	18
Projected increase in jobs by the digital economy (by 2025)	60–65 million
Projected production growth in the agricultural sector due to digitalization (by 2025)	70x

x, indicates present status.
Courtesy: McKinsey Global Institute. (2019). Digital India: Technology to transform a united nation: Report by MGI (144pp).

Information and communication technology (ICT) tools have a wide range of functional utilities as they can be used to find, explore, analyze, exchange, and present information indiscriminately and simply. Selling and buying online is a present trend in which the role of the Internet and communication cannot be dispensed. In today's world, it is a key parameter for economic development. Any farmer needs to make several decisions in limited intervals during crop production until the market. Many of these solutions, which directly affect the economy of production, are objects of digitalization (Rose et al., 2016). For agricultural production activities, farmers need a variety of information at different stages *viz* weather forecasts, a package of practices, including land preparation, variety selection, sowing period, nutrient, pest and disease management, harvesting and marketing (de Silva and Ratnadiwakara, 2008 and Mittal et al., 2010). A report suggests that the world comprises approximately 1.5 billion smallholder farmers who, in turn, account for about 75 % of the world's poorest people (Ferris, Robbins, Best, Seville, & Shriver, 2014).

Agricultural extension services (AES) defined here as services through which the adoption and application of knowledge, technologies, and practices are promoted and deliver positive returns on investment, boost productivity, improve food security and rural livelihoods thereby transforming the global agricultural scenario (Araji, Sim, & Gardner, 1978; Benina et al., 2011). The AES can be a potential means to help smallholders break the cycle of low productivity, vulnerability, and poverty, thereby enabling food security. The awareness and tools regarding modern agricultural practices and their provision to farmers link them to new technology, which provides them with greater access to financial and marketing solutions. The AES is a common age-old practice done through an extension officer visiting a farmer face-to-face or group and conducting farmers' field school (Stringfellow, Coulter, Lucey, McKone, & Hussain, 1997). However, AES is limited by lack of extension personnel, up-to-date information regarding market access, expertise, timeliness, information storage, widening gap between traditional and modern technologies (Fig. 8.1). Therefore, digitalization can be critical in overcoming such limitations through the utilization of various ICT tools, especially for

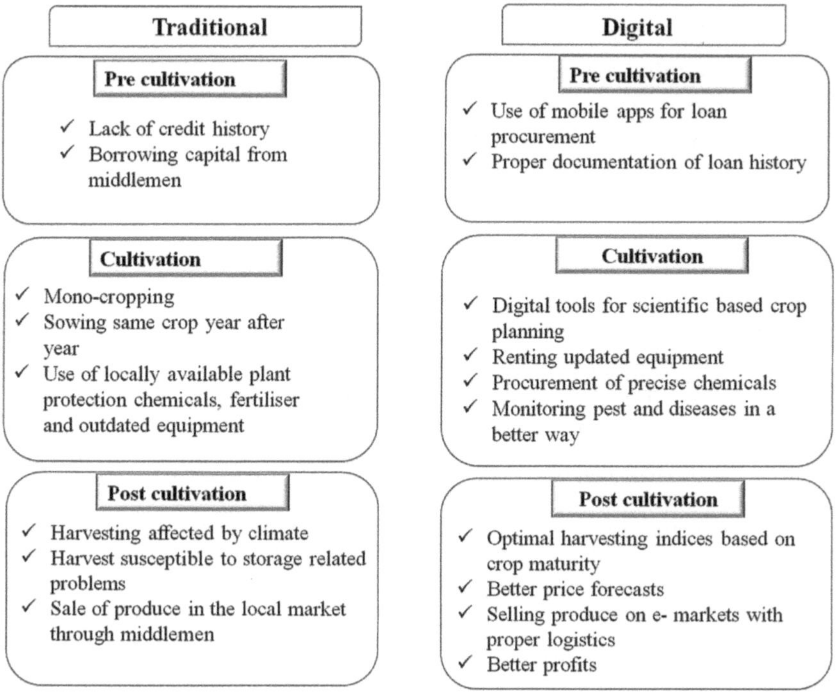

FIGURE 8.1 Comparison between traditional and digital extension services. *(Courtesy: McKinsey Global Institute. (2019). Digital India: Technology to transform a united nation: Report by MGI (144pp)).*

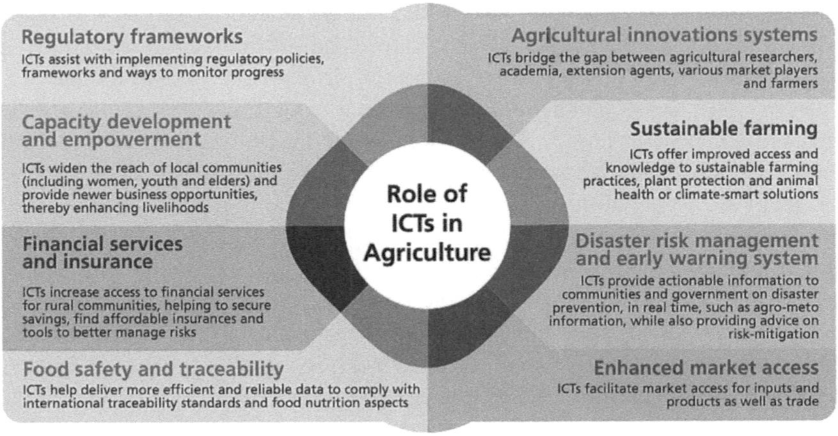

FIGURE 8.2 Role of information and communication technology in agriculture. *(Courtesy: FAO-ITU, National Agriculture Strategy).*

smallholder farmers (Baumuller, 2018; Fabregas et al., 2019). The main goal of the extension is to bring the agricultural technologies obtained through research from lab-to-land and also to collect the feedback of problems faced by the farmer on-field and report it back to the research system.

In order to fulfill this purpose, the latest information and knowledge of the subject is a must. There are many reasons one can come up with in order to understand the information delay between farmers and agriculture researchers (Fig. 8.1).

ICT has played a substantial role in terms of improving food and nutritional security, poverty reduction, as well as climate change resilience, which has added for supporting smallholder farmers (Source: UNFCC https://unfccc.int/resource/mfc2017/project.html?p=project-16). In India, various methods and plans have been implemented by a joint venture of the agriculture department and government through e-technology. With the advent of the ICT revolution, there have been rapid improvements regarding the augmentation of agricultural and rural development all the way through better information and communication processes (Fig. 8.2) (FAO-ITU, National Agriculture Strategy).

Implementation of ITC can directly impact farming systems by providing solutions that ensure informed and quality decisions at the grower's end, which will have a positive impact on the way agriculture and allied activities are conducted (Table 8.2).

8.2 Digital extension services in India

The Indian government has taken key initiatives in order to promote ICT in agriculture, which include National e-Governance, various touch screen

TABLE 8.2 Utilization of information communication technology and tools in DES.

Extention function	TV and video	Cell phones	Feature and smart devices	Computer and internet
Diagnose problems	Visuals are very helpful as "seeing is believing." even better if combined with ways to receive feedback.	Some potential if farmers can call or text in and sufficient expertise is available.	Additional potential to a simple cell phone as it enables web or App access to special diagnostic tools.	Good comprehensive diagnostic tools are available
Collect information		Can use for data collection.	Good for data collection with GPS.	Some potential if Internet available.
Raise aware of general opportunities or needs; convince farmers to try something new	Visuals are usually very helpful as "seeing is believing"	Is an option if users are registered to receive such messages (SMS)	Is an option if users are registered to receive such messages (SMS, email)	Is an option if users are registered to receive such messages (email)
Provide specific information needed for change. What is involved? What are the benefits/ Demonstrate or train?	Good option as "seeing is believing"	Potential if farmers can call or text in and sufficient expertise is available	Additional potential to a simple cell phone as it enables web access and plays videos.	Good option for intermediaries to seek information and videos.
Facilitate access to credit and inputs	Can be used to inform of available services, but one-way communication	Mobile banking and negotiate directly with the suppliers	Mobile banking and negotiate directly with the suppliers	Mobile banking and negotiate directly with the suppliers

Continued

TABLE 8.2 Utilization of information communication technology and tools in DES.—cont'd

Extention function	TV and video	Cell phones	Feature and smart devices	Computer and internet
Link farmers to markets		Access to price information (call in, subscription)	Can bring potential buyers and producers together; access price information	Can bring potential buyers and producers together; price info.
Collect and respond to farmer feedback	Good if producers can call or text and sufficient expertise is available	Some potential if farmers can call or text in and sufficient expertise is available	Good option for intermediaries to seek information (if optimized for smart devices)	Good option for intermediaries to seek information
Assist with business planning	Some potential		Simple farm management "Apps"; record-keeping	Farm management tools; record-keeping

Courtesy: Bell, M, & Payne, J. (2011). ICT options in relation to extension functions. *MEAS*. Original found at www.meas-extension.org/resources/ict.

Kiosks (like Warana), National e-Governance Plan in Agriculture (NeGP-A), Krishi Vigyan Kendras (KVK's), Kisan Call Centers, Agri-Clinics, Common Service Centers, mKisan, Kisan TV, Indian Farmers Fertiliser Cooperative Limited (IFFCO). Agri-portal, e-Choupal, aAqua, Rice knowledge management portal (RKMP), Village knowledge canters (VKCs), Village resource centers (VRCs), and various other applications (Nagesh and Saravanan, 2019). Mobile applications play an important and effective role in agricultural extension. Many agriculture-based applications are being designed every year across the globe to provide substantial solutions to small, medium, and large scale farmers regarding various challenges faced throughout the crop cycle up to the marketing of the produce (Table 8.3). In India, also, many apps are being developed for use in DES (Table 8.4 and Fig. 8.3).

One such application (app) extending digitalized service for farmers is eSolution against Agricultural Pests (eSAP) (Fig. 8.4) (https://play.google.com/store/apps/details?id=com.tene.eSAP). It enables the field user to register farmers, identifying problems related to crop and to estimate the magnitude of the problem, as well as to provide farmers with an effective solution and to

TABLE 8.3 List of agricultural apps and their utility among selected countries.

Application name	Utility
	Africa
iCow	It gives information on milk productivity, poultry, eggs, crops and soil fertility. https://www.icow.co.ke/
Rural eMarket	It provides market related information such as price, product availability in different geographical areas to help farmers take decisions. http://rural-emarket.com/en/
Esoko	It collects and sends out market data using simple text messaging by providing automatic and personalized price alerts, buy and sell offers to the farmers. https://esoko.com/
M-Shamba	It is an interactive platform that gives information regarding crop cultivation practices. https://m-shamba.net/
	USA
Sirrus	It helps the farmers to be updated with information regarding cropping, record keeping and also posts current weather conditions. https://apps.apple.com/us/app/sirrus/id684978309
FarmLog	It provides market related information and also assists the farmer to view specific soil maps which gives first-hand information on fertilizer application in the field. https://farmlogs.com/
FarmEdge	It gives updated information on cultural operations, record and track on the farm storage which helps to realise greater profits. https://www.farmersedge.ca/
Farm At Hand	It is a farm management app which attends to information regarding field operations from seed to sale with more precision. https://www.farmathand.com/
	Australia
Elders weather	It gives updated weather data which helps to plan out important cultural operations. https://www.eldersweather.com.au/
Trap view	This is an automated pest monitoring app which helps to mitigate pest damage. https://www.trapview.com/v2/en/

Continued

TABLE 8.3 List of agricultural apps and their utility among selected countries.—cont'd

Application name	Utility
Feralscan mapping	It is used to record the location of pest animal in a specific area, record the problems they cause, and control the actions they have to undertake. https://play.google.com/store/apps/details?id=com.invasiveanimals.feralscan_pest_mapping&hl=en_IN&gl=US
Holland	
Agren soil calculator	It estimates soil erosion and economic cost, and helps the farmers to implement soil conservation practices like crop rotation, tillage system. https://tracxn.com/d/companies/agrentools.com
Tiger-sul	It is a fertilizer calculating app which helps the farmer to plan balanced fertilizer application. https://www.tigersul.com/
Brazil	
AgroBrazil	It provides ready information on day-to-day deals related to agricultural commodities through which farmers can realize greater profit margins. https://play.google.com/store/apps/details?id=com.hypelabs.agrobrazil&hl=en_IN&gl=US
Agritempo	It provides meteorological information and is endowed with features like weather maps and drought index and also supports the recommendation of the Agricultural Zoning Climate Risk. https://www.agritempo.gov.br/agritempo/index.jsp

follow up with the farmer. In order to digitalize agriculture extension, the government of Karnataka, India, has taken the initiative by adopting eSAP. This has enabled various organizations in Karnataka to work independently in one instance. The app is being managed by the University of Agricultural Sciences (UAS), Raichur, in collaboration with other state agricultural universities. This app also works offline in English and Kannada languages. Is utility, it has played a major role in providing information on managing diseases, pests, nutritional disorders, and weeds that affect crop health in a damaging way. Additionally, the user can also recommend the best production practices to the farmers.

TABLE 8.4 List of selected agricultural apps in India.

Sl. No.	APP name	URL
1.	eSAP	https://play.google.com/store/apps/details?id=com.tene.eSAP
2.	Meghdoot	https://play.google.com/store/apps/details?id=com.aas.meghdoot
3.	IFFCO Kisan	https://play.google.com/store/apps/details?id=com.IFFCOKisan
4.	Indian Satellite weather	https://play.google.com/store/apps/details?id=com.shahul3d.indiasatelliteweather
5.	Kisan Suvidha	https://play.google.com/store/apps/details?id=in.cdac.bharatd.agriapp
6.	Agricultural Business	https://play.google.com/store/apps/details?id=com.AgriculturalBusiness3dsp
7.	Kheti Badi	https://play.google.com/store/apps/details?id=com.app.khetibadi
8.	ICAR Technologies	https://play.google.com/store/apps/details?id=gov.krishi.icar.technologyrepository
9.	Modern Kheti	https://play.google.com/store/search?q=modern%20kheti&c=apps
10.	Fertilizer calculator	https://play.google.com/store/apps/developer?id=Dr.+Vishwanath+Koti
11.	Agri Live	https://play.google.com/store/apps/details?id=agri.live
12.	Agri App	https://play.google.com/store/apps/details?id=com.criyagen
13.	Farm Bee Pomegranate Expert	https://play.google.com/store/apps/details?id=com.rml.Activities
14.	Agriculture dictionary	https://play.google.com/store/apps/details?id=com.ermilogic.dat
15.	Organic Farming	https://play.google.com/store/apps/details?id=com.vasithwam.apps.organicfarming
16.	Ag Mart	https://play.google.com/store/apps/details?id=info.agmart
17.	Kisan Yojana	https://play.google.com/store/apps/details?id=com.purplechai.admin.kissanyojnaapp
18.	Zero Budget Natural Farming	https://play.google.com/store/apps/details?id=com.oyepages.zbnf

Continued

TABLE 8.4 List of selected agricultural apps in India.—cont'd

Sl. No.	APP name	URL
19.	Digital Mandi	https://play.google.com/store/apps/details?id=com.maswadkar.digitalmandi
20.	AgroIndia	https://play.google.com/store/apps/details?id=kisan.helper.mandibhaw

8.3 Usage of Apps in DES in India

8.3.1 Meghdhoot

The Megdhoot app (https://play.google.com/store/apps/details?id=com.aas.meghdoot) was developed with combined efforts of the India Meteorological Department (IMD), Indian Institute of Tropical Meteorology (IITM), Pune, India and the Indian Council of Agricultural Research (ICAR), New Delhi. It provides information on location, crop, and livestock specific weather-based agro advisories in 10 different languages (Fig. 8.5).

8.3.2 Indian Farmers Fertiliser Cooperative Limited (IFFCO)

Are one of India's most prominent cooperative societies, and its web sites are https://www.iffco.in and IFFCOkisan app (https://play.google.com/store/apps/details?id=com.IFFCOKisan). It is extended globally like Jordan India Fertilizer Co. Jordan, Kisan International Trading, Dubai, Oman India Fertiliser Company, Oman, Industries Chimiques du Senegal, Senegal, and IFFCO CANADA, Canada (Fig. 8.6).

8.3.3 Kheti Badi

Kheti Bada app (https://play.google.com/store/apps/details?id=com.app.khetibadi) is developed for providing information on farm product from all over the Indian agriculture markets for agriculture traders and farmers to get the online prices in India (Fig. 8.7).

8.3.4 Agri live

Agri Live app (https://play.google.com/store/apps/details?id=agri.live) is developed by SathvTech Information Services Private Limited, Bangalore, India. It is available in regional languages such as Hindi, Bengali, Punjabi,

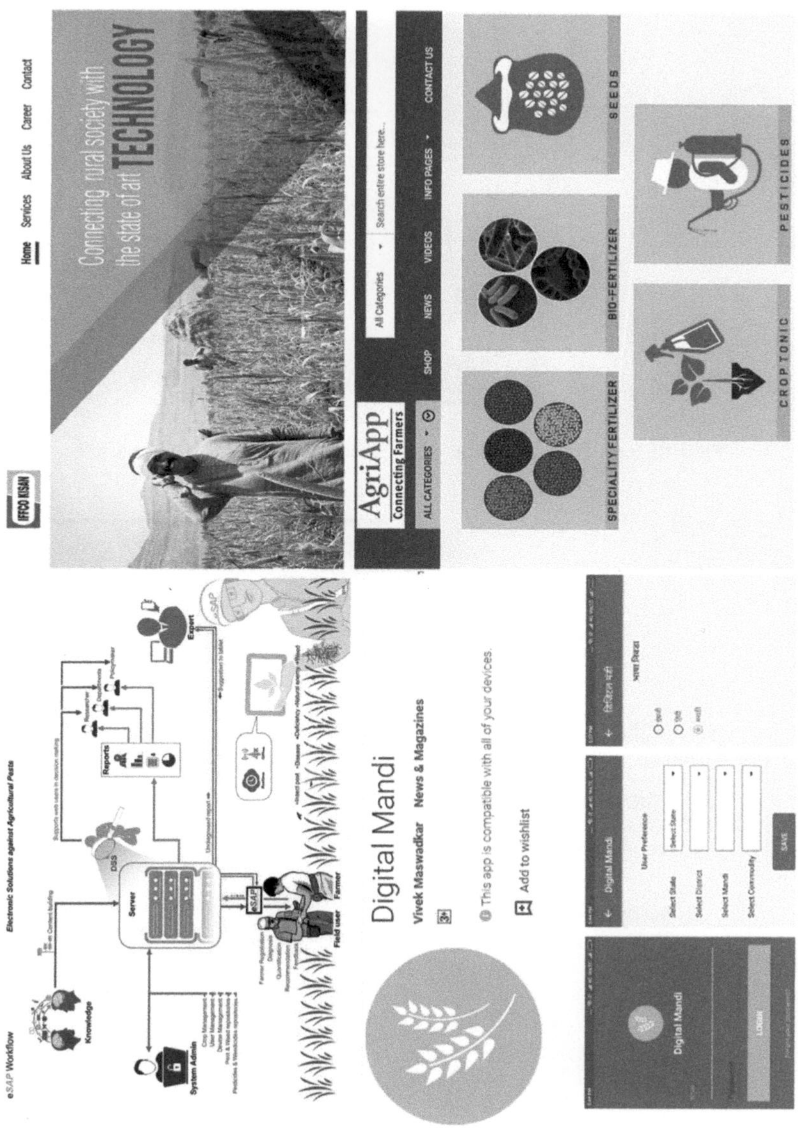

FIGURE 8.3 Selected apps for agriculture purposes in India. *(Courtesy: Images from different apps developers).*

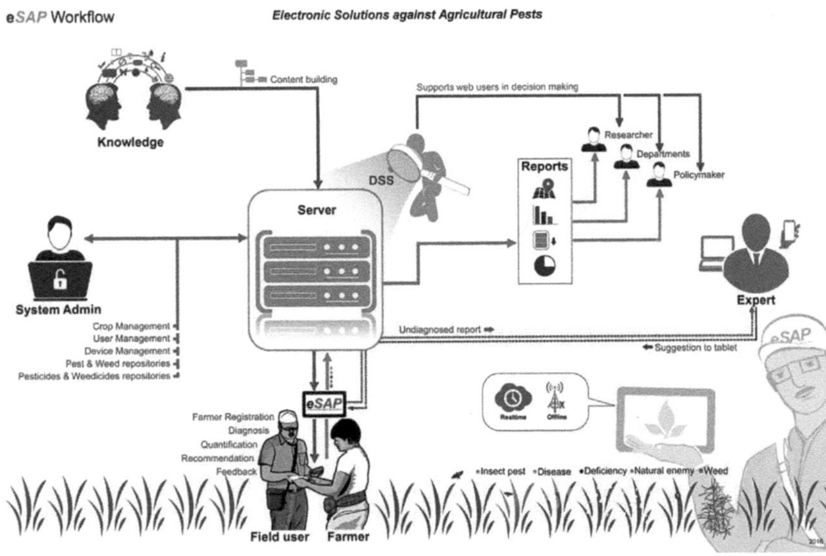

FIGURE 8.4 Overview of the eSAP process. *(Courtesy: eSAP developer).*

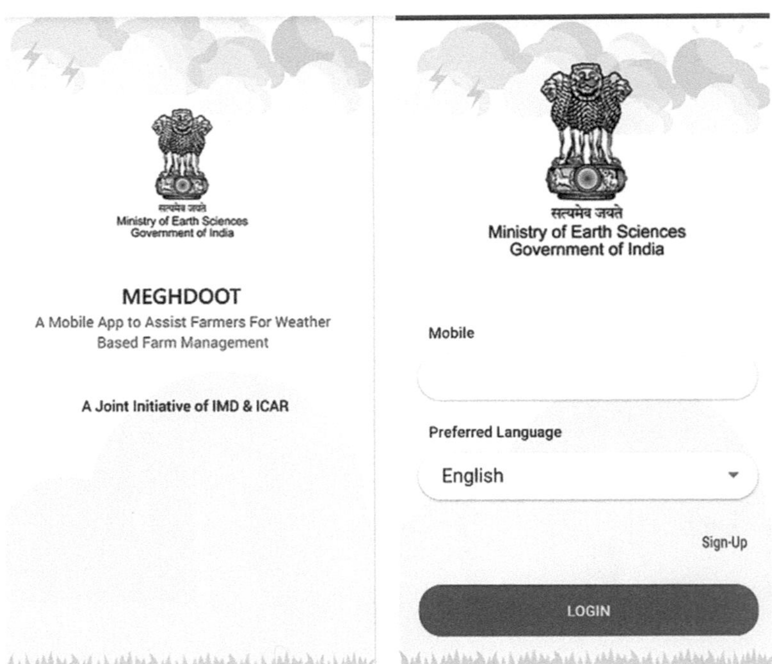

FIGURE 8.5 Overview of meghdhoot. *(Courtesy: IMD, IITM Pune and ICAR New Delhi).*

FIGURE 8.6 Overview of Indian Farmers Fertiliser Cooperative Limited (IFFCO). *(Courtesy: IFFCO).*

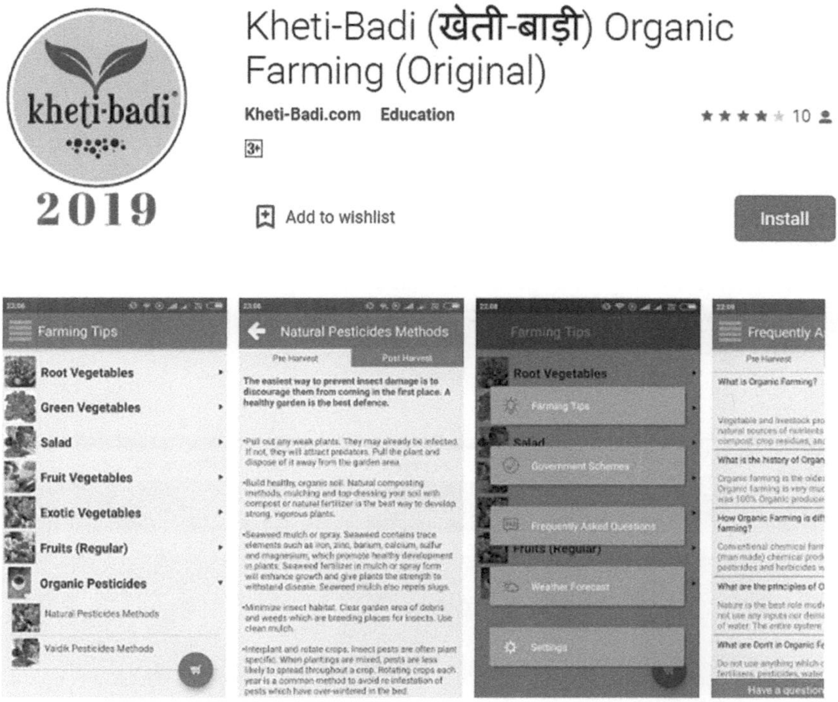

FIGURE 8.7 Overview of Kheti Badi. *(Courtesy: Kheti Badi app developers).*

Marathi, Kannada, Tamil, Malayalam, and Telugu. Its main goal is to generate site a crystalline agriculture market system by ensuring steady demand and uniform distribution of agricultural commodities such as grains, Pulses, Fruits, Vegetables, Flowers, Seeds, Spices, Woods, Livestock and Aqua culture (Fig. 8.8).

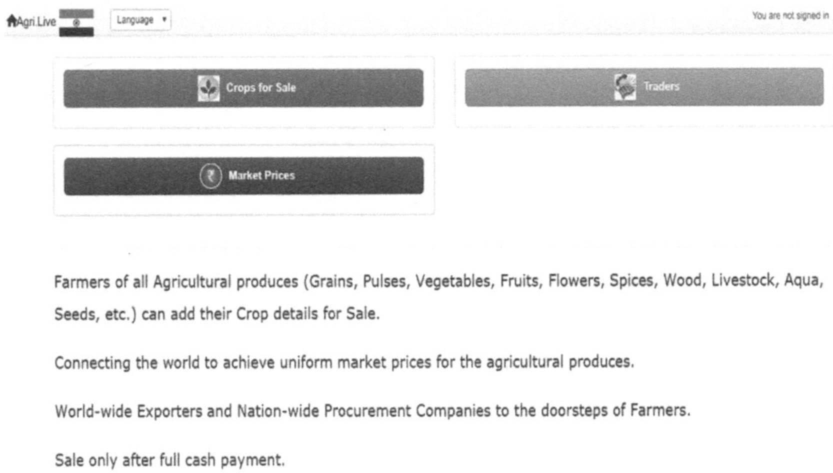

FIGURE 8.8 Overview of agri live. *(Courtesy: Agri live app developers).*

8.3.5 Agri app

Agri app (https://play.google.com/store/apps/details?id=com.criyagen) is developed by Pace Wisdom Solutions private limited, Bangalore, India. It provides detailed information on the package of practice, chats with field officers, calls to subject experts, Available in English, Hindi, and Kannada languages, Videos related to agriculture practices, and news related to agriculture aspects (Fig. 8.9).

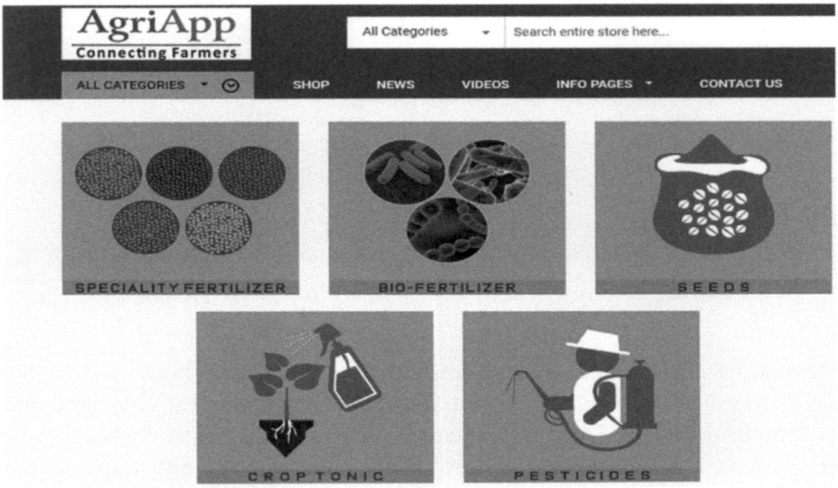

FIGURE 8.9 Overview of agri app. *(Courtesy: Agri app developers).*

8.3.6 Agriculture dictionary

Agriculture dictionary app (https://play.google.com/store/apps/details?id=com.ermilogic.dat) is developed by ERMILOGIC, Greece. This app is a storehouse of terms related to Agriculture, Crops, Fruit, Farming, Animal Husbandry, Livestock, Products, Agricultural Machines, and Engineering, Agricultural Policy, European Union policies, Weather, Environment, Economics, Information and Communication Technologies, which helps in guiding farmers, students, researchers of agriculture (Fig. 8.10).

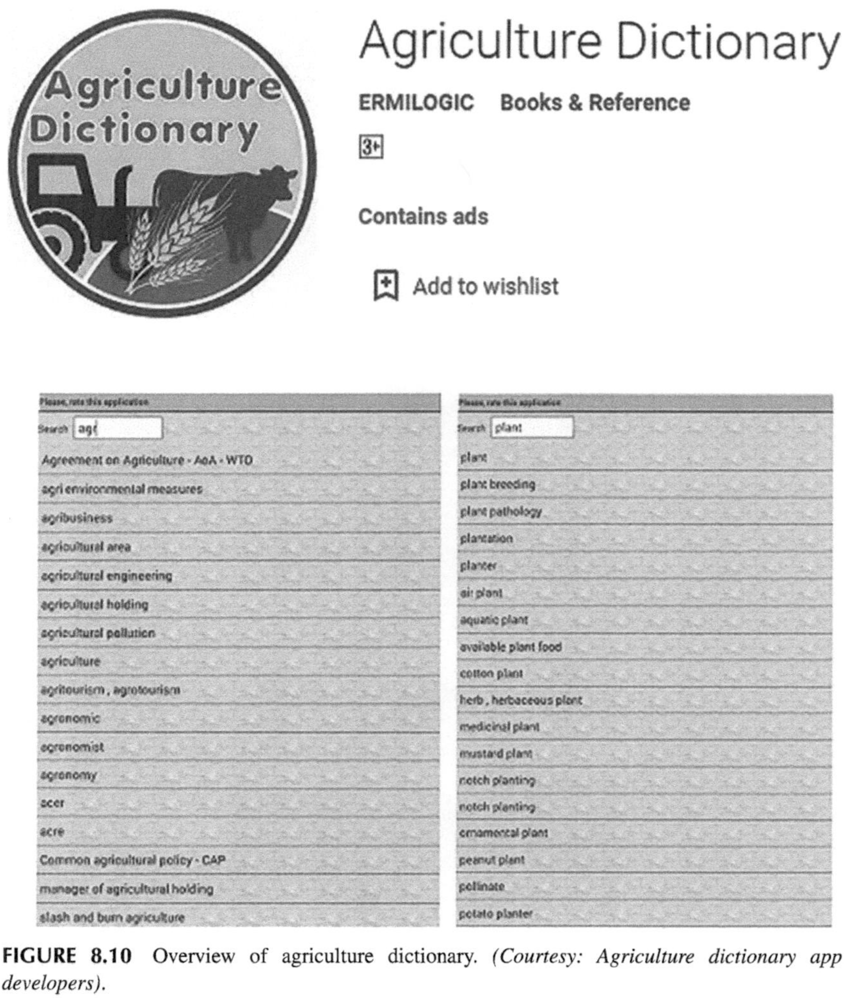

FIGURE 8.10 Overview of agriculture dictionary. *(Courtesy: Agriculture dictionary app developers).*

8.3.7 Ag mart

Agmart app (https://play.google.com/store/search?q=agmart%20app&c=apps) is developed by M/S. AgMart Agro Services Pvt. Ltd., Raichur-584101, Karnataka. App provides the latest information on agri market rate updates to the farmers and buyers (Fig. 8.11).

8.3.8 Zero budget natural farming

Zero budget natural farming app (https://play.google.com/store/apps/details?id=com.oyepages.zbnf) is developed with the combined effort of Dipak

FIGURE 8.11 Overview of Ag mart. *(Courtesy: AgMart app developers).*

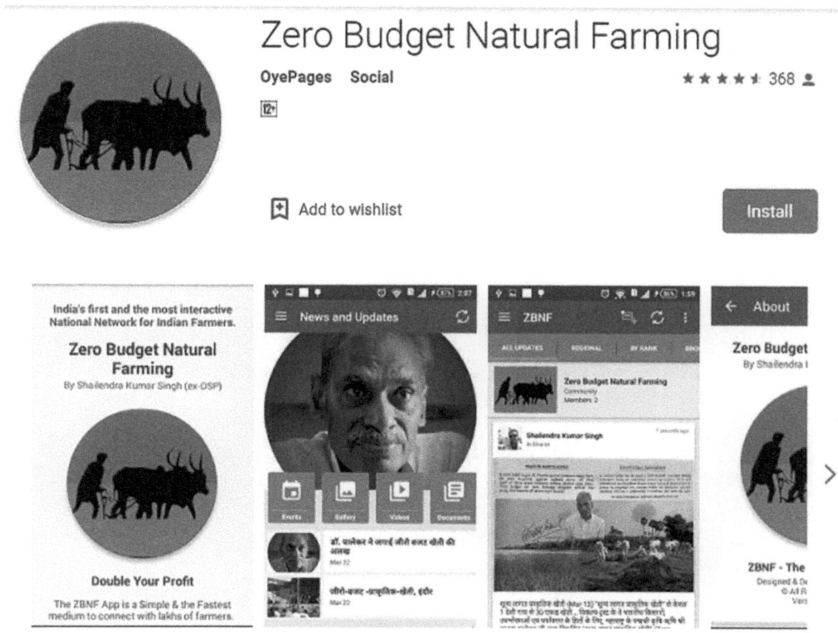

FIGURE 8.12 Overview of zero budget natural farming. *(Courtesy: Zero budget natural farming app developers).*

Mortale and Keshav Rahegaonkar with the support of Subhash Palekar, an agriculture specialist. The app provides the information about Zero Budget Natural Farming (Fig. 8.12).

8.3.9 Digital mandi India

Digital mandi India app (https://play.google.com/store/apps/details?id=com.maswadkar.digitalmandi) helps the farmers to be updated with the latest Indian agricultural commodities mandi price from various states and districts. Features of the app browse through category wise commodities, states, mandi price, favorite commodity-specific to state, copy of mandi price and share option, Sync data from the Agmarket.nic.in an Indian government portal (Fig. 8.13).

8.3.10 Agro India

Agro India (https://play.google.com/store/apps/details?id=kisan.helper.mandibhaw) is an agriculture news app which was developed by MindWings Software Pvt. Ltd., Kholapur, India (Fig. 8.14).

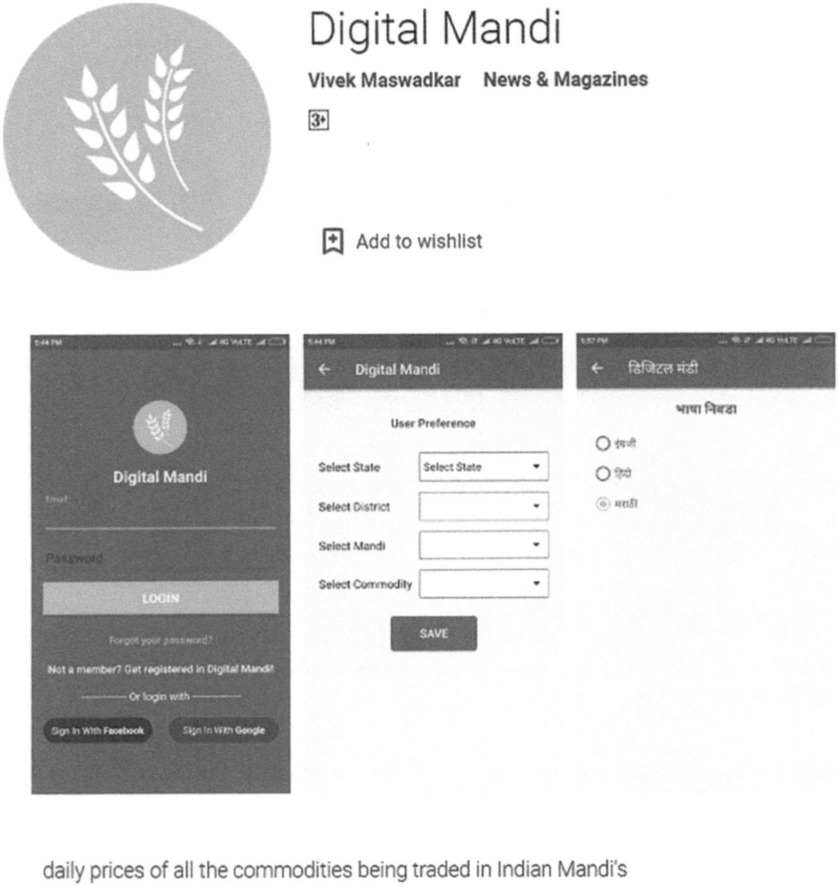

FIGURE 8.13 Overview of digital mandi India. *(Courtesy: Digital mandi India app developers).*

8.3.11 Fertilizer calculator

Fertilizer calculated (https://play.google.com/store/apps/developer?id=Dr.+Vishwanath+Koti), it was developed by ICAR-Central Coastal Agricultural Research Institute, Goa, India. It helps the user to calculate the amount of fertilizers to be applied for a desired area and crop based on soil fertility (Fig. 8.15).

8.4 Case study

8.4.1 Understand the knowledge, perception, and utilization of ICT tools by farmers in India

The study was conducted in Yarzarvi, Tadasaluru, and Budhigoppa villages of Belagavi district of Karnataka, India, during the year 2019. The objective of

FIGURE 8.14 Overview of Agro India. *(Courtesy: Agro India app developers).*

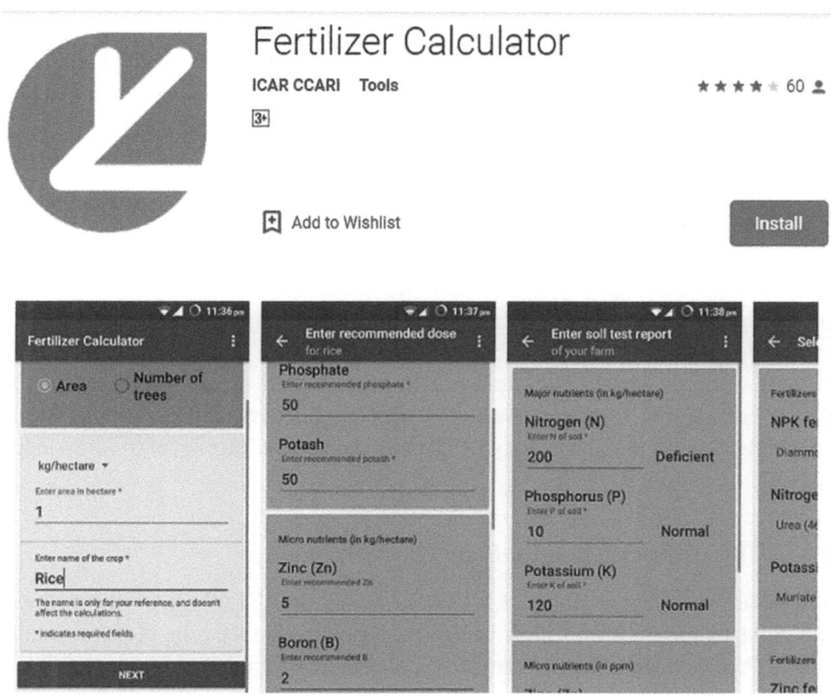

FIGURE 8.15 Overview of fertilizer calculator. *(Courtesy: Fertilizer calculator App developers).*

the study was to understand the knowledge, perception, and utilization of ICT tools by farmers. The total sample size for the study was 120 farmers. Data was collected by interviewing farmers with the help of a structured interview schedule developed for the study. Data collected from farmers was scored, tabulated, and analyzed using Microsoft Excel for frequency, percentage, mean, and Standard Deviation (SD). Based on some of the scores, the farmers were categorized into three groups based on mean and standard deviation. Results obtained from the study presented below;

8.4.2 Extension contact

An insight into Table 8.5 indicates that nearly 50 percent (49.17 %) of farmers belonged to high extension contact followed by medium (32.50 %) extension contact. Due to less availability of extension staff to farmers and the ratio of farmers to extension personnel noticed to below. A study conducted by Vishwatej during 2013 in the Belgaum, District, Karnataka, found that 40.71 % of the farmers had medium extension contact.

8.4.3 Perception of farmers towards ICT tools

More than 50 (51.66 %) percent of farmers belong to the high perception category toward ICT tools, and 31.67 % of farmers belonged to the medium perception category (Table 8.5). The study conducted by Ajayi, Alabi, and Okanlawon (2018) reported that the majority (84.7 %) of the farmers had indifferent perceptions toward ICT use (Table 8.5).

8.4.4 Knowledge of farmers about the subject area covered under ICT tools

As per the study, nearly 50 (48.34 %) percent of the farmers had high knowledge, followed by 32.86 % of farmers who had a medium level of knowledge about ICT tools (Table 8.5). Subhashsingh, Bharat, and Rai (2010), in their study, revealed that most of the farmers had great knowledge of the Kisan call center. A study conducted by Nagalakshmi and Narayanaswamy (2011) reported that the majority of farmers have a medium level of knowledge of ICTs.

8.4.5 The utility of ICT tools by the farmers

It was observed from Table 8.5 that 47.50 % and 31.67 % of the farmers belonged to medium and high utility categories, respectively. Kafura, Islam, Safiul, Prodhan, and Dipanwita (2016) observed that television was the most

TABLE 8.5 Knowledge, perception, and utilization of ICT tools by farmers.

Extension contact		
Category	**Frequency**	**Percentage**
Low (<12.12)	22	18.33
Medium (12.13–16.14)	39	32.50
High (>16.15)	59	49.17
	Mean = 14.14 SD = 4.74	
Perception of farmers toward ICT tools		
Level of perception	**Frequency**	**Percentage**
Low (<34.87)	20	16.67
Medium (34.88–40.28)	38	31.67
High (>40.29)	62	51.66
	Mean = 37.58 SD = 6.38	
Knowledge of farmers about subject area covered under ICT tools		
Knowledge level	**Frequency**	**Percentage**
Low (<32.85)	22	18.33
Medium (32.86–36.03)	40	33.33
High (>36.04)	58	48.34
	Mean = 34.45 SD = 3.75	
Utility of ICT tools by the farmers		
Level of usefulness	**Frequency**	**Percentage**
Low (<20.53)	25	20.83
Medium (20.54–24.19)	57	47.50
High (>24.20)	38	31.67
	Mean = 22.37 SD = 4.32	
Usefulness of ICT tools by the farmers		
Level of usefulness	**Frequency**	**Percentage**
Less useful (<18.55)	30	25.00
Moderate useful (19.42–21.92)	58	48.33
Most useful (>20.80)	32	26.67
	Mean = 19.68 SD = 2.65	

preferred ICT tool followed by mobile phones, radio, and computers in Ghazipur district of Bangladesh, revealing that mobile-based apps hold great potential as extension tools to provide the farmers with necessary information. As per the study conducted by Meena, Sharma, and Aishwarya (2011) reported that the majority of farmers expect information on wide yielding varieties, plant protection practices, and market information.

8.4.6 The usefulness of ICT tools by the farmers

A total of 48.33 % and 26.67 % of farmers belonged to a moderate and highly useful category, respectively (Table 8.5). Similar observations were reported by Kafura et al. (2016) in Bangladesh, where the level of education, income, and innovativeness had a significant positive relationship with the extent of use of ICT tools while age and farming experience had a significantly negative relationship. Raghuprasad, Devaraja, and Gopala (2012) reported that more than two-fifth (40.83 %) of the farmers were moderately favorable toward ICT tools, while 31.67 % were least favorable, and 27.50 % were most favorable. Level of education, size of landholding, income, economic motivation, scientific orientation, and participation in extension activities found to have a positive and significant effect on the utilization of ICT tools by farmers in Bangalore rural, Chikkaballapura and Kolar districts of Karnataka. A study conducted to understand the Effectiveness of the mobile agri-advisory service extension model in Bihar, Haryana, Jharkhand, Madya Pradesh, Uttar Pradesh, and Rajasthan states of India (Kansiime, Alawya, Allenb, Subharwalc, Jadhavd et al., 2019) using a database CABI's Direct2Farm (D2F) (https://platform.cabi.org/projects/our-impact/d2f/). Total 400,000 farmers registered to the database and active. Their study found that around 73 % of farmers get information from fellow farmers and 58 % from government extension service providers. The study concluded that in a brief period, CABI-D2f mobile service reached more farmers and benefited in adopting new information as compared with the conventional extension approach.

8.4.7 Constraints faced by farmers in using ICT tools

Majority of farmers (87.5 %) feel clarification of doubts was the major constraint followed by lack of reliable and useful online content (81.66 %), difficulty in understanding technical words (79.17 %), lack of awareness about ICTs (65.83 %) and lack of adequate skills to use ICTs (46.66 %) as the other major constraints (Table 8.6). During 2018, Naik conducted a study in Andhra Pradesh, India found that high cost of gadgets, lack of sufficient skills, lack of uninterrupted power supply, language barriers of ICT gadgets, poor connectivity of Internet in villages, low level of education, insufficient training and practical exposure toward ICTs, poor Internet

TABLE 8.6 Constraints faced by farmers in using of ICT tools.

Sl. No	Constraints	F	%
1.	Clarification of the message is difficult if any doubt arises	105	87.50
2.	Difficult to understanding of technical words	95	79.17
3.	Lack of practical exposure regarding technologies receive through ICT tools	102	85.00
4.	Lack of awareness about different ICTs	79	65.83
5.	Lack of reliable and useful content online	98	81.66
6.	Lack of adequate skill to use ICTs	56	46.66

F, Frequency; %, Percentage.

amenities and economic condition of rural people to be the main constraints faced by farmers in utilizing the ICT tools. Similar inferences were reported by Lokeswari (2016) in Tamil Nadu. Dhaka and Chayal (2010) also reported lack of adequate ICT skills, lack of awareness about ICT, and insufficient regional specific information are other major constraints in utilizing ICT tools. Nagalakshmi and Narayanaswamy (2011) revealed that problems in infrastructure, as the constraints for widespread ICT usage in rural areas.

8.5 Success stories of ICT tools in India

8.5.1 e-NAM (National Agriculture Market)

National Agriculture Market (https://enam.gov.in/web/) is one online portal launched by the Ministry of Agriculture and Farmers' Welfare to enable agriculturalists, dealers, consumers, exporters, and processors with a common stage for trading supplies. It aims at optimizing the demand-supply chain by streamlining the various aspects of agricultural marketing and integrating the existing APMC mandis for maximum symmetry of information between the buyers and sellers. One of the many advantages of e-NAM is transparency in online auctions and payments, thereby facilitating real-time price discovery. In fact, online payments in e-NAM had crossed the 500-crore mark, thereby strengthening the true meaning of "Digital India." 942 FPOs (Farmers Produce Organizations) were registered, and 1,28,015 registered traders conduct trade on eNAM. A total of 585 markets were integrated across 16 states and two union territories in India. The e-NAM trade includes cereals (25), of oilseeds

(13), fruits (29), vegetables (40), spices (14), and other miscellaneous (29). The process of trading starts with farmer registration and lot generation at the gate entry followed by sampling and assaying of the produce that is then traded online through proper bid management and auction weighing sale agreement. The bid is then sealed by online payment through challan, cheque, or Internet banking. Various types of eLearning videos are available on the website to facilitate a better understanding of various steps in the trading procedure like sampling and assaying of produce, gate entry and weighment, sale agreement, and bill procurement. eNAM also facilitates trade (commodity wise annual cash prize) related and payment (rebate and reduction of mandi fee) related incentives. The eNAM has played a vital role in improving livelihood of the farmers by helping them realize higher returns for their produce in many success stories across Gujarat, Madhya Pradesh, and Nizamabad. APMC mandis connected with eNAM also have better logistics, thereby improving the trade of commodities. Most of the beneficiaries attribute the success of eNAM to updated information and transparent trade process. eNAM also facilitates a mobile application that can be downloaded from the play store and contains a farmers module, a traders module, and a stakeholder module. Regular training programs are also conducted for the farmers to improve their adaptability to various digital platforms and digitalized ICT tools.

8.5.2 Bhoomi project in Karnataka

It is developed and executed by the National informatics center, India. With the aim of improving integrated governance and to improve the access of citizens to information through ICTs, the government of Karnataka has implemented a Project Bhoomi (http://landrecords.karnataka.gov.in/bhoomiweb/) as a part of smart governance where 20 million computerized land records are made available benefitting over 6.7 million farmers of the state as it eliminates the various kinds of delays, harassments, and bribes that would be involved in the land records were to be procured from the village accountants in the traditional method. This fact permits them to take some of the information such as Rights, Tenancy, and Crops from kiosks (Bhoomi centers) located in taluka offices.

8.5.3 Gyandoot project

One of the pioneer examples of how e-government services and e-democracy are working in cooperation to bring about a positive change in the farming community is the Gyandoot project, which was implemented in rural drought-prone areas of Madhya Pradesh, India. Thirty-eight telekiosks, owned by the government, came into being in focal areas like markets and major roads connecting to the villages in Dhar District (Cecchini and Raina, 2004) These places are characterized by a population of which 60% live below poverty (Jafri, Dongre, and Tripathi, 2002). The functioning of the kiosks brings into light a very interesting picture. Each of them approximately provides services for 25 to 30 villages through cybercafes instead of using LAN (Local Area Network) and other such technologies. The e-government services that are housed within the cybercafes include news related to employment, price information that is regularly updated, domicile and income certificate applications, and some other services related to loans. One of the interesting features of the Gyandoot project is that it has an online rural newspaper service that regularly keeps the citizens updated with all the necessary information related to local politics and agriculture (World Bank, 2017).

8.5.4 Lifelong learning for farmers

The Commonwealth of Learning Has come up with an app known as Lifelong Learning for Farmers (L3F), which is featured by an Open and Distance Learning for Development, initiated in Commonwealth countries (Balasubramanian and Daniel, 2010). The L3F initiative has been supported by universities, banks, as well as marketing agencies. This has been implemented by using open and distance learning, integrating two main aims of strengthening the learning process, which is self-sufficient amongst men and women who hail from farming background. The main objective herein is to bring forth farmer's skills and knowledge onto a platform where they can be enhanced to the maximum capacity. In India, it has its major influence in Tamil Nadu. VIDIYAL is an Indian NGO that has efficiently utilized L3F for promoting community banking amongst 5000 women who are organized into self-help groups (SHGs). During 2008, around 300 women from the SHGs could learn various aspects of sheep and goat rearing through open and distance learning (World Bank, 2017).

8.5.5 Agropedia

The Consortium for Agricultural Knowledge Management was initiated under the World Bank-funded NAIP in India and has been active since 2008. It is centered around a highly advanced online content aggregation system known as Agropedia. This system functions by delivering and exchanging information through a Web portal and mobile networks that are approachable by phones with restricted data capability. It has a subsidiary platform known as Agrilore, which aids in efficient agriculture extension learning process. It is characterized by the information that is spread over 10, 000 pages containing all the necessary information pertaining to 10 important crops available in four different languages and have got nearly 2000 registered users with great expertise. In 2009—10, under two cultivation seasons, the consortium was able to contact mobile phones of 27,000 farmers in four different languages and recorded around 2.2 million texts well as voice transactions made possible through 687 specific messages (World Bank, 2017).

8.5.6 iKISAN

It is a web-based wide-ranged Agri portal developed by Nagarjuna Fertilisers and Chemicals Limited, Andra Pradesh, India, and it can be accessed using the URL www.ikisan.com. It provides comprehensive content on Fertilizers, management of crops, pest, water, soil, agri-machinery, pesticides, diagnostics of pest and nutrition, weather forecast, information on the market, insurance, dairy and poultry, trade channel partners.

8.5.7 e-Choupals

e-Choupal (https://www.itcportal.com/businesses/agri-business/e-choupal.aspx) is one of the web platform in rural India that was implemented during 2000. These services reach out to more than four million farmers growing a wide variety of crops, including rice, pulses, soybean, coffee, wheat in over 40,000 villages through 6500 kiosks across 10 states (Madhya Pradesh, Haryana, Uttarakhand, Karnataka, Andhra Pradesh, Uttar Pradesh, Rajasthan, Maharashtra, Kerela, and Tamil Nadu).

8.6 ICTs into agricultural extension among selected countries

Globally, different approaches incorporate the utilization of ICTs into the agricultural extension to enable the better transfer of technology and to improve knowledge dissemination at the grass-root level. A super platform is an effective digital agricultural solution that lists not only multiple different services for farmers in one platform but also brings complementary services into the picture that enables them to strengthen each other, which, in turn, creates an added value. By having a comprehensive approach for service provision, it has been able to link farmers to market and also to the broader areas of financial advice and other services, which reduces the role of intermediaries, which ensures the creation of immediate economic value.

8.6.1 Africa

GODAN (Global Open Data for Agriculture and Nutrition) action was a 3-year project underneath the Center for Agricultural and Rural Cooperation ACP-EU (CTA), Wageningen, whose prime objective was to strengthen and develop the efficiency of potential data users like researchers, ICT professionals, journalists, and policymakers so as to get a good understanding regarding the practical utility of open data, mainly to tackle the challenges faced by agriculture. One of the best approaches adopted to deliver training and capacity building was innovative face-to-face workshops, which experienced a positive wave of response, which, in turn, paved the way for more such workshops. A four-week e-learning course was supported to reach huge open data users. Monthly webinars were also provided to reach a large number of audiences, which, in turn, encouraged professionals and practitioners to engage in the GODAN Capacity Development Working Group in order to share their experiences. The project assessed the impact of dealings, considering both the capability and the effectiveness of individuals on the open data (Mesengezi, 2019). CTA's Data4Ag project has been very effective in implementing digital solutions in Africa by concentrating on one of the key contributors to change; the farmer organization. It has supported seven initiatives across the continent and has got 120,000 farmers registered. The digital-based solutions included farmer registration and drone trials, which, in turn, have been able to meet with the resilience and sustainability of livelihoods of Ugandan coffee farmers of the National Union of Coffee Agribusinesses and Farm Enterprises (NUCAFE) (Addison, 2019).

A super platform is an effective solution for addressing problems related to agriculture in a digital sense, and it lists not only multiple different services for farmers in one platform but also brings complementary services into the picture that enables them to strengthen each other, which, in turn, creates an added value. By having a comprehensive approach for service provision, it has been able to create a platform for linking farmers to the market chain, as well as the broader areas related to financial advice and other such services, which reduces the role of intermediaries, which ensures the creation of immediate economic value. Some of the super-platform solutions supporting agriculture are Digital Okam, Digi farm, Econet, Agrikore, Farm to market al.liance, and other platforms (Addom, 2019; Rambaldi, 2019).

One of the key components of CTA's flagship project, "Scaling-up Climate-smart Agricultural Solutions for Cereal and Livestock Farmers" offers an agricultural solution integrated with climate-smart technology for smallholder farmers; the weather-based index insurance (WBI), which began in August 2017. The project emphasized increasing the food security and income for 140,000 rural smallholder households who faced challenging consequences as a result of changing climates within maize-livestock-based farming systems; one of the keys enabling technology in this project is the mobile phone, which has been used to carry out digital registrations and farmer profiling. The weather and agricultural advisory message were provided by e-extension services, which were made possible through mobile phones usage. The impact of this project was 154,578 farmer profiles had been digitized, 54% men (83,796), and 46% women (70,782). A total of 59,046 smallholder farmers (45% of women) were made aware of and trained in WBI. Among them, 21,293 farmers who were trained had shown a great amount of interest by subscribing to a bundled WBI product for which they would be paying a monthly fee (52% of these insured farmers were women) (Oluyede & Mariam, 2019).

e-Granary is a new platform raised by the Eastern Africa Farmers Federation (EAFF) in Uganda, as well as Eastern Africa in order to improve the way in which farmers engage in business-related activities. This platform mainly seeks to use mobile services in order to fetch a better piece for the produce of the farmers (World Bank, 2017).

8.6.1.1 Kenya

Kenya is a country that houses around five million farmers comprising of small landholders, as well as big industrial agriculturalists. AkiraChix, a female team of developers, came up with a mobile service known as M- Farm, which made it possible to connect farmers with each other leading to the development of Kenya's agriculture sector. This service also has a connection with the government and many NGOs so that an efficient connection could be established with farmers (World Bank, 2017).

Kilmo Salama, known as "safe farming" in Kiswahili, is a crop insurance program that is being operated by UAP Insurance, Safaricom, and Syngenta Foundation. Its main functioning is to deliver crop insurance in rural Kenya to small and marginal farmers by using mobile phones. The insurance is given to farmers when they buy seeds and fertilizers from registered vendors. This system mainly relies on climatic information from weather stations located in agricultural regions. So when hard times prevail, with the aid of M-PESA mobile money services, the insured customers are paid automatically by this service. Thus, it truly demonstrates how efficiently the farmers are protected and efficiently mitigated from risks associated with crop loss (World Bank, 2017).

8.6.1.2 Ghana

Mark Davis, the manager of Ghana's largest ICT center, BusyLab, as well as a successful British Technology entrepreneur, set up Esoko, which began initially as TradeNet, under the private initiative, in 2005. It is a service related to market information and provides all the necessary information regarding prices of agricultural commodities, as well as creates a virtual marketplace that is accessible through mobile phones and the Internet for buyers and sellers. It has emerged as the most successful ICT based agricultural service in Africa, which operates in about nine different African countries. The website of Esoko is http://www.esoko.com/, and all its activities revolve around the usage of mobile phones. The four key services provided by Esoko are live market feeds, direct SMS marketing and extension, scout polling, and online profiling and marketing. Real-time market information can be exchanged by participants throughout agricultural value chains. Using the simple feature of bulk text messaging the associations and governments can share critical information with thousands of farmers and this keep them updated with all the necessary details pertaining to market-related information like current demands, seed and fertilizer outlets, and market prices. It is a form of bulk texting wherein important information is shared by government and associations to many using this efficient and simple mobile features. Users can find relevant resources and many members who would want to buy and sell various kinds of agricultural products. Esoko also offers services like training and strategy sessions, which enables one to use this platform, as well as provide customer services for farmer groups (World Bank, 2017).

Farmerline was launched in 2013 to help the farmers increase the quality of their life in Ghana. It has reached over 5000 small-scale farmers across rural Ghana and also more than 2 lakhs through partner organizations in Cameroon, Nigeria, and Sierra Leone. Its impact was the considerable increase in farmers' yields (up to 55 %), as well as income (up to 44 %). Voice messaging technology is one of its key features to make communication channels more efficient in order to empower women, and thereby, realizing greater crop yields and good returns, create an easy marketing network that would boost the marketing skills of the women and more importantly, make the right decisions regarding their health. Women Advancing Agriculture is an initiative by Farmerline in Ghana that has increased its gender equality by making information more available to women community. This initiative functions by sending educational voice messages to the mobile phones of women agricultural workers. This process creates awareness in them of the trending agricultural practices, market prices, and weather forecasts together with information related to family planning, maternal health, and financial literacy. The prime importance is given not only to increase women farmers' productivity but also to mitigate their unequal access to all these rich information (World Bank, 2017).

8.6.1.3 Tanzania

The Tanzanian government had launched several important policies which recognizes the importance of ICT tools for effective spreading of information in agriculture through mobile phone introduction (URT, 2013). Digital Early Warning Network is a mobile service initiative by the Great Lakes Cassava Initiative in Tanzania's Lake Zone, wherein farmers from 10 districts had been trained in all the skills pertaining to identify symptoms of cassava brown streak disease and cassava mosaic disease. The farmers use their mobile phones in order to send messages to researchers regarding disease incidence and, in turn, receive suitable control measures from them (World Bank, 2017; Berta et al., 2020). In sub-Saharan Africa the model for the rural development was promoted through sustainable intensification (SI), and their test with 97 farmers in Tanzania found that farmers actively involved with the service to access agricultural advice and extension agents were able to answer questions with reduced workload compared to conventional communication channels. The study highlighted how User-Centered Design could be used to develop information services for complex and resource-restricted smallholder farming contexts.

8.6.1.4 Botswana

The Livestock Identification Trace-Back System, developed by Inala Identification Control (IIC) in South Africa, has emerged as one of the largest networks of innovative forms of ICT for animal husbandry revolving around 300 million cattle. This system utilizes radio-frequency identification (RFID), which meets many purposes include the beef import requirements for the European Union (EU), which is the main terminal for 80 % to 90 % of Botswana's beef exports. This system also provides veterinary services and also aims for livestock health (World Bank, 2017).

8.6.1.5 Uganda

The Technical Center for Agricultural and Rural Cooperation (CTA) with partnership from various organizations like the Alliance for a Green Revolution in Africa (AGRA), aWhere Inc., the East Africa Farmers' Federation (EAFF), Environmental Analysis and Remote Sensing (EARS) Earth Environment Monitoring (EARS-E2M), the eLEAF Competence Center (eLEAF), and Mercy Corps, Uganda, has implemented Market-Led, User-Owned ICT4Ag-Enabled Service (MUIIS) that supports the extension and advisory services for the farmers using the power of setallite data. The MUIIS project functions by making available all the necessary information regarding crop management and climatic risks to the farming community. One of its striking features is the proof-of-concept function, which aims to increase the farmers' productivity by about 40 % through satellite data-enabled extension and advisory services (World Bank, 2017).

8.6.2 Asia

8.6.2.1 Bangladesh

An SMS Gateway developed by FAO has been very crucial in reducing the spread of H5N1 HPAI avian influenza through poultry farms across Bangladesh with an efficient mechanism of interaction between the farmers and concerned experts. This fact has supported small scale farmers on a huge scale by providing sustainable resilience. In addition to this, reports regarding animal disease outbreaks, as well as the exchange of other types of information, take place at a faster pace in an efficient manner (FAO, 2016).

Services offered by Bangladesh Institute of ICT in Development (BIID) (https://www.biid.org.bd/).

- e-Krishok is an initiative that mainly centers its services around ICT-enabled agricultural products and services, and it includes a portal that is both advisory and informative in function as a network of local information centers connects it. It integrated the Farmbook business planning solution, the Zero Cost phone line, and extension into one big pool of information.
- An Irrigation Scheduling Application is being developed, which is being collaborated by University of Twente (the Netherlands) and CIMMYT.
- In October 2015, Zero Cost Extension and Advisory Service was launched in association with the Katalyst, a development agency funded by multiple donors and Bangladesh Seed Association (BSA). The main principle driving this service is that it is of prime importance to easy access to advice, as well as to promote the use of quality agricultural inputs. Zero Cost indicates that the extension and advisory services are being offered free of cost for the farmers through mobile phones (World Bank, 2017).

8.6.2.2 Turkey

The government of Turkey, through the Agriculture Directorate of Kastamonu Province, initiated a project for local weather forecasting, which had a collaboration with international donors. This project has focused on increasing productivity by monitoring climatic conditions, diseases, and pests, keeping in view the microclimatic conditions that are favorable for them. In one case study related to frost damage, it has been reported that not even a single farmer out of 500 suffered crop loss from frost, whereas those farmers, to whom the service was not extended to, did experience the damage (World Bank, 2017).

8.6.2.3 Malaysia

A Web-based GIS provides information on precision farming in the Sawah Sempadan rice-growing area. This interactive system allows farmers to explore the information related to rice cultivation in their area. It is also very cost-effective as it uses open-source software. Farmers have the advantage to print information regarding fertilizers, historical data related to yield per plot in the previous season. This fact helped the farmers in analyzing and devising the best strategy for the next growing seasons (World Bank, 2017).

8.6.2.4 Japan

Akisai cloud is one such database that has been field-tested by Fujitsu Group for food and agriculture. Fujitsu's service is very peculiar and is designed to provide a pillar of support to all aspects regarding agricultural management and research. Cloud's computing being a Saas-based solution, is extended to support various activities related to agriculture. This initiative has come up with measurable results leading to notifying changes in companies' working patterns, as well as training of farmers (World Bank, 2017).

8.6.2.5 China

Construction and Popularization of Agriculture Info-Service System, a project based in Anhui Province, China, set a perfect example of how ICT has enhanced farmer organizations' access to knowledge. It was launched by the Hefei Institute of Physical Sciences of the Chinese Academy of Sciences and is being funded by the World Bank. It is characterized by three main features: an Internet portal, information assistants, and information dissemination models. Its main target is the specialized farmer's cooperatives, which is a pivotal force for agricultural development in China (World Bank, 2017).

SOUNONG is an Internet search engine, developed by the Institute of Intelligent Machines, which assembles information and makes it available to the farmers' cooperatives in a meaningful manner. It maintains multilateral collaboration with China's official agricultural websites, which have higher user rates and are more authorized in promoting information. This engine governs over the activities of nearly 7000 websites per day, integrating nearly all of China's agricultural data. These sites cater to information regarding prices of wholesale farm products, prices in 9000+ markets, and prices for 20,000 types of agricultural commodities. Information is also made available from a number of databases regarding climate, crop species, and pest and disease diagnostics (World Bank, 2017).

8.6.3 Europe

8.6.3.1 Italy

The Open System for Tractor's autonomous Operations (STRATOS) has demonstrated the application of information and communications technology (ICT), in particular ISOBUS and wireless communication technologies, to agricultural applications by developing an open System for tractor's autonomous Operations. The STRATOS supports collecting information on soil and cultivation. The tractor is fully equipped with five sensors: humidity, temperature, soil pH, barometric pressure; and, acceleration (Zuniga-Soto, 2018).

8.6.3.2 Denmark

Modern agricultural machinery has degenerated soil quality with traffic. The Preparing for the EU Soil Framework Directive by optimal use of information and communications technology across Europe (PredICTor) project had as an online decision support tool in order to deal the risk of compaction for an intended field traffic situation with the help of Terranimo that is a web-based computer model which has much scientific application and has been primarily designed for farmers, agricultural contractors, consultants, and enforcement authorities (Zuniga-Soto, 2018).

8.6.3.3 Germany

3D-Mosaic is a support system designed for precision management of orchards using the concept for a decision support system (DSS), which aims at optimizing the efficiency of inputs, as well as mitigating the environmental necessities of fruit production. The concept has been very well demonstrated with stakeholders to ensure an integrated approach using robotics, sensors, geostatistics, plant physiologists, and horticulturists in managing orchards. It has made possible for the scientific community to practice precision horticulture (Zuniga-Soto, 2018).

8.6.4 North, Central and South America

8.6.4.1 Chile

Mobile Information Project (MIP) is a system that uses software evolved by DataDyne, a non-profit organization (US-based), and has played a prominent role in delivering necessary agricultural information from the Internet directly to farmers in Chile. This software functions by organizing related content from the Internet into ready news feeds, and these are transmitted to the farmers using SMS. The plus point of this system is that it can function over slow networks with intermittent connectivity and works on low-cost mobile phones, which cost only 15–20$ in Chile (World Bank, 2017).

8.6.4.2 Mexico

A nitrogen sensor developed by The International Maize and Wheat Improvement Center (CIMMYT) and tested on 174 wheat plots in Mexico's Yaqui Valley, under collusion with the State of Sonora, Oklahoma State University, and Stanford University. It is a hand device operated by an infrared sensor that traps light in order to amount for the biomass, as well as the chlorophyll content. These matters indicate the amount of nitrogen and fertilizers required by a plant. The main advantage associated with this technology is that it not only brings down the costs but also reduces the risk of environmental damage (World Bank, 2017).

8.6.4.3 Caribbean

The Caribbean Farmers Network (CaFAN) has come up with the fact that the farming community in the Caribbean region prosper very much by functioning as clusters that are either geographically or thematically created (Greene, 2010). Through such close working, or with those who share a common interest, can set up a cluster to share their experiences regarding technical information, devising a plan for new market demands, as well as improve their promoting and bargaining skills. CaFAN has encompassed a 30 member organization, which together comprises half a million farmers in and around 12 countries. One key principle behind creating clusters is that clusters overcome membership boundaries. Farmers use computer and Internet services like Skype, email, and the CaFAN website to communicate with each other. Text messages make it possible to communicate with farmers directly and also make it possible to share product information across the farming community (World Bank, 2017).

8.6.4.4 Dominican Republic

Digital orthophoto quads (DOQs) is a system that not only provides digital maps but also makes it possible to create water databases that are very important for an efficient irrigation system. This database provides all the necessary information related to sparsely and heavily irrigated locations, drainage problems, water use statistics, and even issues related to salinity. An example quoted in this regard is that of a database that was highlighted in a program to improve users' management of irrigation systems (PROMASIR) in the Dominican Republic, which was supported by the Inter-American Development Bank and Utah State University. The DOQs were combined with other necessary information like information on property ownership. This fact has enabled water users to get information about other water users, obtain estimates of irrigation water demands, observe property boundaries, as well as review monthly crop and water statistics. Users are made available with more authentic information to be used when their infrastructure is being updated, as well as tackling potential maintenance problems. Such systems' major benefit is that it helps in overcoming conflicts over water as all the users have access to the same precise information, such as plot size and price information (World Bank, 2017).

8.6.4.5 Jamaica

The Rural Agricultural Development Authority (RADA) of Jamaica has come up with the Agricultural Business Information System (ABIS), which has become its flagship ICT effort. ABIS gives a more efficient business outlook to agriculture by skillfully managing agricultural information. It is the principal source of information for the agricultural sector, as well as advisory services. The data contains all the necessary information related to registered farmers, stakeholders, technical information, livestock, crops, as well as farm practices. It keeps track of various activities on the field like managing pests and diseases, weather forecasting, buyers' and seller's chain-links of agricultural commodities, and the production is monitored with the help of its Agricultural Resource Planning Tool (World Bank, 2017).

8.6.4.6 United States of America

In the United States of America, John Deere, the agricultural firm, has shown the utility of remote sensing, big data, and crowdsourcing can be integrated to provide farmers with the best possible means to increase efficiency and productivity. The entire process is summed up as follows: The data is collected from the farmers' sensors by the company's online portal. The collected data, from thousands of farmers, are pooled and then sourced with external data sets such as weather data. The system is featured with an advanced technology that can predict when machinery is likely to break and notify a nearby parts distributor immediately to replace that part (World Bank, 2017).

8.6.5 Australia

In Australia, the next big wave of productivity in agriculture is expected to occur due to better utilization of ICTs, including digital technologies that simplify the electronic capture, processing, storage, and information exchange. About 96 % of Australian farmers used ICT tools, and 95 % had an Internet connection. While vegetable and other cereal-based farms used GPS-enabled know-hows, dairy farms utilized electronic ID and herd supervision tools (Duffy and Jackson, 2018).

Mobile applications like eLocust3 are being used globally to monitor the desert locust, which is one of the most notorious pests. This application combines information, communication, and satellite technologies into a unified monitoring and detection system, thereby significantly reducing the severity of locust plagues in Asia and Africa. Sousa, Nicolay, and Home (2016) reported that utilization of video calling features could improve the broadcasting of knowledge, especially in the case of the farmer to farmer communication.

8.7 General challenges in using ICTs

Inconsistent power supply, network fluctuations, high cost of ICT infrastructure, rural farmers' low income, rural areas require policies to boost ICT development and need essential skills to use the technologies (Agwu & Mercy, 26AD). Although there was a positive outcome from the study conducted by Kansiime et al. (2019), they also reported some of the limitations such as women, older persons, and small-scale farmers were less likely to be benefitted by ICTs, especially mobile-based information owing to less smart-mobile ownership and literacy issues. Using media, information management, and ICT are not being utilized properly. Extension workers cannot transfer all the available technology from the lab to the land. The agricultural information literacy is lagging far behind the horizon, which becomes a big hurdle in updating farmers with the latest information. This, in turn, has become a major reason for poor technological knowledge for both farmers and village-level extension personnel.

8.8 Conclusion

If the major constraint, i.e., clarification of doubts is effectively addressed, the perception, utility, and usefulness of ICT tools can be explored and exploited to the full extent. Digital platforms that permit a two-way flow by ideas and technologies in a cost-effective manner play an important role in the transfer of technologies effectively to reach the grass-root level. The digitalization of agriculture will prove to be a paradigm shift in farming and food production over the coming years. Potential associated challenges like differences in accessibility are being addressed through improved connectivity, which is particularly beneficial to smallholder farmers who faced a digital divide in terms of farm productivity, economic, and social integration. The diffusion of DES in rural India is increasing with advances in technologies and governing policies. DES can change the ideas, activities, and knowledge of the farmers, thus empowering them to adopt proper actions as and when needed. Agriculture has come a long way and established several breakthroughs in terms of production and productivity. Hence, the use of DES in agricultural extension is essential and impactful to improve supply chain management.

References

Addison, C. (2019). GODAN action: Digital capacity building. *ICT Update, 93*, 14–15.

Addom, B. (2019). Super platforms- going beyond bundling digital solution. *ICT Update, 93*, 8–9.

Agwu, A. E., & Mercy, N. U. (26 September 2019). Challenges and opportunities for ICT adaptation in agricultural extension. *ICT Update Newsletter.* https://ictupdate.cta.int/en/article/challenges-and-opportunities-for-ict-adoption-in-agricultural-extension-sid03b7c751a-f2db-48c7-a5ed-40344e00e00a.

Ajayi, A. O., Alabi, O. S., & Okanlawon, B. I. (2018). Knowledge and perception of farmers on the use of information and communication technology (ICT) in Ife-Central Local Government Area of Osun State: Implications for rural development. *Journal of Agricultural Extension and Rural Development, 10*(3), 44–53.

Araji, A. A., Sim, R. J., & Gardner, R. L. (1978). Returns to agricultural research and extension programs: An ex ante approach. *American Journal of Agricultural Economics, 60*, 964–968. https://doi.org/10.2307/1240129

Balasubramanian, K., & Daniel, J. (2010). Knowledge transfer for a horticultural revolution: The lifelong learning for farmers model. In *Paper presented at the 28th international horticultural congress, lisbon.*

Batchelor, S., Scott, N., Manfre, C., & Edwards, D. (2014). Is there a role for mobiles to support sustainable agriculture in Africa?. In *2nd International conference on ICT for sustainability* (pp. 272–280). Stockholm, Sweden: ICT4S.

Baumüller, H. (2018). The little we know: An exploratory literature review on the utility of mobile phone-enabled services for smallholder farmers. *Journal of International Development, 30*(1), 134–154.

Bell, M., & Payne, J. (2011). ICT options in relation to extension functions. *MEAS*. Available at www.meas-extension.org/resources/ict.

Benina, S., Nkonyaa, E., Okechob, G., Randriamamonjya, J., Katoa, E., Lubadec, G., et al. (2011). Returns to spending on agricultural extension: The case of the National Agricultural Advisory services (NAADS) program of Uganda. *Agricultural Economics, 42*, 249–267. https://doi.org/10.1111/j.1574-0862.2010.00512.x

Berta, O., Jonathan, S., Carlos, F. Q., Van de Gevel, J., Happy, D., Majuto, G. M., et al. (2020). User-centred design of a digital advisory service: Enhancing public agricultural extension for sustainable intensification in Tanzania. *International Journal of Agricultural Sustainability*. https://doi.org/10.1080/14735903.2020.1720474.

Cecchini, S., & Raina, M. (2004). Electronic government and the rural poor: The case of Gyandoot. *Information Technologies and International Development, 2*(2), 65–75.

Dhaka, B. L., & Chayal, K. (2010). Farmers' experience with ICTs on transfer of technology in changing agri-rural environment. *Indian Res. J. Ext. Educ., 10*(3), 114–118.

Duffy, N., & Jacbkson, T. (2018). *Information and communication technology use in Australian agriculture: A survey of broadacre, dairy and vegetable farms, research report*. Australia: Department of Agriculture and water resources.

Fabregas, R., Kremer, M., & Schilbach, F. (2019). Realizing the potential of digital development: The case of agricultural advice. *Science, 366*, 6471.

FAO. (2016). *SMS Gateway: Improving animal health through information and communication technologies*. Information factsheet.

FAO. (2017). *The future of food and agriculture – trends and challenges* (Rome).

Ferris, S., Robbins, P., Best, R., Seville, D., Shriver, J., & Wei, E. (2014). *Linking smallholder farmers to markets and the implications for extension and advisory services*. Brief #4 MEAS.

Greene, J. (2010). The bridge between farm and market. *ICT Update*. http://ictupdate.cta.int/en/Regulars/Perspectives/The-bridgebetween-farm-and-market.

http://landrecords.karnataka.gov.in/bhoomiweb/.

http://www.esoko.com.

https://www.digitalindia.gov.in/.

https://www.un.org/development/desa/en/.

www.ikisan.com.

https://www.icow.co.ke/.

http://rural-emarket.com/en/.

https://m-shamba.net/.

https://apps.apple.com/us/app/sirrus/id684978309.

https://farmlogs.com/.

https://www.farmersedge.ca/.

https://www.eldersweather.com.au/.

https://www.trapview.com/v2/en/.

https://play.google.com/store/apps/details?id=com.invasiveanimals.feralscan_pest_mapping&hl=en_IN&gl=US.
https://tracxn.com/d/companies/agrentools.com.
https://www.tigersul.com/.
https://play.google.com/store/apps/details?id=com.hypelabs.agrobrazil&hl=en_IN&gl=US.
https://www.agritempo.gov.br/agritempo/index.jsp.
https://play.google.com/store/apps/details?id=com.aas.meghdoot.
https://play.google.com/store/apps/details?id=com.IFFCOKisan.
https://play.google.com/store/apps/details?id=com.app.khetibadi.
https://play.google.com/store/apps/details?id=com.criyagen.
https://play.google.com/store/apps/details?id=com.ermilogic.dat.
https://play.google.com/store/search?q=agmart%20app&c=apps.
https://play.google.com/store/apps/details?id=com.oyepages.zbnf.
https://play.google.com/store/apps/developer?id=Dr.+Vishwanath+Koti.
https://platform.cabi.org/projects/our-impact/d2f/.
https://enam.gov.in/web/.
https://www.itcportal.com/businesses/agri-business/e-choupal.aspx.
https://www.farmathand.com/.
https://play.google.com/store/apps/details?id=agri.live.
https://www.biid.org.bd/.
https://unfccc.int/resource/mfc2017/project.html?p=project-16.
https://play.google.com/store/apps/details?id=com.tene.eSAP.
https://play.google.com/store/apps/details?id=com.maswadkar.digitalmandi.
https://www.itu.int/en/ITU-D/Statistics/Documents/publications/misr2016/MISR2016-w4.pdf.

Jafri, A., Dongre, A., & Tripathi, V. (2002). Information communication technologies and governance: The Gyandoot experiment. In *ODI working paper 160*. London: Overseas Development Institute (ODI).

Kafura, R., Islam, M., Safiul, I., Prodhan, F. C., & Dipanwita. (2016). Use of ICT as extension tool by the farmers of Gazipur district in Bangladesh. *Indian Research Journal of Extension Education, 16*(2), 1–5.

Kansiimea, M. K., Alawya, A., Allenb, C., Subharwalc, M., Jadhavd, A., & Parr, M. (2019). Effectiveness of mobile agri-advisory service extension model: Evidence from Direct2Farm program in India. *World Development Perspectives, 13*, 25–33.

Lokeswari, K. (2016). A study of the use of ICT among rural farmers. *International Journal of Communication Research, 6*(3), 232–238.

McKinsey Global Institute. (2019). *Digital India: Technology to transform a united nation* (p. 144pp). Report by MGI.

Meena, M. L., Sharma, N. K., & Aishwarya, D. (2011). Role perception about information communication technology among farmers. *Journal of Communication Studies, 29*(1), 98–105.

Mesengezi, C. (2019). GODAN action: Digital capacity building. *ICT Update, 93*, 12–13.

Mittal, S., Gandhi, S., & Tripathi, G. (2010). *Socio-economic impact of mobile phone on Indian Agriculture. ICRIER Working paper No.246*. International Council for Research on International Economic relations.

Nagesh, N. S., & Saravanan, R. (2019). *Impact of ICTs on agriculture growth and development case studies from Karnataka Region, Discussion Paper 9, MANAGE-Centre for Agricultural Extension Innovations, Reforms and Agripreneurship*. National Institute for Agricultural Extension Management (MANAGE), Hyderabad, India.

Nagalakshmi, G., & Narayanaswamy, B. K. (2011). Perception, awareness, attitude and knowledge of extension personnel about information communication technologies. *Mysore Journal of Agricultural Sciences, 45*(2), 421–426.

Naik, J. B. (2018). *A study on ICT tools usage by the farmers in Anantapur district of Andhra Pradesh*. M. Sc Thesis (Agric. Extension). India: Acharya N.G. Ranga Agricultural University.

Oluyede, A., & Kadzamira, M. (2019). Weather-based index insurance: A climate-smart agricultural solution for smallholder farmer. *ICT Update, 93*, 16–17.

Raghuprasad, K. P., Devaraja, S. C., & Gopala, Y. M. (2012). Attitude of farmers towards utilization of information communication technology (ICT) tools in farm communication. *Research Journal of Agricultural Sciences, 3*(5), 1035–1037.

Rambaldi, G. (2019). Drone-based services taking off to transform Africa's agriculture. *ICT Update, 93*, 10–11.

Rose, D. C., Sutherland, W. J., Parker, C., Lobley, M., Winter, M., Morris, C., et al. (2016). Decision support tools for agriculture: Towards effective design and delivery. *Agricultural Systems, 149*, 165–174.

de Silva, H., & Ratnadiwakara, D. (2008). *Using ICT to reduce transaction costs in agriculture through better communication: a case-study from Sri Lanka*. LIRNEasia.

Sousa, F., Nicolay, G., & Home, R. (2016). Information Technologies as a tool for agricultural extension and farmer to farmer exchange: Mobile one video use in Mali and Burkina Faso. *International Journal of Education and Development Using Information and Communication Technology, 12*(3), 19–36.

Stringfellow, R., Coulter, J., Lucey, T., McKone, C., & Hussain, A. (1997). *Improving the access of smallholders to agricultural services in sub-Saharan Africa: Farmer cooperation and the role of the donor community Natural Resource Perspectives 20*. Overseas Development Institute.

Subhashsingh, P., Bharat, M., & Rai, D. P. (2010). Sustainable models of Information technology for agriculture and rural development. *Indian Research Journal of Extension Education, 10*(1), 20–23.

United Republic of Tanzania (URT). (2013). *National Agriculture Policy, Ministry of Agriculture Food Security and Cooperatives Dar Es Salaam*. Tanzania October. http://www.faoilo.org/fileadmin/user_upload/fao_ilo/pdf/ICA_MLW_and_TZ/NATIONAL_AGRICULTURAL_POLICY-2013.pdf.

Vishwatej, R. (2013). *Awareness, accessibility and utilisation pattern of Information and communication technology (ICT) projects by farmers of Belgaum district*. M.Sc. Thesis. Dharwad, Karnataka (India): Univ. Agric. Sci.

World Bank. (2017). *ICT in agriculture: Connecting smallholders to knowledge, networks, and institutions*. Updated Edition. Washington, DC: World Bank. https://doi.org/10.1596/978-1-4648-1002-2.

Zuniga-Soto, E. (2018). *Highlights of projects funded by the ICT-AGRI ERA-NET (2009-2014) and ICT-AGRI-2 ERA-NET (2014-2018*.

Chapter 9

Social acceptability of radical food innovations

Arnout R.H. Fischer, Ellen J. Van Loo
Wageningen University, Marketing, and Consumer Behaviour Group, Wageningen, The Netherlands

9.1 Introduction

The food market has long been hinging on two opposing stories: convenient and processed food versus authentic, traditional foods. Society has been fascinated by the idea of convenient, high tech foods, such as the food pills for consumers illustrated by the space-age Jetsons cartoon originating in the 1960s, which, although nightmarish to many, still captures interest (Belasco, 2000). Another, probably better known, imaginary high-tech food production system is Star-Trek's food replicator that can create any kind of traditional food and beverage by request. The food replicator is an adaptation of food-synthesizers from the Original Star-Trek series, which created cubes of nutrient-rich foods, not unlike the Jetson's food pills. The change from the original cubes of nutrients toward more traditional foods (according to the Star-Trek show a series of improvements between the 22nd and 24th centuries; Belasco, 2000) already indicates that society may not be ready for food pills or food cubes. This links to the second story where consumers and society view good food and farming practices in a romantic, traditional perspective (e.g., Mazzacano D'Amato & Falzon, 2015) of free-ranging animals and flowering alpine meadows.

This dual story puts pressure on food innovation systems, which likely influences the success and failure of radical innovations in agri-food businesses. However, even the romantic view on farming practices is likely the outcome of previous radical innovations in the food supply. Perhaps the most societally radical innovation has been the domestication of plants and animals, thought to have occurred at the end of the Pleistocene. The more predictable food stocks supported the permanent colonization of suitable habitats and the creation of static, permanent, and broader communities, which lead to radical changes in human societies (Diamond, 2002). Through

Food Technology Disruptions. https://doi.org/10.1016/B978-0-12-821470-1.00002-1

innovations that introduced more complex preparation methods like cooking and fermentation, humans gained access to otherwise inedible or indigestible foods and food ingredients (Hillman & Davies, 1990). These innovations radically changed the process of food production, preparation, and storage, and hence, changed the dietary patterns of humans. Not only in prehistoric times but also in historical times, important innovations radically changed society. The introduction of the moldboard plow in early medieval Europe allowed the European population to farm fertile but challenging to manage lands, such as those containing heavy clay, at manageable input of labor. It has been argued that the introduction of these plowing systems played a pivotal role in Western European recovery from the period of scarcity and population reduction after the collapse of the Roman Empire. The increased food production allowed the growth of cities from about the 10th century CE (Gimpel, 1992). Wind and watermills further provided energy sources required for these cities to flourish and develop economies based on mechanical power sources, making them completely different than those in the classical antiquity.

Radical innovations have continued to influence agricultural practices into modern times. The introduction of mechanized farming in the 19th century increased food production, and reduced labor demand in agriculture resulted in the movement of populations from rural to urban environments. In the early 20th century, the development of internal combustion engines led to the development of tractors and other mechanization of agriculture. At the same time, chemical fertilizers and pesticides were developed, which heralded the "green revolution" (Evenson & Gollin, 2003). The innovations during the green revolution resulted in crops being produced consistently in volumes unparalleled in history. The consistent crop quality and sufficiency reduced the probability of famines and hugely increased yields on arable lands; thus, breaking the then-current Malthusian predictions about the food capacity of the earth. The green revolution allowed for much of the world population growth in the 20th century.

The efficiency advocated and achieved by introducing mechanized agriculture, and consistent crop support resulted in food becoming a commodity market, where large volumes of very similar resources are pooled and shipped around the world with small profit margins. To increase profit margins, identification of a specific niche or creation of a brand to distinguish products from competitors has been an ongoing effort by food producers. Food has become one of the markets where brands and brand image provide the most added value. The brand name Coca-Cola for example, originated in 1886, and through consistent advertising campaigns, has been associated with the happy American family life. The creation of Christmas advertisements dressing Santa Claus in Coca-Cola company's colors, red and white, has even shaped the American, and later global image of Santa Clause (Mckay, 2008). The Coca-Cola brand is considered a highly valuable asset of the Coca-Cola company (Fischer & Himme, 2017), with the value of the name alone being estimated between 50 and 80 billion US$ in 2010 (depending on sources and estimation methods). More recently, the primary production of agriculture has also started to diversify and trying to create specific niche markets for their specific products outside commodity streams (Meulenberg & Viaene, 2005), for example, by creating branded vegetables and fruits such as the Zespri kiwi fruit. This approach requires radical reorganization of resource streams as niche products need to be kept separate from commodities.

The radical and often transformative innovations that agricultural production and marketing have gone through has not always been unopposed. The Luddite movement of the late 18th and early 19th century included farmworker protests voicing their concern about changes in traditional employment and ways of living driving farm laborers to seek employment in industries (Sale, 1996). These protests were, in the light of labor conditions and pollution of the major cities at the time of the industrial revolution, not necessarily unwarranted. Nowadays, the term Luddite remains in use as a generally derogatory term to refer to individuals or groups that oppose innovation. While famines and the world wars of the early 20th century raised the need for innovation to ensure food security, by the time the high-tech utopia as fantasized in the Jetsons and Star-Trek came to the fore, societal concerns were once again gaining attention. Concerns about the harmful effects of agrochemicals on human health and nature increased signaled by the publication of "the silent spring" by Rachel Carson in (1962). Carson used the metaphor that the increase in pesticide use resulted in fewer bird sounds heard in agricultural areas. This book is considered as a landmark of environmentalists' movement in the latter half of the 20th century. More recently, concerns have been expressed about the replacement of natural ecosystems such as the replacement of natural rainforest by monocultures of soy or palm crops. Other societally controversial innovations in the agri-food sector have led to societal protest, to the extent that these technologies suffered delayed introduction and were introduced at much more limited scales than initially envisaged. For example, the introduction of irradiated foods relatively close in time to the widespread destruction of crops as far away as Sweden (~1500 km) to counter the radiation by pollution after the 1986 Chernobyl accident resulted in general public rejection of irradiation of food (for example, see: Bruhn, 1995). The development and introduction of genetically modified foods and crops where all primary benefits accrued to multinationals, pesticide producers, and other actors in the production chain resulted in significant protests and stringent regulations initially in Europe and later elsewhere (Frewer et al., 2013, among others). The food industry has been trying to buy in on the trend for natural rather than high tech food in society. Images of grandmothers preparing traditional foods or rainforests where traditional local agriculture provides sought after resources are some of many romanticized claims of authentic, low tech, traditional production in food. Brands like Unilever's Bertolli or Associated British Food's Patak gain much of their value from the attention to traditional products and are advertised with direct reference to traditional agriculture and authentic preparation methods. Other brands have also been increasingly capitalizing on the growing societal demand for more authentic products (Askegaard & Brogård, 2016). Over the last decades, such authenticity branding has played an essential role in de-commodifying food markets. This increased attention to authenticity has been an incremental process making it more difficult for radical innovations that cannot be aligned with images of authenticity to enter the market.

While radical innovations in food production systems have been ongoing since the dawn of time, in the current era, resistance to further acceptance and adoption of such technologies in the production of foods remains substantial. Nevertheless, radical innovations in the food system are emerging. Internet technology allows consumers and citizens to access information and sometimes even video footage of production systems, Internet marketing allows for new ways of ordering foods and keeping track of consumer stocks, and despite some concerns, consumers do not seem reluctant about these innovations.

Before providing an overview of the ways how society deals with disruptive and radical innovations, first, there is a need to clarify some of the terms in use. Much of the literature on the acceptance of innovations and new products or services based on technological innovations is about acceptance. However, the word "acceptance" is used in various ways (Fischer & Reinders, 2016; Ronteltap, Fischer, & Tobi, 2011). In marketing and product innovation literature, the term "acceptance of a novel product" is generally used to indicate that the innovation has become part of society and is actively being used (Rogers, 1962/1995). In the public understanding of science and risk analysis literature, "acceptance of an innovation" relates more to tolerance (cf. Taebi, 2017) of the innovation, which is not so much acceptance in daily behavior, but more a tacit or psychological acceptance indicating that people do not categorically reject the innovation (see, e.g., the studies of Bredahl, 2001; Schenk et al., 2011; Siegrist, 2000). Regardless of whether acceptance is considered actual use of an innovation or the psychological acceptance of that innovation, in both cases, the underlying assumption is that there is an innovation that enters society, and only after its introduction society accepts or rejects the innovation. Alternatively, to put it more directly, the burden of accepting the innovation lies with society. This fact may easily lead to "blaming" society for not accepting an innovation deemed brilliant in the mind of the developers of that innovation. This approach to public or societal acceptance, prevalent in innovation acceptance literature, foregoes the fact that to be accepted, an innovation needs to be acceptable (Taebi, 2017). Although the linguistic difference between acceptance and acceptability seems small, we should not underestimate the consequence of the subtle difference. Acceptability, in contrast to acceptance, is not an act of society, but a built-in property of the innovation. It cannot be achieved by merely informing or educating consumers, but must be built into the fabric of innovation. The burden of creating a societally acceptable innovation lies with developers, not the public. How to create such acceptable innovations can still be informed by the broader literature on acceptance to identify what society is sensitive about, but then, this needs to be translated into the innovation itself.

The challenge is to align with the demand by consumers and society for foods and food services that fit in their modern lifestyles on the one hand and align with the demand for natural and authentic products and production chains on the other hand. It is essential to consider societal drivers, general scientific knowledge about consumer and societal response to innovations, and then reflect upon some current innovations. This chapter aims to provide such an overview.

Therefore, we will briefly introduce some drivers that influence consumers in 2020 to make choices, change choices, or decide on the desirability of innovations in the food market. Reference to a specific time frame is related to the trend that, especially in developed countries, high levels of food security and food safety combined with rising welfare levels resulted in more focus on health and sustainability since the 1990s (Kambewa, 2007). The emergence of the Internet combined with the rapidly decreasing cost of sensors to measure personal health and nutritional needs and the emergence of new generations of the millennials and Generation Z, for whom sustainability is central to their lifestyle, has opened the food market for disruptive developments, discussed throughout this book.

9.2 Consumers in the early 2020s

To determine societal acceptability, we need to consider that what may have been acceptable to our parents' parents may not be acceptable to us. The view of what is acceptable slowly changes in society, and this is often related to the time in which one is raised, i.e., the cohort one belongs to (Buss, 1974). In the early 2020s, the **Millennial** generation (born between approximately 1985 and 2000) has become a dominant force in the marketplace. The millennials, next to **generation X**, the generation preceding the millennials with birth years starting from 1965 and the **baby-boomers** (born ~1945–1965) complement (the majority of) the adult population. The next generation, **"Generation Z"** (born from about 2000), is starting to mature and gaining influence. The delineation between cohorts differs between definitions and is not strict. Those born in 1965 and 1966 are often classified in different cohorts but (obviously) have more in common than those born in 1966 and 1984 (classified in the same cohort).

Nevertheless, considering cohorts as distinct can provide some insights into how society responds to innovations. To understand how society will respond to innovations, considering these cohorts in the light of technology generations and the years when people develop their character, worldview, and personality (roughly until their mid-20s) would be useful. In terms of technology generations, baby-boomers were raised in the precomputer age and were exposed to computers only in later life. While baby-boomers may have embraced computers and the Internet, this is not recognizable from their youth. Generation X can be considered the first computer generation raised during the emergence of programmable (home) computers and to whom computers are a visible part of life (Dimock, 2019). Millennials are the first Internet generation (Dimock, 2019) raised when the Internet was ubiquitous, and people started to be connected from a young age (Sackmann & Winkler, 2013). Baby-boomers were raised during Cold War tensions between the USSR and the West. However, the baby-boomers experienced the thriving economies and countercultures of the 1960s and watched the first human walking on the moon. Generation X was reaching adulthood during the recession in the 1980s and adopted a more cynical approach to society (Fishman, 2016). Generation X was also growing up when the Cold War ended, and the Berlin wall fell. The next cohort, Millennials, grew up during the booming economy of the 1990s, but also experienced the 9/11 attacks, the subsequent start of the war on terror and the 2008 economic crisis (Dimock, 2019). Millennials are shown to be the least quality and most prestige focused of the current cohorts (Moore & Carpenter, 2008), but the millennial cohort also contains substantial subgroups who focus on ethical consumption and cause-based purchasing (Bucic, Harris, & Arli, 2012). It remains hard to capture the identity of the newest generation, "Generation Z," as that generation is still developing and seems rather diverse. Generation Z is the generation that is becoming adult in a world where Internet is ubiquitous from a young age, and while their views are in many respects similar to these of millennials, they are demanding more government intervention in societal issues compared to previous cohorts (Parker, Graf, & Igielnik, 2019). How the substantial government interventions, including forced lockdown of many countries, the struggling health care systems, and the dramatic increase in distance learning and working from home during the 2020 COVID-19 pandemic will influence this generation's worldview is yet unknown. However, it makes sense to assume this will be an essential part of their formative experiences.

Population cohorts and the changing consumer positions influence more short-term consumer trends. Such trends can give insight into what societal priorities are relevant for current innovations to be societally acceptable. Ten global food trends for 2020 were identified by market research agency Innova (Williams, 2019). About six out of the ten main trends involve technologies or food contents that are a disruptive departure of traditional food markets. Consumers increasingly demand **transparency** about the provenance of ingredients, and they want to be able to verify this, which requires accurate and specific tracking and tracking technologies and may require redesign or even novel production methods. The consumer is **moving away from animal products to plant**-based alternatives, even if they are not vegans or vegetarians, which requires the development of high-quality alternatives to animal products. Society expects **companies to invest in sustainability**, reduce food and plastic (packaging) waste, which requires radical changes in the way how products and packaging are designed, and how waste is dealt with. The society also places a burden on consumers to better manage their food supply and limit household food waste. Individual consumers demand **foods that are healthy, convenient, and provide pleasure.** This trend, related to health, paves the way for a new generation of functional foods and the growth in demand for supplements, e.g., powdered protein drinks. Increased attention for pleasurable experience and convenience may further increase the demand for ready-to-eat meals, but also meal boxes and a shift to digital marketing of food. Increased focus on **named known ingredients,** as mistrust in artificial additives remains, society calls for clean labels with few ingredients which they know, can recognize and interpret their relevance (also see Lion, Willems, Fischer, & van Trijp, 2020). **Increased personalization or limited releases** asking for foods tailored to individual preference or even individual nutritional profile, which requires made to order foods and keeping track of personal preferences and needs.

9.3 Societal responses to radical innovation

9.3.1 Generic individual and cultural influences

Several theories have been developed to assess consumer response to innovations (Ronteltap et al., 2011). Among the most frequently used is the **Theory of Planned Behavior** (TPB) (Ajzen, 1991), which is a further development of the **Theory of Reasoned Action** (TRA) (Fishbein & Ajzen, 1975). The Theory of Planned Behavior posits that behavior follows from a behavioral intention (plan) to conduct that behavior; under the condition that people think they have, and they also actually have the control to engage in that behavior. Behavioral intention is determined by attitudes, social norms, and perceived behavioral control. Attitudes are summary evaluations in terms of an overall positive or negative opinion about an attitude object. Attitudes are based on beliefs about how product properties either positively or negatively contribute to the overall product performance. Social norms depend on beliefs and evaluations about social approval of conducting the behavior by relevant peers such as family, colleagues, and friends. "I have perceived behavioral control results from beliefs and evaluations of the likelihood that the behavior is under the control of the individual" (Ajzen, 1991). Perceived behavioral control has two roles in the model. Perceived behavioral control contributes to the formation of an intention: "If I do not believe I can do it, I will not form an intention." The second role relates to whether intention results in actual behavior: "If at the moment of action I do not believe I can do it, I will not follow my intention and will not execute the planned behavior."

While the Theory of Planned Behavior was not developed to investigate novel products or disruptive innovations, it is among the most frequently used theories and is applied straightforwardly to novel foods including potentially disruptive innovations (e.g., Saba & Vassallo, 2002) such as insect-eating

(Menozzi, Sogari, Veneziani, Simoni, & Mora, 2017). The Theory of Planned Behavior can accommodate the study of innovations because it includes the personal opinion (attitude) of the person forming an intention, but also includes their assessment of peer approval (or peer pressure) to engage in the behavior and an assessment of the likelihood they actually can engage in the behavior. This fact allows people to position innovation adoption behavior in the social context. The versatility and general applicability of the Theory of Planned Behavior also have downsides. Most importantly, while within the Theory of Planned Behavior intentions are generally well predicted, mainly by attitudes, the actual behavior is much less predictable, resulting in the still unresolved attitude-behavior gap (e.g., Vermeir & Verbeke, 2006; Yamoah & Acquaye, 2019). There are some reasons for this attitude-behavior gap. The first reason is that the Theory of Planned Behavior is often used for unplanned, automatic, or habitual behaviors. These unplanned behaviors are considered to be outside the scope of the theory (Verplanken & Orbell, 2003). A second reason is that in many cases, general attitudes, social norms, and perceived behavioral controls are used, which because of their general nature, are unlikely to predict specific behaviors well (Kaiser, Schultz, & Scheuthle, 2007). Kaiser further argues that if attitudes were to be measured at the same level of specificity, compatible with the behavior, the prediction of behavior improves a lot. In the case of disruptive innovations, however, we argue that consumer response to a large extent has to do with general information at first and that, particularly in the food domain, much of the behaviors are nonplanned. Therefore, the application of the Theory of Planned Behavior for disruptive innovations in food should be applied with care.

Another frequently used model to explain the response to innovations is the **Technology Acceptance Model** (TAM) (Davis, 1989). The Technology Acceptance Model shares many properties with the Theory of Planned Behavior. Like the Theory of Planned Behavior, it predicts the use of technology from attitudes and intentions toward using that technology. An essential difference between the Technology Acceptance Model and the Theory of Planned Behavior lies in the determinants of the attitudes. The Technology Acceptance Model was developed based on the first generations of programmable consumer products entering mass markets in the 1980 (most infamously VCR (video cassette recorder) systems). Before the 1980s, such complex systems were mostly reserved for professional experts or at best-dedicated hobbyists; hence, expert knowledge to operate the interface could be assumed. Once programmable VCRs entered the market, novice users became operators. To predict the kind of interfaces that would be accepted, the Technology Acceptance Model strongly focusses on Perceived Usefulness (i.e., does the apparatus deliver what I want of it) and Ease of Use of the technology. Since its original conception, the Technology Acceptance Model has been extended to a **TAM-2** and **TAM-3** with additional predictors of perceived usefulness and perceived ease of use (Venkatesh & Bala, 2008;

Venkatesh & Davis, 1996, 2000; Venkatesh, Thong, & Xu, 2012). TAM and its variants are widely used to explain the adoption of electronic devices and services with some success (Scherer, Siddiq, & Tondeur, 2019). Elements of the Technology Acceptance Model formed the core of a comprehensive model: the **Unified Theory of Acceptance and Use of Technology** (UTAUT) (Venkatesh et al., 2012; Venkatesh, Morris, Davis, & Davis, 2003). UTAUT aims to predict not only the original acceptance of new technology but also its continued use. It brings in social norms, demographics, and elements from other technology adoption models together with ideas of TAM. This fact resulted in a model with several dozens of independents variables. While this abundance of predictors necessarily improves expressive power, using UTAUT in practice results in having to measure so many variables to make practical use often infeasible. Besides this practical limitation, there can also be some doubt whether the addition of this many variables may lead to predictions that fit the observed data patterns quite well, but does not help in prioritization of prominent issues, or even worse, the abundance of predictors may obscure fundamental knowledge gaps related to core elements of the deliberation process (Bagozzi, 2007).

Whereas the TPB and TAM set the individual decision-maker at a central position, disruptive innovations unavoidably also contain a group dynamic. A central theory in this field is that of **Diffusion of Innovation** (Rogers, 1962/1995). This theory, originating in the 1960s, was based on how innovations diffused in society, and the initial studies included disruptive food technologies such as the mechanization of agriculture. Within the diffusion of innovations approach, different types of adapters are identified. *Innovators* are a small group of people or organizations that adopt an innovation when it is first available. At this stage of the innovation, it remains uncertain whether the innovation will succeed or fail, and innovators take some risk in embracing it. From innovation development, it would be useful to involve such innovators in the development process to ensure the innovation fits their expectations. *The lead user approaches* aim to bring in such early end-user input to codevelop products or services with the earliest users (von Hippel, 1986). Arguably, such lead users would most fit the innovator group or at least a second group of people adopting an innovation. This second group to embrace innovation is labeled as early adopters. *Early adopters* buy into the mature innovation, and if they appreciate the innovation, they can become opinion leaders. The emergence of vloggers and product placement in their Internet broadcasts capitalizes on the opinion leaders of this new generation of celebrities (Lin, Bruning, & Swarna, 2018). The next group to adopt are *early and late majorities*, at which stage the innovation becomes commonplace, followed by *laggards* who are among the last to adopt the innovation. While the studies by Rogers and the subsequently developed mathematical models to describe the diffusion across the different user groups (Bass, 1969) describe innovators as about 2.5% of the market, early adopters about 13.5% and both majorities roughly 34% leaving 16% for the laggards, there are some issues in using the distinction between user groups to predict the diffusion of innovation. The main issue is that generally, the transitions between innovators, early adopters, early and late majorities, and laggards are determined after the market is saturated. At the moment of an evolving market, it is generally not possible to estimate whether the end-users buying into a product is an early adopter (predicting rapid growth of future uptake) or early or even late majorities (predicting a decreasing growth in demand). This fact is because the market potential of innovation may not be known at first. Hopefully, it is more than a tiny niche group, but almost certainly it will fall shy of the population as a whole. As a consequence, much of the research on innovation diffusion takes a reflective approach, describing specifics of an innovation trajectory. While this is very relevant in understanding what happened afterward, the predictive power of the findings from such an approach for future success or failure of innovations is limited (Claudy, Garcia, & O'Driscoll, 2015).

Nevertheless, the diffusion of innovations has identified five determinants that can be used to predict whether an innovation is likely to be successfully adopted in the market: (i) relative advantage, i.e., delivering a benefit to the user over preceding technologies, (ii) compatibility, i.e., fitting with values, experiences, and needs of potential adopters, which is where radical innovations often face a challenge, (iii) complexity, more or less the opposite of ease of use, the central element in the TAM model introduced previously, (iv) trialability, i.e., the possibility to experiment with innovation before actually committing to buying and implementing it in the long run and (v) observability by others. Observability by others relates to social norms, similar to TPB and UTAUT. In diffusion of radical innovation, observability by others is frequently associated with a social identity, i.e., "do I want to transmit my identity as being innovative." Identity is one of the drivers of the radical changes in the U.S. beer market with the emergence of microbreweries (Rao, 2009; Verhaal, Khessina, & Dobrev, 2015), or in reducing the attitude-behavioral gap in sustainable consumption (Van Dam & Fischer, 2015).

Despite the identification of laggards and innovators in the diffusion of innovations literature, it is infamously difficult to characterize these consumers a priori. Consumers who are innovators for one technology may be laggards for another. Therefore it is hard to distinguish the adopter types based on demographic or psychographic profiles.

However, some personality characteristics may indicate systematic differences in innovation uptake between individuals. Perhaps the most frequently mentioned personality trait in the acceptance of new foods is **food neophobia** (Pliner & Hobden, 1992). Food neophobia is a personality trait where people consistently favor known foods and are unwilling to adopt new foods. Food neophobia is shown to be an essential predictor for adopting foreign foods, and also predicts positive response to novel foods (Barrena & Sánchez, 2013; Siegrist, 2008), new health products (Schickenberg, Van Assema, Brug, & De Vries, 2008) and new protein sources such as insects

(Hartmann & Siegrist, 2016; Tan, Fischer, van Trijp, & Stieger, 2016; Verbeke, 2015). The character trait of food neophobia has been further specified as **food technology neophobia**, which is the dislike to use new technologies to create foods (Cox & Evans, 2008), which does not focus on new foods alone but also includes an aversion to (new) processing technologies. Classification of consumers on food neophobia may help innovators to assess who in society may reject what food innovations (Henriques, King, & Meiselman, 2009). We should probably accept that extremely neophobic people may not accept any innovation. At least identifying how many and what kind of people may object can help prioritize further innovation. Other personality characteristics influence response to innovations. For example, variety seeking (e.g., Van Trijp, 1994) with people who consistently look for variety in their foods probably being more positive about innovations. Personal value orientations (e.g., Schwartz, 1994) can also influence innovation opinions, with people who have strong conservative values (tradition, security, conformity) more likely to be opposed to innovations, while those scoring high on openness to change (stimulation, self-direction, hedonism) more likely to support innovations. Evidence of the effect of such values on general consumer innovativeness general showed moderate to small effect sizes (Steenkamp, Ter Hofstede, & Wedel, 1999). The influence of values on innovative choice has been further taken up in consumer decisions to engage in eco-innovations (Jansson, Marell, & Nordlund, 2010), including the adoption of more radical innovations such as electronic cars (Barbarossa, De Pelsmacker, & Moons, 2017).

9.4 Radical food innovations today – societal issues

So far, we have introduced some generic points of view on consumers in the early 2020 and theoretical models that can shed light on the adoption of disruptive innovations in food products, food production, and food supply chain management.

9.4.1 Novel protein sources

One radical ongoing innovation involves the introduction of novel foods and ingredients. Most notably we see a consumer and societal trend toward more demand for high-quality proteins in asking for foods with specific, functional, health benefits (Martins, Morales, Barros, & Ferreira, 2017), and the notion that the current, mostly animal-based, production of proteins cannot be sustainable (Aiking, 2011). This latter realization has instigated the search for alternative protein sources, which includes the introduction of plant-based meat analogs (Hoek, van Boekel, Voordouw, & Luning, 2011; Van Loo, Caputo, & Lusk, 2020), insects (Van Huis et al., 2013), cultured meat (van der Weele & Tramper, 2014), algae (Ścieszka & Klewicka, 2019), duckweed (de Beukelaar, Zeinstra, Mes, & Fischer, 2019), and legumes.

From a consumer point of view, perhaps the most radical change has been that meat is no longer seen as a necessary daily part of the meal. The rise of flexitarianism (Dagevos, 2015) and low meat diets (de Gavelle et al., 2019) have influenced the societal position that a meal without meat is only half a meal. The idea of what a complete meal is and should be is being redefined in society. In line with this development, several national dietary guidelines (e.g., Sweden) and nonofficial guidelines (e.g., LiveWell) call to moderate or even limit meat consumption, especially consumption of red and processed meat (Gonzalez Fischer & Garnett, 2016).

Innovative protein sources themselves also hold radical, innovative elements for society. By far, the most impactful has been the increasing popularity of plant-based meat analogs. Meat analogs are plant-based meat products created in such a way to mimic original meat in preparation and taste as much as possible. This fact allows consumers to replace animal proteins with alternatives without adjusting behavior. Brands such as Beyond Meat, the Vegetarian Butcher (Ingenbleek & Zhao, 2019) and the Impossible Burger (Heffernan, 2017) create products that look like the originals. In the case of the impossible burger it even contains plant-based juice similar to those in "real" burgers. Impossible Whopper even became part of the Burger King menu across the USA in 2019. The disruptive element in this innovation may lie not so much on the consumer side but in the supply, where plant-based analogs may threaten the current business models of meat processing companies (Keefe, 2018). To deal with such changes, companies are investing in meat alternatives, with Unilever buying the Vegetarian Butcher in 2018 (Tziva, Negro, Kalfagianni, & Hekkert, 2019) and gaining the European Burger King contract for plant-based burgers in 2019.[1] Dutch-based meat processor Vion converted one of its beef processing plants into a plant-based site (Fortune, 2019). With the trend of plant-based meat alternatives, there is an ongoing debate to what extent using meat-like names (e.g., burger, schnitzel) for plant-based products will remain legally allowed under European and member state laws (Seehafer & Bartels, 2019). In 2017, following complaints by the dairy industry, the EU regulated that dairy words like "milk," "cheese," "butter," "yogurt," and "cream" were banned for use by nondairy products, and this inspired the meat industry for its cases. This observation may be not so clear cut, as even for dairy, that has a limited number of names, exemptions were made for long-established products such as cacao butter, coconut milk, and peanut butter. In the USA several states no longer allow the use of "meaty" words for products not produced by raising livestock (Watson, 2019) while Seehafer and Bartels (2019) argue that customary use of meat-like names (such as burger) will remain allowed in the EU, as well as other naming strategies as long as

1. The red juice (a heme protein: leghemoglobin) of the impossible burger was created with genetically modified yeast not approved for the EU market; hence, the EU contract could not be granted to Impossible Foods.

consumers are not misled by the product names. Of notice is, that at the time of writing (early 2020), the demand of plant-based meat analogs had increased to a level where shortages of processed soy and other legume proteins, required for the manufacturing of some of these products started to limit the market growth (Zhang et al., 2020). This shows that uptake of innovation may have disruptive effects further along in the production chain.

Cultured meat, in vitro meat, synthetic meat, lab-grown meat, clean meat, fake meat, animal-free meat (probably known under several other names as well), is another product that aims to mimic meat. Cultured meat is produced by growing meat substitutes from actual animal stem cells (Post, 2012). However, cultured meat has run into some challenges. In order to be a relevant, sustainable meat alternative, the growth medium needs to shift from bovine serum to growing medium that does not need such serum. While the development up until 2020 has been promising, nonanimal growth media remained inferior to those containing serum (Kolkmann, Post, Rutjens, van Essen, & Moutsatsou, 2019). Another issue is the legal status of cultured meat. In Europe, it is considered a novel food which requires extensive testing before consumption is allowed (Seehafer & Bartels, 2019). Also, while the cost of cultured meat has come down dramatically since its introduction as a €250,000 hamburger in 2013, in 2019, costs were still at least €200 per kilo (Purdy, 2019). It will still be a distant future where every village has its cultured meat plant (van der Weele & Tramper, 2014), or where consumers grow their meat in their bio processor on the kitchen counter. From a consumer perspective, cultured meat seems desirable (Bryant & Barnett, 2018; Verbeke, Sans, & Van Loo, 2015; Wilks & Phillips, 2017). However, consumers show a tough time in considering it as either meat or meat substitute (Bekker, Fischer, Tobi, & van Trijp, 2017; Bekker, Tobi, & Fischer, 2017). The problem consumers have with fitting in cultured meat into their view on meat suggests that adoption of cultured meat may require a (partial) redefinition of what meat is.

Another new protein type in the Western countries that has received much attention is that of eating insects. Hailed initially as a very promising meat replacer (e.g., Van Huis et al., 2013), insect-eating by humans has shown to be somewhat less prominent than hoped for. This is mainly due to western consumers exhibiting neophobia (Verbeke, 2015) or even disgust response when confronted with insects (Hartmann, Shi, Giusto, & Siegrist, 2015), especially when recognizable insects are mixed with existing foodstuffs (Hartmann & Siegrist, 2016). It is not only this initial dislike that consumers have to overcome. When comparing insect-eating between countries where insect-eating is part of the local cuisine, it becomes clear that consumers from these countries consider specific insects in the context of specific recipes, contribution to dishes and preparation methods (Tan et al., 2015) and show a more balanced view on specific insects compared to western populations (Hartmann et al., 2015). People from a culture where insect-eating is uncommon to have problems even recognizing insects as food (Tan et al., 2015), and have difficulties imagining how to prepare them (Tan et al., 2016). Recognition of insects as food seemed to be largely insect species-specific rather than culture-specific. Insects unknown food in insect-consuming cultures used to insect consumption were also not recognized as foodstuffs (Tan et al., 2016), while there is some evidence that those exposed to insect tastings in a culture without insects in their cuisine start to recognize specific insect species as edible (Fischer & Steenbekkers, 2018). Many of the studies on consumer acceptance of eating insects focus on willingness to try (once), but it is, in the end, the repeated consumption that would make this innovation successful, and willingness to repeatedly use insects in food seems low (Tan et al., 2016). Early introductions of insect products in the supermarket have largely failed arguably because of the high price, inferior taste, and the difficulty fitting them into existing routines (House, 2016), as well as the placement among vegetarian products. All these factors may have contributed to a "passive rejection" where people may have been willing to try but were not ready to include it in their diets. These new burgers with insects have long been withdrawn from retail, showing that ignoring changes consumers have to make in adopting an innovative product in their daily life is not trivial.

A similar issue may occur when trying to bring aquatic crops such as algae and duckweed products in the market. Initial consumer research on acceptance of duckweed as human food suggests that use in salads or applications where greenery is expected as a condiment would make acceptance much more likely than heavily processed applications (de Beukelaar et al., 2019). Algae are an exciting class of innovative protein sources. Macroalgae such as seaweed are associated with specific cuisines, most notably sushi (Birch, Skallerud, & Paul, 2019), and maybe introduced as such. Microalgae, such as spirulina, may enter the market as functional food ingredients (Grahl, Strack, Weinrich, & Mörlein, 2018), particularly in food drinks, or for specific groups of vulnerable people (Santos, Freitas, Moreira, Zanfonato, & Costa, 2016). Spirulina as high protein supplements is now becoming more common among specific groups with high motivation to gain the functional aspects (sporters, vegetarians) or those interested in food innovation in general (Moons, Barbarossa, & De Pelsmacker, 2018). However, in the light that only a minority of the population reports using functional foods regularly (Sparke & Menrad, 2009), it remains to be seen to what extent these functional food applications of spirulina are a start of further diffusion of spirulina supplements to the broader food market, or whether (in terms of diffusion of innovations) the entire markets consists of these highly motivated niches.

Herein, we identified a number of relevant issues to end-users, which shows that a genuinely radical approach inputting new food products on the market often runs into trouble (e.g., insect foods). We also argue introductions that allow consumers to evolve toward a new food consumption pattern (e.g., meat analogs in plant-based meat substitutes) are more likely to be successful. For more radical innovations to be successful, it seems, therefore, that a more incremental rather than radical or disruptive innovation approach is the best way to get in the market.

9.4.2 High tech and precision agriculture

While mechanized agriculture is now commonplace, fully automated and precision agriculture is emerging. This has substantial effects on how farms operate and may impact the societal view on farming. An example that is being adopted across Europe and Northern America is that of Automatic Milking Systems or Robot Milkers (Butler, Holloway, & Bear, 2012). Such systems can milk dairy cattle without (real-time) involvement of the farmer. The cow moves into the robot system whenever she wants to be milked, is recognized by the system, and receives tailored feed while being milked. An automatic analysis of milk quality is then used to adjust feed composition for the next milking session. The role of the farmer has become that of a controller who has to check the robot generated data to ensure their cows are healthy and receive the best possible feed. There could, however, be a catch to this approach, as continuing automatization of animal husbandry could be considered as further objectifying the animals by removing the remaining contact with a caring human (cf. Bos, Bovenkerk, Feindt, & Van Dam, 2018). In the case of the milking robots, these initial concerns were incorporated in the final development of the Robot Milker of a Dutch company.

By ensuring the cow chooses when to enter the robot, the autonomy of the cow was established as a more critical increase in animal welfare than the reduced contact with the farmer was detrimental for animal welfare (Driessen & Heutinck, 2015). Society is readily accepting this more natural behavior as desirable even if facilitated by high-tech, and hence, assumedly nonnatural (Rozin, 2005) robot milking. This early example of mechanized agriculture is only the beginning, however. More all-encompassing automation is expected in agriculture. Current studies are ongoing to develop fully autonomous greenhouses (Ko, Ryuh, Kim, Suprem, & Mahalik, 2014), drone surveillance (Mogili & Deepak, 2018), robot weeders (Hussmann, Knoll, Meissner, & Holtorf, 2019) and fully automated harvesters that can find their way across fields (Rahman, Ishii, & Noguchi, 2019). All these development will generate big data that has to be processed and interpreted to maximize the benefits of these systems and to integrate data streams from different systems. These studies have focused to a large extent to radical changes in the production chain, most dominantly the position of the farmer that is rapidly changing from manual laborer to a process operator (Bramley & Ouzman, 2019). Given the amount and complexity of the data, the farmer needs the support of data companies for the advanced analyses needed. This fact raises the issue of data ownership, i.e., to what extent does the farmer keep a say over their data, and to what extent can the data company use it freely for their profit (Fraser, 2019); a discussion reminiscent of the data ownership concerns that emerged with (on the face of it) free of charge social media platforms (Veletsianos, Johnson, & Belikov, 2019).

Similarly, drone surveillance raises privacy concerns among people living within the range of such drone surveillance (Ahirwar, Swarnkar, Bhukya, & Namwade, 2019; Oltvoort, de Vries, van Rompay, & Rosen, 2019). Given the relatively small scale agriculture in Europe, this almost certainly will be an issue for agricultural drone surveillance. Similar analogies can be drawn between fully automated harvester and self-driving cars (by early 2020 neither being allowed on public roads) (Suganuma, 2019). A fully automated harvester while contained within a field may be acceptable, although the killing of wildlife may raise some protests. Strong guarantees would be needed that such harvesters cannot leave the designated field and that while traveling to and from the fields, it will ensure the safety of other road users (Borenstein, Herkert, & Miller, 2019). Again, the relatively small scale of European agricultural plots is likely to make these issues salient in that continent.

Precision agriculture is a radical innovation that will have a significant impact on primary food producers (farmers) but may also raise some societal concerns. These are most likely due to an aversion against increasingly high-tech agriculture leading to (further) objectivation of animals and the deprecation of rural areas as a naturally perceived environment. Also, privacy concerns and bystander safety need to be dealt with in developing fully automated systems.

9.4.3 Smart in–home appliances

The Internet was initially developed for the transfer of data. To date, most of the Internet still is about transferring data, making online orders, engaging in games, audio or video streams, social media, or otherwise communicating with peers or stakeholders. Increasingly the Internet is also becoming integrated into physical products. Doorbells, alarm systems, smart television, and many other in–home appliances now communicate with the Internet, transfer data between each other and external data analysis services. This Internet of things (IoT) may also start affecting food consumption in society. A long-standing promise has been the smart fridge, which in theory, should be able to provide recipes, keep track of stock, and prompt users on their smartphone to replenish specific products, keep track of use-by dates, and provide nutritional advice based on currently stocked foods. Consumers perceive the smart fridge functionality as moderately useful and particularly like keeping track of use-by dates and nutritional advice (Rothensee, 2008, pp. 123–139). The function of keeping track of use-by dates is (for about 20 years already) a technological challenge that requires food to communicate use-by dates to the fridge. This can either be by unique barcodes that report use-by dates (Keluskar, no date) or Radio-frequency identification (RFID) that transmit these use-by dates (Ferrero, Vakili, Giusto, Guerrera, & Randazzo, 2019).

Given that neither is the default in the food market and given the low margins in food retail, such expensive product-specific tags are unlikely to be implemented in retail any time soon. Therefore, at this moment in time, the only way to have smart fridges keeping track of use-by dates is by having consumers attach them to their groceries before storing. Individually labeling each product with smart-fridge readable use-by dates will add a considerable burden on consumers for each shopping trip. Not surprisingly, the current generations of high-end smart fridges do not keep track of use-by dates. Some of the higher end fridges contain cameras that allow the user to visually see the inside of the fridge, and thus, track contents while shopping, assuming that the static camera has an unobstructed view on all relevant items.

Another technology that through information technology could change how consumers deal with food stocks and production at home is 3D printing of food. 3D food printing is a technique in which a food product is constructed in layers from cartridges with preprocessed contents. In theory, a 3D printer could create many shapes of foods with a range of tastes by mixing essential ingredients. In that way, the 3D food printer holds promises similar to the Star-Trek replicator we started this chapter with (Desai 2019). Imagine just having to purchase the compatible cartridge and download a recipe from the Internet, and award-winning food appears in the domestic kitchen. In practice, the current state of the art is much more limited, and 3D food printing is mainly used for professional applications. 3D printing sees some applications in custom made sweets (mainly chocolate and sugar work, and some biscuit shapes). For these products, 3D printing allows the creation of complex or highly personalized shapes that cannot be made at all or not at any reasonable cost in other ways (Liu, Zhang, Bhandari, & Wang, 2017). Another application of 3D printing is to support the appetite of the elderly who cannot eat solid foods. In the past, the solution was to provide these people with unappealing pureed foods, resulting in a reduced appetite. By creating more appealing shapes through 3D printing, the purees in the shapes of recognizable foodstuff, appetite, and with that quality of life of the elderly can be improved, while maintaining the softness of the puree (Liu et al., 2017). These applications are, however, far from the domestic kitchen. Costs are too high, 3D printing does not fit routine food preparation behaviors, and prefilled cartridges are not available while creating the fillings of cartridges in the domestic kitchen increases complexity and effort needed for food preparation rather than reducing it.

One issue that is evolving around the emergence of the Internet of things in general and in—home appliances specific is that of cybersecurity. This issue has recently moved high on the agenda for the Internet of things and needs considerable attention from developers of smart appliances and data services that collaborate with them (Guhr, Werth, Blacha, & Breitner, 2020). So far, it seems that smart appliances have a much lower security level than, for example, computers. This poses several risks. First, through the weak protection of smart appliances, criminals may gain access to the home network. Second, and not unimportantly, mere access to the data streams from, e.g., the fridge to suppliers or smartphones may inform others about the exact contents of your fridge in real-time. This fact may sound trivial, but do you want burglars to be able to track when a fridge is not restocked for some time, which may indicate a holiday. Do you want your health insurance or employer to have access to your fridge contents, allowing them to keep track of the health of your diet or the amounts of, e.g., beers consumed in your household?

Internet of things and smart hardware has the potential for radical changes in domestic food usage. The current versions of these appliances do, however, only offer limited functionality of this kind. The security of these appliances, their data streams, and the privacy sensitivity of the associated data are only just being incorporated into the designs. There seems to be much promise, but actual adoption may take some time.

9.4.4 Transparency in food supply chain through traceability by blockchain

New data-driven technologies are being applied in the food sector, such as blockchain technology, allowing for storage and sharing of data along the food supply chain in a transparent and unmodifiable way. While a traditional database is often owned by one organization or person and susceptible to manipulation, blockchain technology allows the sharing of a decentralized database across a whole network where everyone can contribute, and data cannot be altered. While blockchain was initially mainly associated with financial services, the blockchain's decentralized database for sharing, storing, and securing data offers innovative food quality traceability and transparency systems to the food supply chain, in particular in tracking individual products and their ingredients to their specific origins. The traceability information can relate to any of the six traceability elements identified by Opara (2003) in Karlsen, Dreyer, Olsen, and Elvevoll (2013) (product, process, genetic, input, disease/pest, and measurement traceability). For example, the provider adds information about the crops, the producer on pesticide use, fertilizer use, and other information about the crop cultivation, the distributor on details on shipping, storage conditions, and the retailer add details on expiration dates. The consumer can then access all of the information for the individual products, for example, through scanning a QR (quick response) code.

Several useful foods tracing applications along the supply chain include (1) fast and specific responses to foodborne illness outbreaks, (2) food integrity and authenticity, and avoiding food fraud, (3) building consumer trust in the food supply.

Fast and specific responses to foodborne illness outbreaks. Blockchain technology may contribute to a transformation in food safety prevention. In Europe, every year, over 23 million people suffer from a foodborne illness from eating contaminated food (WHO, 2017), mainly caused by Norovirus, Campylobacter, and Salmonella poisoning. In order to reduce foodborne illnesses, it is key to act quickly when food contamination outbreaks occur. When using the existing food tracing system, it can take some time to identify the specific source of the outbreak, which can cause further illnesses and often leads to overly broad product recalls resulting in unnecessarily large amounts of food being lost. For example, during an *E. coli* outbreak linked to Romaine lettuce, the U.S. CDC (2018) (Center for Disease Control and Prevention) warned U.S. consumers not to eat any Romaine lettuce and restaurants and retailers not to sell it. This resulted in a huge loss of Romaine lettuce. With adequate recordkeeping and data sharing, the source of food contaminations could be located quicker, leading to improved public safety, and less food wasted by retailers and farmers. In 2017, a group of large food producers (e.g., Nestlé, Unilever, Tyson) and retailers (e.g., Walmart, Kroger) started to integrate IBM's blockchain Food Trust into their supply chain to increase food traceability, and increasingly more fast-moving consumer goods (FMCG) companies are joining. One of the largest retailers (e.g., Walmart Inc.) requires their suppliers of lettuce, spinach, and other greens, as well as the suppliers of these suppliers (farmers, logistics companies, and business partners) to join the food-tracking blockchain system (Nash, 2018). This led to increased product traceability and a targeted recall. By supporting food safety during a recall through its blockchain records, it thus also creates consumer trust (Galvez, Mejuto, & Simal-Gandara, 2018).

Food integrity and authenticity - preventing food fraud. Blockchain applications in the food supply chain may help to fight food fraud. Blockchain technology provides an unchangeable, transparent, and traceable record along the food chain. As a result, blockchain systems can help to assure food integrity and help tackle problems such as fraud, counterfeit, and mislabeling along the food supply chain. Thus blockchain technology can play a role in managing the reported loss of public and consumer confidence in food production and the food industry. In the blockchain system, all information on the product can be stored from the start of the supply chain to the retail such as pesticides use, details about the livestock, feed, and importantly once stored cannot be changed. Hence, in a food supply chain controlled with blockchain technology, fraud and contamination can be detected more quickly, and even more, the source of the fraud and contamination can easily be identified. This may help to prevent food fraud regarding specific food certifications, (accidental) misdescribing and mislabeling, and adulteration such as the 2013 horse meat scandal, where substantial quantities of horse meat were entered into beef production chains. If the origin of the adulterated meat had been stored using blockchain registers, the companies who were first to enter these horse meats into the supply chain could likely have been identified much faster.

Building consumer trust in food supply from farm to fork. Blockchain systems allow food businesses to store all data from farm to fork in a reliable and transparent system, resulting in radical innovations in the way food chains are managed. If companies make this information available to consumers, this may help to build a trust relationship with consumers (Xiong, Dalhaus, Wang, & Huang, 2020). By transparently sharing information about the individual product, guaranteed by blockchain registration, companies can build a better reputation for their company and products (Xiong et al., 2020). Already, some producers or retailers provide QR codes that consumers can scan to access product traceability information across the entire supply chain. This includes information about the route the product has taken from the farm to the shelves. For consumers, blockchain technology ensures reliability and transparency of such information on food production and addresses their potential concern about food safety and food quality. There is some evidence that a food traceability system can facilitate consumer confidence that strict requirements for food labeling claims are met. For example, Charlebois and Haratifar (2015) reported that Canadian consumers would appreciate a traceability system for organic milk. While 4% of consumers indicate to always purchase organic milk under current regulation without a traceability system for consumers, 27% would always and 53% would consider purchasing traceable organic milk (Charlebois & Haratifar, 2015). Consumers' preferences for identifiable information also differs based on the food category (Liu, Li, Steele, & Fang, 2018). Sander, Semeijn, and Mahr (2018) reported a positive relationship between the presence of the blockchain traceability for meat and consumers' quality perception and purchase decision. Several studies indicate that Chinese consumers are willing to pay a premium for traceable food (Jin, Zhang, & Xu, 2017; Liu, Gao, Nayga, Snell, & Ma, 2019; Zhang, Bai, & Wahl, 2012); there are, however, no studies whether this also holds for other countries. So far, no research evaluated real consumers' purchase behavior or nonhypothetical willingness to pay for traceable food products by observation or auction experiments. For traceability for fish, consumers have positive attitudes about traceability data and indicate that they mainly want to use it to verify the origin and producer claims (Rodriguez-Salvador & Dopico, 2020).

Food chain actors have high hopes blockchains data will be read, and thus, increase consumer and societal confidence in their productions. Whether this reassurance of food quality and safety is valued and will be read by consumers remains uncertain. Liu et al. (2019) reported that only 1.2% of all consumers accessed the information provided by a QR code on a food package, although more than half of the participants accessed the QR code when provided with a smartphone with QR scanning software preinstalled. Liu et al. (2019) argue that convenient and fast access is essential for consumers to retrieve information. Studying the consumer acceptance of a QR code for food traceability information using the TAM model, Kim and Woo (2016) similarly found that the consumers' attitude toward using the QR code and subsequent scan intention is affected by the expected information provided, as well as the perceived usefulness, and perceived ease of use.

Blockchain is a promising technology for the food supply sector, but it faces some challenges for the blockchain to be implemented along the food supply chain (Kamilaris, Fonts, & Prenafeta-Boldu, 2019). It can only be successful when all actors along the supply chain participate. At this moment in time, however, many food industry companies still use stand-alone record keeping, and no history of shared registers of product lines, such as blockchain, exists in the food sector (Aaron, 2018). Setting up the infrastructure for blockchain and ensuring commitment to the system for all actors is, therefore, an important hurdle to the implementation of blockchain to ensure transparency. While for consumers, the innovation to access product history may not seem very large, the changes within the supply required to ensure this are disruptive.

9.4.5 Personalized dietary advice

The dietary habits in the developed world are leading to a high prevalence of noncommunicable diseases such as obesity, type 2 diabetes, and cardiovascular diseases due to the high consumption of saturated fat, added sugar, and salt and insufficient fruit and vegetables (WHO, 2018). Consequently, there is a need to change toward a healthier dietary pattern. Most free dietary advice is population-level, or at least population subgroup level recommendations, and follow a one-size-fits-all approach, such as "eat at least 200 g of vegetables daily" and "limit salt intake to 6 g daily" (WHO, 2015). Population-level dietary advice does not effectively address individual needs. Instead, a personalized nutritional approach specifies dietary recommendations tailored to the individual and has the potential to be more effective in changing dietary behavior (Celis-Morales et al., 2017; Celis-Morales, Lara, & Mathers, 2015; Ferguson et al., 2016). Personalized dietary advice is tailored to one's specific individual needs based on genotype or physiological characteristics, phenotype, and current nutritional status (Gibney & Walsh, 2013; Livingstone et al., 2016). In addition, an individual's lifestyle and personal preferences may be considered to formulate personalized dietary advice by personalized nutrition services. While to date, there is no consensus on a definition of personalized nutrition, experts agree that "the goal of personalized nutrition is to advance human health and wellbeing by tailoring nutrition recommendations and interventions to individuals or groups of individuals with similar traits" (Bush et al., 2020, p. 5).

There are three important stages in the information exchange process for personalized nutrition services (Berezowska et al., 2014 Fig. 9.1), contributing to the consumer acceptance of personalized nutrition services. First, consumers need to be willing to disclose the requested personal, potentially sensitive

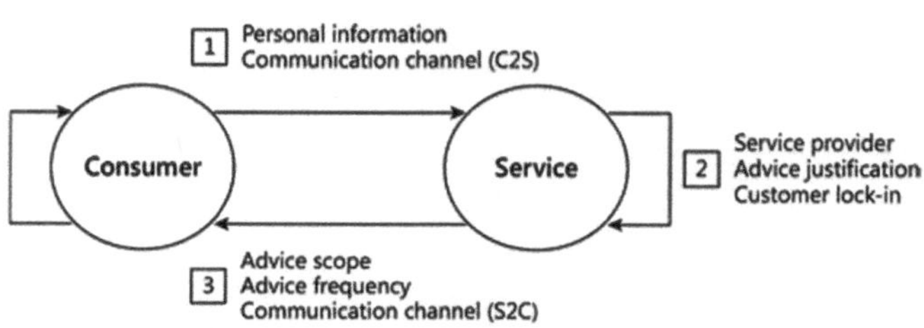

FIGURE 9.1 Schematic overview of the personalized nutrition information exchange process and its attributes. *Berezowska, A., Fischer, A. R. H., Ronteltap, A., Kuznesof, S., Macready, A., Fallaize, R., et al. (2014). From: Understanding consumer evaluations of personalised nutrition services in terms of the privacy calculus: A qualitative study.* Public Health Genomics, 17*(3), 127–140.*

information to the service provider through the chosen communication channel. Second, the personalized nutrition service provider will develop personalized dietary advice, followed by communicating the advice to the consumer, where the consumer needs to understand and trust the given advice. The consumer adoption of the advice will depend on the trade-off between the perceived privacy risk and the perceived personal benefit (Berezowska et al., 2014; Berezowska, Fischer, Ronteltap, Van der Lans, & Van Trijp, 2015).

Consumer acceptance of personalized advice may differ depending on the specific personalized nutrition service design attributes, and thus, the personalized nutrition design attributes need to be carefully selected. A wide range of personalized nutritional services exists (Ronteltap, Van Trijp, Berezowska, & Goossens, 2013). Specifically, the personalized nutrition service approaches differ in terms of the type of personal information disclosed (e.g., dietary intake data, phenotypic information, genotypic information, lifestyle), the type of business model and service provider (e.g., employer, dietitian, fitness club, consultant), target group, and communication channels (e.g., personal contact, app, Website, e-mail) (Berezowska et al., 2014, 2015; Ronteltap et al., 2013). The details of the advice itself will have an impact on consumer acceptance, and this advice can differ in term of the type of personalized offer or advice scope (e.g., personal diet plan, personal coach, personalized shopping list, whether or not the nutritional advice is combined with advice on physical activity), advice frequency, the message framing, as well as the information presentation (Berezowska et al., 2014, 2015; Nguyen et al., 2017; Ronteltap et al., 2013). The consumer response and willingness to adopt varies based on the personalized nutrition service design attributes and the implications of these design attributes for the information exchange process. For example, Berezowska et al. (2014) reported that consumers prefer an expert or dietitian to be the personalized nutrition service provider rather than the government or the employer.

The implemented personalized nutrition service attributes influence intention to adopt as they relate to the perceived personalized benefit, i.e., the degree to which the advice is personally relevant, and the perceived risk of privacy loss as personal and perhaps sensitive health information is shared. A highly relevant and personalized recommendation is more likely to lead to a perceived personal benefit but may also require the sharing of sensitive information, leading to perceived privacy risk. Berezowska et al. (2014) argue that consumers trade-off the perceived personal benefit and the perceived privacy risk called the risk-benefit calculus, which is related to the intention to adopt personalized advice. To encourage consumer adoption, it essential to reduce the perceived privacy risk and increase perceived personal benefit of personalized nutritional advice (Berezowska et al., 2014; Reinders, Bouwman, van den Puttelaar, & Verain, 2020). Berezowska et al. (2015) further suggested to clearly communicate the personal benefit of personalized advice over non-personalized advice to enhance its effectiveness. The perceived privacy risk perceptions should be minimized by transparently sharing how the personal data will be protected (Berezowska et al., 2015; Reinders et al., 2020). However, the intention to adopt personalized advice depends on more than the trade-off between the risk and benefits perception of the information disclosure.

Personality traits, knowledge, and values should be taken into account when developing the personalized advice and how to communicate it to optimize its effectiveness (Ronteltap, van Trijp, Renes, & Frewer, 2007). For example, Berezowska, Fischer, & van Trijp (2017) suggest that the personalized dietary recommendation takes one's level of self-determination into account. For individuals with high levels of autonomous motivation, i.e., the decision to eat healthily is self-determined, one should focus on the benefits, while for individuals with high levels of controlled motivation, risk-reduction properties of such a service should be emphasized. Ambivalent feelings, which occur when one has both negative and positive evaluations, may negatively impact the intention to adopt the personalized nutrition advice, so directly countering risks with more benefits, resulting in higher risk and benefit perception, and hence, higher ambivalence may not be the way forward (Reinders et al., 2020). Consumers' concerns need to be taken seriously, and the advice should minimize potential ambivalent feelings.

9.5 Digital food purchases and food ordering

Digital technologies are transforming the food shopping landscape. Recently food online to off-line (O2O) e-commerce is growing with increasingly prevalent online food shopping and online meal ordering platforms (Roh & Park, 2019). Consumers can thus shop online for (1) groceries, which in the content of goods is closely related to a traditional shopping trip, (2) meal kits delivered to their homes which provides conveniently portioned food allowing faster preparation (3) prepared, ready to eat, meals. These different online food delivery services cater to consumer demand for convenient food purchases in different ways:

(1) *Online grocery shopping and delivery.* E-commerce and online shopping have boomed in the last two decades, allowing them to have any products at our fingertips without having to leave the house, with only a matter of a few clicks. While online shopping for groceries is not yet as common compared with the other product categories such as clothes and travel (Eurostat, 2020), it is proliferating. Research by FMI and Nielsen (2018) predicted that 70% of U.S. consumers would be purchasing groceries online by 2024, illustrating the "rising digital grocery landscape. " In Europe, the adoption of online grocery shopping sometimes referred to as e-groceries, is also rising quickly. In 2019, 27% of the e-shoppers in the EU-28 purchased food online (Eurostat, 2020). This fact is amounting to substantial monetary revenue, with recent research predicting that 10 leading global online grocery market will account for €200 billion in sales.

The most important underlying motivations to shopping for e-groceries are convenience (e.g., reduction in effort, ordering at any moment of the day at any place, access to a wider product assortment), saving time (Seitz, Pokrivčák, Tóth, & Plevný, 2017), and the often more favorable prices (IRI, 2019). Online food shopping differs from online shopping for other products (Mortimer, Fazal e Hasan, Andrews, & Martin, 2016) as many food products are perishable, and consumers value freshness of food products. This is also reflected in the barriers to online food shopping, such as lack of trust and lack of control and concern about product quality since you cannot select your own products. While online food shopping faces some challenges, consumers are increasingly finding their way to shop for groceries away from the brick-and-mortar stores. New companies or companies initially not involved in food sales enter the e-grocery market (e.g., AmazonFresh, AmazonPantry, Picnic, Instacart). At the same time, traditional brick-and-mortar grocery chains have now created online grocery shopping environments, and allow for both home delivery, as well as pick-up points ("click-and-collect" options) where consumers collect their order.

More so than in other sales channels, many consumers were reluctant to change their habitual grocery shopping behavior and were not ready to leave their trusted brick and mortar grocery shop, in part because they were unfamiliar with shopping online for food. There are, however, indications this may have changed during the 2020 COVID-19 outbreak. With rising concerns to enter physical stores due to the risk for contamination and the government lockdown policies, many consumers adopted online grocery shopping resulting in a surge in e-groceries. This caused, both in the U.S. and Europe, large delivery waiting times, and consumers sometimes even being placed on waiting lists. Grocery stores needed to expand their online operations and had to look for additional employees to assist with the rapid expansion of online orders. While we do not know what will happen once the food retail and the society have recovered from this pandemic, many sources are predicting that this pandemic may have catalyzed the digital transformation, including for the transition to online grocery shopping and grocery delivery as consumers may have become more familiar with online groceries and may continue to shop online for groceries after the pandemic (Goddard, 2020; Hobbs, 2020).

(2) *Subscription meal boxes.* Meal kits were originally offered as subscription meal boxes (e.g., Blue Apron, Marley Spoon, HelloFresh) and include recipe cards and preportioned ingredients to prepare home-cooked meals. Consumers subscribe to meal box plans to save time but still experience a home-cooked meal with the help of the easy recipes on recipe cards. The meal boxes thus save time and reduce the mental and physical efforts since consumers do not have to shop for groceries nor have to engage in meal planning (Hill & Maddock, 2019; Khan & Sowards, 2018). Meal kits are seen as a convenient food, which lessens the dinner related tasks, but still gives consumers the feeling of "cooking from scratch" (Hertz & Halkier, 2017). This compares to traditional convenience foods which are often ready-to-eat meals, giving consumers the feeling of guilt for not cooking a meal from catch (Costa, Schoolmeester, Dekker, & Jongen, 2007). While delivery of meal kits may offer time-saving, it may lack some flexibility as consumers precommit to the number of meal kits each week, which may not allow last-minute change of plans. This lack of flexibility may prevent consumers from remaining a meal kit subscriber, and meal kit providers are developing strategies to deal with this (e.g., a pause option to delay delivery). Meal kits have impacted food retail, which now also offers meal kits in their brick and mortar stores.

(3). *Meal delivery services.* While in the past, meal delivery was often a service provided by few restaurants, most notably pizza chains, with Pizza Hut being the first to allow online ordering, it has now expanded tremendously. Digital meal ordering is becoming increasingly prevalent, and the global online food delivery market was estimated at $35 billion and expected to rise to $365 billion by 2030 (Business Insider, 2018). Digital meal orders can be placed on a mobile app, Website, or with a text message. More recently, companies specialized in meal delivery, the so-called third-party meal delivery services, have emerged that offer a restaurant-to-consumer platform, allowing restaurants without their own preexisting delivery service to home-deliver. This online-to-offline (O2O) meal delivery market is multiplying, and many players are active in this market (e.g., Uber Eats, Takeaway, Just Eat, Deliveroo, GrubHub) currently competing for a market share in the food delivery space. Ray, Dhir, Bala, and Kaur (2019) reported that convenience, societal pressure, customer experience, delivery experience, search of restaurants, quality control, listing, and ease-of-use all influence the consumers' use of food delivery apps, with performance expectancy being significant for the intention to use online food delivery systems (Gunden, Morosan, & DeFranco, 2020). Digital meal ordering may impact consumer diets, for better or worse. Some academics are advocating to study how to incorporate behavioral change techniques into the digital meal ordering platform to assist consumers in choosing healthier meal options (Stephens, Miller, & Militello, 2020).

These three O2O food delivery services target convenience-seeking consumers, and convenience-oriented consumers are more likely to adopt such delivery services (Roh & Park, 2019). Nevertheless, some concerns remain as consumers with a higher moral obligation are more hesitant to adopt online food services (Roh & Park, 2019).

While still developing, online groceries and food ordering collect customer data and through its flexible online environment could be used to create tailored product offerings, personalized suggestions to help consumers make healthier dietary decisions and apply targeted advertising. This fact may further change how consumers plan and purchase foods in the future, but may also lead to privacy concerns and reluctance with consumers associated with that.

9.6 Radical food innovations and society – what next

Considering the disruptions caused by current innovations in food science and technology, a few issues can be identified that will facilitate or hinder its societal acceptance.

The move toward a palette of sustainable protein sources to (partially) replace, and hence, reduce the sustainable impact of animal husbandry will not be as easy as sometimes thought. Across many studies, a general willingness to consider eating less meat, for trying alternatives, even less than apparent protein alternatives such as insects, or high tech produced cultured meat, is observed. However, when we look closer, we also realize this does not translate directly into a shift toward eating less meat. While flexitarian diets are now considered more "normal," people still consume about the same amount of meat. While people are willing to try insects once, they do not consider them relevant alternatives to meat in their regular diet. Duckweed foods that brought duckweed to the meal as a salad were most acceptable, but salads are not regular meat replacements in meals. Cultured meat, in particular, formed a puzzle to consumers as to whether it is meat or something else. This suggests that the disruptive element in these innovations is not as much in a deliberate willingness to try the new products, but more in a reluctance, or inability to change long-standing interpretations and patterns of the daily meal. Breaking such habits is, because of their unconscious nature, often more challenging than merely convincing consumers to try, and that they are willing to try shows an opening nonetheless.

The second group of disruptive innovations in food production revolves around the convergence of microelectronics, information technology, and food technology. While there may be less impact on actual food properties (except 3D printed foods), the idea that food production is even further industrialized in precision agriculture may upset some among the public. More importantly, there are ethical concerns around such high tech information applications. Who is responsible for autonomous robots if accidents occur? Who safeguards privacy around drone surveillance or even more importantly about the fridge contents kept by a smart fridge, shopping lists in digital marketing applications, and who controls the storage, interpretation of health profiles that may include DNA profiles in personalized nutrition services? These concerns have been changing the way we position ourselves about social media but are now also encroaching on food supply chain issues. Besides concerns, part of the solution may lie in information technology, however. Blockchain technology may make supply chains more transparent, and thus, present reassurance to the public. Mobile phones are now ever-present, and many new innovative portable biosensors to be linked with our personal devices are being developed. Together with the development of personalized self-learning algorithms, digital nutritional expert systems providing personalized nutritional advice are likely to exist soon (as outlined in more detail in Michel & Burbidge, 2019) and may contribute to a reliable, robust food supply.

Food technologies related to resilient and robust supply chains have become even more urgent over the time we spent writing this chapter. When we started writing this chapter in early 2020, supply chains of food and medical supplies were assumed to be stable and sufficient if not abundant in developed countries. While we are finalizing this chapter only a few weeks later, hoarding had emptied supermarket food (and toilet paper) supplies, and essential medical supplies across the world became depleted. Important production sites have been offline for extended periods, and international travel has mostly stopped. The ripple effect showing the vulnerability of our highly optimized interdependent high tech production systems may trigger new ways in how we deal with food, food technology, and organizing supply chains and information streams around it. A food retail system based on a just-in-time approach is efficient in regular times, but may cause many problems and is vulnerable during a crisis as we currently see during the COVID-19 pandemic. Currently, we can only see short-term problems arising in the food supply chain, and the long term effects on consumer demand and consumer perceptions are unknown. However, the pandemic's effects are radical and disruptive, with potentially lasting changes in food production and retail. In recovering from the pandemic, public opinions of innovations discussed in this book may rapidly change for the better or worse and may create a changed outlook on food production in general. While we do not precisely know how culture may change, we can nevertheless expect that general principles in which neophobia, diffusion of innovation, personal deliberations, and the impact of (sub) cultures on acceptance of disruptive innovations influence society will remain the same and will help us interpret what comes next.

Acknowledgments

The Dutch 4TU Federation supported the writing of this chapter through the Pride and Prejudice project.

References

Aaron, C. (2018). *Romaine, blockchain, and the future of food safety.* FSR Magazine.

Ahirwar, S., Swarnkar, R., Bhukya, S., & Namwade, G. (2019). Application of drone in agriculture. *International Journal Curr. Microbiological Applied Science, 8*, 2500–2505.

Aiking, H. (2011). Future protein supply. *Trends in Food Science & Technology, 22*(2), 112–120. https://doi.org/10.1016/j.tifs.2010.04.005

Ajzen, I. (1991). The theory of planned behavior. *Organizational Behavior and Human Decision Processes, 50*(2), 179–211.

Askegaard, S., & Brogård, D. (2016). 'Authentic food'and the double nature of branding. In *The practice of the meal* (pp. 39–54). Routledge.

Bagozzi, R. P. (2007). The legacy of the technology acceptance model and a proposal for a paradigm shift. *Journal of the Association for Information Systems, 8*(4), 244–254.

Barbarossa, C., De Pelsmacker, P., & Moons, I. (2017). Personal values, green self-identity and electric car adoption. *Ecological Economics, 140*, 190–200. https://doi.org/10.1016/j.ecolecon.2017.05.015

Barrena, R., & Sánchez, M. (2013). Neophobia, personal consumer values and novel food acceptance. *Food Quality and Preference, 27*(1), 72–84.

Bass, F. M. (1969). A new product growth model for consumer durables. *Management Science, 15*(5), 215–227.

Bekker, G. A., Fischer, A. R. H., Tobi, H., & van Trijp, H. C. M. (2017). Explicit and implicit attitude toward an emerging food technology: The case of cultured meat. *Appetite, 108*, 245–254. https://doi.org/10.1016/j.appet.2016.10.002

Bekker, G. A., Tobi, H., & Fischer, A. R. H. (2017). Meet meat: An explorative study on meat and cultured meat as seen by Chinese, Ethiopians and Dutch. *Appetite, 114*, 82–92. https://doi.org/10.1016/j.appet.2017.03.009

Belasco, W. (2000). *Future notes: The meal-in-a-pill.*

Berezowska, A., Fischer, A. R. H., Ronteltap, A., Kuznesof, S., Macready, A., Fallaize, R., et al. (2014). Understanding consumer evaluations of personalised nutrition services in terms of the privacy calculus: A qualitative study. *Public Health Genomics, 17*(3), 127–140.

Berezowska, A., Fischer, A. R. H., Ronteltap, A., Van der Lans, I. A., & Van Trijp, H. C. M. (2015). Consumer adoption of personalised nutrition services from the perspective of a risk-benefit trade-off. *Genes and Nutrition, 10*(6), 2–14.

Berezowska, A., Fischer, A. R. H., & van Trijp, H. C. M. (2017). The moderating effect of motivation on health-related decision-making. *Psychology and Health, 32*(6), 665–685. https://doi.org/10.1080/08870446.2017.1293055

de Beukelaar, M. F. A., Zeinstra, G. G., Mes, J. J., & Fischer, A. R. H. (2019). Duckweed as human food. The influence of meal context and information on duckweed acceptability of Dutch consumers. *Food Quality and Preference, 71*, 76–86. https://doi.org/10.1016/j.foodqual.2018.06.005

Birch, D., Skallerud, K., & Paul, N. A. (2019). Who are the future seaweed consumers in a western society? Insights from Australia. *British Food Journal, 121*(2), 603–615. https://doi.org/10.1108/BFJ-03-2018-0189

Borenstein, J., Herkert, J., & Miller, K. W. (2019). Autonomous vehicles and the ethical tension between occupant and non-occupant safety. *Computer Ethics-Philosophical Enquiry (CEPE) Proceedings, 2019*(1), 6.

Bos, J. M., Bovenkerk, B., Feindt, P. H., & Van Dam, Y. K. (2018). The quantified animal: Precision livestock farming and the ethical implications of objectification. *Food Ethics, 2*(1), 77–92.

Bramley, R., & Ouzman, J. (2019). Farmer attitudes to the use of sensors and automation in fertilizer decision-making: Nitrogen fertilization in the Australian grains sector. *Precision Agriculture, 20*(1), 157–175.

Bredahl, L. (2001). Determinants of consumer attitudes and purchase intentions with regard to genetically modified foods - results of a cross-national survey. *Journal of Consumer Policy, 24*(1), 23–61.

Bruhn, C. M. (1995). Consumer attitudes and market response to irradiated food. *Journal of Food Protection, 58*, 175–181.

Bryant, C., & Barnett, J. (2018). Consumer acceptance of cultured meat: A systematic review. *Meat Science, 143*, 8–17. https://doi.org/10.1016/j.meatsci.2018.04.008

Bucic, T., Harris, J., & Arli, D. (2012). Ethical consumers among the millennials: A cross-national study. *Journal of Business Ethics, 110*(1), 113–131. https://doi.org/10.1007/s10551-011-1151-z

Bush, C. L., Blumberg, J. B., El-Sohemy, A., Minich, D. M., Ordovás, J. M., Reed, D. G., et al. (2020). Toward the definition of personalized nutrition: A proposal by the American nutrition association. *Journal of the American College of Nutrition, 39*(1), 5–15. https://doi.org/10.1080/07315724.2019.1685332

Business Insider. (2018). *UBS: Online food delivery could be a $365 billion industry by 2030 — here are the winners and losers from that 'mega trend'*. Retrieved from https://www.businessinsider.com/ubs-online-food-delivery-death-of-kitchen-note-2018-7?international=true&r=US&IR=T.

Buss, A. R. (1974). Generational analysis: Description, explanation, and theory. *Journal of Social Issues, 30*(2), 55–71. https://doi.org/10.1111/j.1540-4560.1974.tb00715.x

Butler, D., Holloway, L., & Bear, C. (2012). The impact of technological change in dairy farming: Robotic milking systems and the changing role of the stockperson. *Journal of the Royal Agricultural Society of England, 173*, 1–6.

Carson, R. (1962). *The silent spring*. Boston, MA: Houghton Mifflin.

(CDC), C. f. D. C. a. P. (2018). *CDC food safety alert:* E. coli *outbreak linked to romaine lettuce*. Retrieved from https://www.cdc.gov/media/releases/2018/s1120-ecoli-romain-lettuce.html.

Celis-Morales, C., Lara, J., & Mathers, J. C. (2015). Personalising nutritional guidance for more effective behaviour change. *Proceedings of the Nutrition Society, 74*(2), 130–138. https://doi.org/10.1017/S0029665114001633

Celis-Morales, C., Livingstone, K. M., Marsaux, C. F., Macready, A. L., Fallaize, R., O'Donovan, C. B., … Navas-Carretero, S. (2017). Effect of personalized nutrition on health-related behaviour change: Evidence from the Food4me European randomized controlled trial. *International Journal of Epidemiology, 46*(2), 578–588.

Charlebois, S., & Haratifar, S. (2015). The perceived value of dairy product traceability in modern society: An exploratory study. *Journal of Dairy Science, 98*(5), 3514–3525. https://doi.org/10.3168/jds.2014-9247

Claudy, M. C., Garcia, R., & O'Driscoll, A. (2015). Consumer resistance to innovation-a behavioral reasoning perspective. *Journal of the Academy of Marketing Science, 43*, 528–544. https://doi.org/10.1007/s11747-014-0399-0 (prepublished online).

Costa, A. I.d. A., Schoolmeester, D., Dekker, M., & Jongen, W. M. F. (2007). To cook or not to cook: A means-end study of motives for choice of meal solutions. *Food Quality and Preference, 18*(1), 77–88. https://doi.org/10.1016/j.foodqual.2005.08.003

Cox, D. N., & Evans, G. (2008). Construction and validation of a psychometric scale to measure consumers' fears of novel food technologies: The food technology neophobia scale. *Food Quality and Preference, 19*(8), 704–710.

Dagevos, H. (2015). Exploring flexitarianism: Meat reduction in a meat-centred food culture. In *Impact of meat consumption on health and environmental sustainability* (pp. 233–243).

Davis, F. D. (1989). Perceived usefulness, perceived ease of use, and user acceptance of information technology. *MIS Quarterly, 13*(3), 319–340.

Desai, D. (2019). How democratized production challenges society's ability to regulate. In *3D printing and beyond*. Edward Elgar Publishing.

Diamond, J. (2002). Evolution, consequences and future of plant and animal domestication. *Nature, 418*(6898), 700–707.

Dimock, M. (2019). Defining generations: Where Millennials end and Generation Z begins. *Pew Research Center, 17*, 1–7.

Driessen, C., & Heutinck, L. F. (2015). Cows desiring to be milked? Milking robots and the co-evolution of ethics and technology on Dutch dairy farms. *Agriculture and Human Values, 32*(1), 3–20.

Eurostat. (2020). *E-commerce statistics for individuals* (Retrieved from).

Evenson, R. E., & Gollin, D. (2003). Assessing the impact of the green revolution, 1960 to 2000. *Science, 300*(5620), 758–762.

Ferguson, L. R., De Caterina, R., Görman, U., Allayee, H., Kohlmeier, M., Prasad, C., ... Martinez, J. A. (2016). Guide and position of the international society of nutrigenetics/nutrigenomics on personalised nutrition: Part 1 - fields of precision nutrition. *Lifestyle Genomics, 9*(1), 12–27. https://doi.org/10.1159/000445350

Ferrero, R., Vakili, M. G., Giusto, E., Guerrera, M., & Randazzo, V. (2019). Ubiquitous fridge with natural language interaction. In *Paper presented at the 2019 IEEE international conference on RFID technology and applications (RFID-TA).*

Fischer, M., & Himme, A. (2017). The financial brand value chain: How brand investments contribute to the financial health of firms. *International Journal of Research in Marketing, 34*(1), 137–153. https://doi.org/10.1016/j.ijresmar.2016.05.004

Fischer, A. R. H., & Reinders, M. J. (2016). Consumer acceptance of novel foods. In C. M. Galanakis (Ed.), *Innovation strategies in the food industry* (pp. 271–292). London, UK: Academic Press.

Fischer, A. R. H., & Steenbekkers, L. P. A. (2018). All insects are equal, but some insects are more equal than others. *British Food Journal, 120*(4), 852–863. https://doi.org/10.1108/BFJ-05-2017-0267

Fishbein, M., & Ajzen, I. (1975). *Belief, attitude, intention, and behavior: An introduction to theory and research.* Reading, MA: Addison-Wesley.

Fishman, A. A. (2016). How generational differences will impact America's aging workforce: Strategies for dealing with aging Millennials, Generation X, and Baby Boomers. *Strategic HR Review.*

FMI, & Nielsen. (2018). *70% of consumers will be grocery shopping online by 2024*. Retrieved from https://www.nielsen.com/us/en/press-releases/2018/fmi-and-nielsen-online-grocery-shopping-is-quickly-approaching-saturation/.

Fortune, A. (2019). *Vion to convert beef facility into plant-based site*. GlobalMeat new.com. Retrieved from https://www.globalmeatnews.com/Article/2019/10/30/Vion-to-convert-beef-facility-into-plant-based-site?utm_source=copyright&utm_medium=OnSite&utm_campaign=copyright.

Fraser, A. (2019). Land grab/data grab: Precision agriculture and its new horizons. *Journal of Peasant Studies, 46*(5), 893–912.

Frewer, L. J., Brennan, M., Coles, D., Kleter, G. A., Fischer, A. R. H., Houdebine, L. M., Mora, C., Millar, K., Salter, B., & Brennan, M. (2013). Genetically modified animals from life-science, socio-economic and ethical perspectives: Examining issues in an EU policy context. *New Biotechnology, 30*(5), 447–460. https://doi.org/10.1016/j.nbt.2013.03.010

Galvez, J. F., Mejuto, J. C., & Simal-Gandara, J. (2018). Future challenges on the use of block-chain for food traceability analysis. *TRAC Trends in Analytical Chemistry, 107*, 222–232. https://doi.org/10.1016/j.trac.2018.08.011

de Gavelle, E., Davidenko, O., Fouillet, H., Delarue, J., Darcel, N., Huneau, J.-F., et al. (2019). Self-declared attitudes and beliefs regarding protein sources are a good prediction of the degree of transition to a low-meat diet in France. *Appetite, 142*, 104345. https://doi.org/10.1016/j.appet.2019.104345

Gibney, M. J., & Walsh, M. C. (2013). The future direction of personalised nutrition: My diet, my phenotype, my genes. *Proceedings of the Nutrition Society, 72*(2), 219–225. https://doi.org/10.1017/S0029665112003436

Gimpel, J. (1992). *The medieval machine: The industrial revolution of the middle ages*. London: Pimlico.

Goddard, E. (2020). The impact of COVID-19 on food retail and food service in Canada: Preliminary assessment. *Canadian Journal of Agricultural Economics/Revue canadienne d'agroeconomie*. https://doi.org/10.1111/cjag.12243. n/a(n/a).

Gonzalez Fischer, C., & Garnett, T. (2016). *Plates, pyramids, and planets: Developments in national healthy and sustainable dietary guidelines: A state of play assessment* (9251092222). Retrieved from.

Grahl, S., Strack, M., Weinrich, R., & Mörlein, D. (2018). Consumer-oriented product development: The conceptualization of novel food products based on spirulina (Arthrospira platensis) and resulting consumer expectations. *Journal of Food Quality*, 1919482. https://doi.org/10.1155/2018/1919482, 2018.

Guhr, N., Werth, O., Blacha, P. P. H., & Breitner, M. H. (2020). Privacy concerns in the smart home context. *SN Applied Sciences, 2*(2), 247.

Gunden, N., Morosan, C., & DeFranco, A. (2020). Consumers' intentions to use online food delivery systems in the USA. *International Journal of Contemporary Hospitality Management, 32*(3), 1325–1345. https://doi.org/10.1108/IJCHM-06-2019-0595

Hartmann, C., Shi, J., Giusto, A., & Siegrist, M. (2015). The psychology of eating insects: A cross-cultural comparison between Germany and China. *Food Quality and Preference, 44*, 148–156. https://doi.org/10.1016/j.foodqual.2015.04.013

Hartmann, C., & Siegrist, M. (2016). Becoming an insectivore: Results of an experiment. *Food Quality and Preference, 51*, 118–122. https://doi.org/10.1016/j.foodqual.2016.03.003

Heffernan, O. (2017). Sustainability: A meaty issue. *Nature, 544*(7651), S18–S20. https://doi.org/10.1038/544S18a

Henriques, A. S., King, S. C., & Meiselman, H. L. (2009). Consumer segmentation based on food neophobia and its application to product development. *Food Quality and Preference, 20*(2), 83–91.

Hertz, F. D., & Halkier, B. (2017). Meal box schemes a convenient way to avoid convenience food? Uses and understandings of meal box schemes among Danish consumers. *Appetite, 114*, 232–239. https://doi.org/10.1016/j.appet.2017.03.016

Hill, B., & Maddock, S. (2019). (No) time to cook: Promoting meal-kits to the time-poor consumer. In J. Byrom, & D. Medway (Eds.), *Case studies in food retailing and distribution* (pp. 241–253). Woodhead Publishing.

Hillman, G. C., & Davies, M. S. (1990). Measured domestication rates in wild wheats and barley under primitive cultivation, and their archaeological implications. *Journal of World PreHistory, 4*(2), 157–222.

von Hippel, E. (1986). Lead users: A source of novel product concepts. *Management Science, 32*(7), 791–805.

Hobbs, J. E. (2020). Food supply chains during the COVID-19 pandemic. *Canadian Journal of Agricultural Economics/Revue canadienne d'agroeconomie*. https://doi.org/10.1111/cjag.12237. n/a(n/a).

Hoek, A. C., van Boekel, M. A. J. S., Voordouw, J., & Luning, P. A. (2011). Identification of new food alternatives: How do consumers categorize meat and meat substitutes? *Food Quality and Preference, 22*(4), 371–383. https://doi.org/10.1016/j.foodqual.2011.01.008

House, J. (2016). Consumer acceptance of insect-based foods in The Netherlands: Academic and commercial implications. *Appetite, 107*, 47–58. https://doi.org/10.1016/j.appet.2016.07.023

Hussmann, S., Knoll, F. J., Meissner, A., & Holtorf, T. (2019). Development and evaluation of a low-cost delta robot system for weed control applications in organic farming. In *Paper presented at the 2019 IEEE international instrumentation and measurement technology conference (I2MTC)*.

Ingenbleek, P., & Zhao, Y. (2019). The vegetarian butcher: On its way to becoming the world's biggest 'meat'producer? *The International Food and Agribusiness Management Review, 22*(1030–2019-628), 295–308.

IRI. (2019). *Winning ahead online E-commerce in FMCG*. Retrieved from https://www.iriworldwide.com/IRI/media/IRI-Clients/International/E-commerce-POV.pdf.

Jansson, J., Marell, A., & Nordlund, A. (2010). Green consumer behavior: Determinants of curtailment and eco-innovation adoption. *Journal of Consumer Marketing, 27*(4), 358–370.

Jin, S., Zhang, Y., & Xu, Y. (2017). Amount of information and the willingness of consumers to pay for food traceability in China. *Food Control, 77*, 163–170. https://doi.org/10.1016/j.foodcont.2017.02.012

Kaiser, F. G., Schultz, P. W., & Scheuthle, H. (2007). The theory of planned behavior without compatibility? Beyond method bias and past trivial associations. *Journal of Applied Social Psychology, 37*(7), 1522–1544.

Kambewa, E. V. (2007). *Contracting for sustainability: An analysis of the lake victory-EU nile perch chain*. Wageningen, Nl: Wageningen Academic Publishers.

Kamilaris, A., Fonts, A., & Prenafeta-Boldu, F. X. (2019). The rise of blockchain technology in agriculture and food supply chains. *Trends in Food Science & Technology, 91*, 640–652. https://doi.org/10.1016/j.tifs.2019.07.034

Karlsen, K. M., Dreyer, B., Olsen, P., & Elvevoll, E. O. (2013). Literature review: Does a common theoretical framework to implement food traceability exist? *Food Control, 32*(2), 409–417. https://doi.org/10.1016/j.foodcont.2012.12.011

Keefe, L. M. (2018). FakeMeat: How big a deal will animal meat analogs ultimately be? *Animal Frontiers, 8*(3), 30–37.

Keluskar, K. (no date). Fridge-O-Matic.

Khan, S. A., & Sowards, S. K. (2018). It's not just dinner: Meal delivery kits as food media for food citizens. *Frontiers in Communication, 3*(39). https://doi.org/10.3389/fcomm.2018.00039

Kim, Y. G., & Woo, E. (2016). Consumer acceptance of a quick response (QR) code for the food traceability system: Application of an extended technology acceptance model (TAM). *Food Research International, 85*, 266–272. https://doi.org/10.1016/j.foodres.2016.05.002

Kolkmann, A., Post, M., Rutjens, M., van Essen, A., & Moutsatsou, P. (2019). Serum-free media for the growth of primary bovine myoblasts. *Cytotechnology*, 1–10.

Ko, M. H., Ryuh, B.-S., Kim, K. C., Suprem, A., & Mahalik, N. P. (2014). Autonomous greenhouse mobile robot driving strategies from system integration perspective: Review and application. *IEEE/ASME Transactions On Mechatronics, 20*(4), 1705–1716.

Lin, H.-C., Bruning, P. F., & Swarna, H. (2018). Using online opinion leaders to promote the hedonic and utilitarian value of products and services. *Business Horizons, 61*(3), 431–442. https://doi.org/10.1016/j.bushor.2018.01.010

Lion, R., Willems, A., Fischer, A. R. H., & van Trijp, H. (2020). *Exploring consumers perceptions of additives and flavour enhancers in Germany, China and Philippines*.

Liu, R. F., Gao, Z. F., Nayga, R. M., Snell, H. A., & Ma, H. Y. (2019). Consumers' valuation for food traceability in China: Does trust matter? *Food Policy, 88*. https://doi.org/10.1016/j.foodpol.2019.101768

Liu, C., Li, J. Y., Steele, W., & Fang, X. M. (2018). A study on Chinese consumer preferences for food traceability information using best-worst scaling. *PLoS One, 13*(11). https://doi.org/10.1371/journal.pone.0206793

Liu, Z., Zhang, M., Bhandari, B., & Wang, Y. (2017). 3D printing: Printing precision and application in food sector. *Trends in Food Science & Technology, 69*, 83–94. https://doi.org/10.1016/j.tifs.2017.08.018

Livingstone, K. M., Celis-Morales, C., Navas-Carretero, S., San-Cristobal, R., Macready, A. L., Fallaize, R., … Food4Me, S. (2016). Effect of an internet-based, personalized nutrition randomized trial on dietary changes associated with the mediterranean diet: The Food4Me study. *American Journal of Clinical Nutrition, 104*(2), 288–297. https://doi.org/10.3945/ajcn.115.129049

Martins, N., Morales, P., Barros, L., & Ferreira, I. C. (2017). *The increasing demand for functional foods. Wild plants, mushrooms and nuts: Functional food properties and applications.*

Mazzacano D'Amato, P., & Falzon, J. (2015). Why do some consumers prefer organic food? A discourse Analytical perspective. *Journal of Food Products Marketing, 21*(3), 255–273. https://doi.org/10.1080/10454446.2013.843487

Mckay, G. A. (2008). *Consumption, coca-colonisation, cultural resistance–and Santa Claus.*

Menozzi, D., Sogari, G., Veneziani, M., Simoni, E., & Mora, C. (2017). Eating novel foods: An application of the theory of planned behaviour to predict the consumption of an insect-based product. *Food Quality and Preference, 59*, 27–34. https://doi.org/10.1016/j.foodqual.2017.02.001

Meulenberg, M. T. G., & Viaene, J. (2005). Changing agri-food systems in western countries: A marketing approach. In W. M. F. Jongen, & M. T. G. Meulenberg (Eds.), *Innovation in agri-food systems* (pp. 17–51). Wageningen: Wageningen Academic Publishers.

Michel, M., & Burbidge, A. (2019). Nutrition in the digital age - how digital tools can help to solve the personalized nutrition conundrum. *Trends in Food Science & Technology, 90*, 194–200. https://doi.org/10.1016/j.tifs.2019.02.018

Mogili, U. R., & Deepak, B. (2018). Review on application of drone systems in precision agriculture. *Procedia computer science, 133*, 502–509.

Moons, I., Barbarossa, C., & De Pelsmacker, P. (2018). The determinants of the adoption intention of eco-friendly functional food in different market segments. *Ecological Economics, 151*, 151–161. https://doi.org/10.1016/j.ecolecon.2018.05.012

Moore, M., & Carpenter, J. M. (2008). Intergenerational perceptions of market cues among US apparel consumers. *Journal of Fashion Marketing and Management, 12*(3), 323–337. https://doi.org/10.1108/13612020810889281

Mortimer, G., Fazal e Hasan, S., Andrews, L., & Martin, J. (2016). Online grocery shopping: The impact of shopping frequency on perceived risk. *International Review of Retail Distribution & Consumer Research, 26*(2), 202–223. https://doi.org/10.1080/09593969.2015.1130737

Nash, K. S. (2018). Walmart requires lettuce, spinach suppliers to join blockchain. *The Wall Street Journal*. https://blogs.wsj.com/cio/2018/09/24/walmart-requires-lettuce-spinach-suppliers-to-join-blockchain/.

Nguyen, M. H., van Weert, J. C. M., Bol, N., Loos, E. F., Tytgat, K. M. A. J., van de Ven, A. W. H., et al. (2017). Tailoring the mode of information presentation: Effects on younger and older adults' attention and recall of online information. *Human Communication Research, 43*(1), 102–126. https://doi.org/10.1111/hcre.12097

Oltvoort, A., de Vries, P., van Rompay, T., & Rosen, D. (2019). "I Am the eye in the sky—can you read my mind?" How to address public concerns towards drone use. In *Paper presented at the international conference on persuasive technology.*

Opara, L. U. (2003). Traceability in agriculture and food supply chain: A review of basic concepts, technological implications, and future prospects. *Journal of Food Agriculture and Environment, 1*, 101–106.

Parker, K., Graf, N., & Igielnik, R. (2019). Generation Z looks a lot like millennials on key social and political issues. *Pew Research Center, 17.*

Pliner, P., & Hobden, K. (1992). Development of a scale to measure the trait of food neophobia in humans. *Appetite, 19*(2), 105–120.

Post, M. J. (2012). Cultured meat from stem cells: Challenges and prospects. *Meat Science, 92*(3), 297–301.

Purdy, C. (2019). *The first cell-cultured meat will cost about $50.* Quartz. Retrieved from https://qz.com/1598076/the-first-cell-cultured-meat-will-cost-about-50/.

Rahman, M. M., Ishii, K., & Noguchi, N. (2019). Optimum harvesting area of convex and concave polygon field for path planning of robot combine harvester. *Intelligent service robotics, 12*(2), 167–179.

Rao, H. (2009). *Market rebels and radical innovation.* McKinsey Quarterly.

Ray, A., Dhir, A., Bala, P. K., & Kaur, P. (2019). Why do people use food delivery apps (FDA)? A uses and gratification theory perspective. *Journal of Retailing and Consumer Services, 51*, 221–230. https://doi.org/10.1016/j.jretconser.2019.05.025

Reinders, M. J., Bouwman, E. P., van den Puttelaar, J., & Verain, M. C. D. (2020). Consumer acceptance of personalised nutrition: The role of ambivalent feelings and eating context. *PLoS One, 15*(4), e0231342. https://doi.org/10.1371/journal.pone.0231342

Rodriguez-Salvador, B., & Dopico, D. C. (2020). Understanding the value of traceability of fishery products from a consumer perspective. *Food Control, 112*, 107142. https://doi.org/10.1016/j.foodcont.2020.107142

Rogers, E. M. (1962/1995). *Diffusion of innovations.* New York: Free Press.

Roh, M., & Park, K. (2019). Adoption of O2O food delivery services in South Korea: The moderating role of moral obligation in meal preparation. *International Journal of Information Management, 47*, 262–273. https://doi.org/10.1016/j.ijinfomgt.2018.09.017

Ronteltap, A., Fischer, A. R. H., & Tobi, H. (2011). Societal response to nanotechnology: Converging technologies-converging societal response research? *Journal of Nanoparticle Research, 13*(10), 4399–4410. https://doi.org/10.1007/s11051-011-0473-1

Ronteltap, A., Van Trijp, H., Berezowska, A., & Goossens, J. (2013). Nutrigenomics-based personalised nutritional advice: In search of a business model? *Genes and Nutrition, 8*(2), 153–163.

Ronteltap, A., van Trijp, J. C. M., Renes, R. J., & Frewer, L. J. (2007). Consumer acceptance of technology-based food innovations: Lessons for the future of nutrigenomics. *Appetite, 49*(1), 1–17.

Rothensee, M. (2008). User acceptance of the intelligent fridge: Empirical results from a simulation. In *Vol. 4952 LNCS. Lecture notes in computer science (including subseries lecture notes in artificial intelligence and lecture notes in bioinformatics).*

Rozin, P. (2005). The meaning of "natural": Process more important than content. *Psychological Science, 16*(8), 652–658.

Saba, A., & Vassallo, M. (2002). Consumer attitudes toward the use of gene technology in tomato production. *Food Quality and Preference, 13*(1), 13–21. https://doi.org/10.1016/s0950-3293(01)00052-0

Sackmann, R., & Winkler, O. (2013). Technology generations revisited: The internet generation. *Gerontechnology, 11*(4), 493–503. https://doi.org/10.4017/gt.2013.11.4.002.00

Sale, K. (1996). *Rebels against the future: The luddites and their war on the Industrial revolution, lessons for the computer age*. New York: Perseus.

Sander, F., Semeijn, J., & Mahr, D. (2018). The acceptance of blockchain technology in meat traceability and transparency. *British Food Journal, 120*(9), 2066–2079. https://doi.org/10.1108/BFJ-07-2017-0365

Santos, T. D., Freitas, B. C. B.d., Moreira, J. B., Zanfonato, K., & Costa, J. A. V. (2016). Development of powdered food with the addition of Spirulina for food supplementation of the elderly population. *Innovative Food Science & Emerging Technologies, 37*, 216–220. https://doi.org/10.1016/j.ifset.2016.07.016

Schenk, M. F., van der Maas, R., Smulders, M. J. M., Gilissen, L. J. W. J., Fischer, A. R. H., Van der Lans, I. A., et al. (2011). Hypoallergenic food products as a novel approach to alleviate mild food allergy. *Food Quality and Preference, 22*(1), 83–91.

Scherer, R., Siddiq, F., & Tondeur, J. (2019). The technology acceptance model (TAM): A meta-analytic structural equation modeling approach to explaining teachers' adoption of digital technology in education. *Computers & Education, 128*, 13–35. https://doi.org/10.1016/j.compedu.2018.09.009

Schickenberg, B., Van Assema, P., Brug, J., & De Vries, N. K. (2008). Are the Dutch acquainted with and willing to try healthful food products? The role of food neophobia. *Public Health Nutrition, 11*(5), 493–500.

Schwartz, S. H. (1994). Are there universal aspects in the structure and contents of human values? *Journal of Social Issues, 50*(4), 19–45.

Ścieszka, S., & Klewicka, E. (2019). Algae in food: A general review. *Critical Reviews in Food Science and Nutrition, 59*(21), 3538–3547. https://doi.org/10.1080/10408398.2018.1496319

Seehafer, A., & Bartels, M. (2019). Meat 2.0 – the regulatory environment of plant-based and cultured meat. *European Food and Feed Law Review, 14*(4), 323–331.

Seitz, C., Pokrivčák, J., Tóth, M., & Plevný, M. (2017). Online grocery retailing in Germany: An explorative analysis. *Journal of Business Economics and Management, 18*(6), 1243–1263. https://doi.org/10.3846/16111699.2017.1410218

Siegrist, M. (2000). The influence of trust and perceptions of risks and benefits on the acceptance of gene technology. *Risk Analysis, 20*(2), 195–204. https://doi.org/10.1111/0272-4332.202020

Siegrist, M. (2008). Factors influencing public acceptance of innovative food technologies and products. *Trends in Food Science & Technology, 19*(11), 603–608.

Sparke, K., & Menrad, K. (2009). Cross-European and functional food-related consumer segmentation for new product development. *Journal of Food Products Marketing, 15*(3), 213–230. https://doi.org/10.1080/10454440902908142

Steenkamp, J.-B. E., Ter Hofstede, F., & Wedel, M. (1999). A cross-national investigation into the individual and national cultural antecedents of consumer innovativeness. *Journal of Marketing, 63*(2), 55–69.

Stephens, J., Miller, H., & Militello, L. (2020). Food delivery apps and the negative health impacts for Americans. *Frontiers in Nutrition, 7*. https://doi.org/10.3389/fnut.2020.00014

Suganuma, N. (2019). Trends in the development of autonomous vehicles and challenges for deployment in society. *IATSS Research, 43*(4), 242–243.

Taebi, B. (2017). Bridging the gap between social acceptance and ethical acceptability. *Risk Analysis, 37*(10), 1817–1827. https://doi.org/10.1111/risa.12734

Tan, H. S. G., Fischer, A. R. H., Tinchan, P., Stieger, M., Steenbekkers, L. P. A., & van Trijp, H. C. M. (2015). Insects as food: Exploring cultural exposure and individual experience as determinants of acceptance. *Food Quality and Preference, 42*, 78–89. https://doi.org/10.1016/j.foodqual.2015.01.013

Tan, H. S. G., Fischer, A. R. H., van Trijp, H. C. M., & Stieger, M. (2016). Tasty but nasty? Exploring the role of sensory-liking and food appropriateness in the willingness to eat unusual novel foods like insects. *Food Quality and Preference, 48*, 293–302. https://doi.org/10.1016/j.foodqual.2015.11.001. Part A.

Tziva, M., Negro, S. O., Kalfagianni, A., & Hekkert, M. P. (2019). Understanding the protein transition: The rise of plant-based meat substitutes. *Environmental Innovation and Societal Transitions*. https://doi.org/10.1016/j.eist.2019.09.004

Van Dam, Y. K., & Fischer, A. R. H. (2015). Buying Green without being seen. *Environment and Behavior, 47*(3), 328–356. https://doi.org/10.1177/0013916513509481

Van Huis, A., Van Itterbeeck, J., Klunder, H., Mertens, E., Halloran, A., Muir, G., et al. (2013). *Edible insects: Future prospects for food and feed security* (Vol. 171). Rome: Food and Agriculture Organization of the United Nations.

Van Loo, Caputo, & Lusk. (2020). Consumer preferences for farm-raised meat, lab-grown meat, and plant-based meat alternatives: Does information or brand matter? *Food Policy, Article 101931. https://doi.org/10.1016/j.foodpol.2020.101931. In press.*

Van Trijp, H. C. M. (1994). Product-related determinants of variety-seeking behaviour for foods. *Appetite, 22*(1), 1–10.

Veletsianos, G., Johnson, N., & Belikov, O. (2019). Academics' social media use over time is associated with individual, relational, cultural and political factors. *British Journal of Educational Technology, 50*(4), 1713–1728.

Venkatesh, V., & Bala, H. (2008). Technology acceptance model 3 and a research agenda on interventions. *Decision Sciences, 39*(2), 273–315.

Venkatesh, V., & Davis, F. D. (1996). A model of the antecedents of perceived ease of use: Development and test. *Decision Sciences, 27*(3), 451–477.

Venkatesh, V., & Davis, F. D. (2000). Theoretical extension of the technology acceptance model: Four longitudinal field studies. *Management Science, 46*(2), 186–204.

Venkatesh, V., Morris, M. G., Davis, G. B., & Davis, F. D. (2003). User acceptance of information technology: Toward a unified view. *MIS Quarterly: Management Information Systems, 27*(3), 425–478.

Venkatesh, V., Thong, J. Y. L., & Xu, X. (2012). Consumer acceptance and use of information technology: Extending the unified theory of acceptance and use of technology. *MIS Quarterly: Management Information Systems, 36*(1), 157–178.

Verbeke, W. (2015). Profiling consumers who are ready to adopt insects as a meat substitute in a Western society. *Food Quality and Preference, 39*, 147–155. https://doi.org/10.1016/j.foodqual.2014.07.008

Verbeke, W., Sans, P., & Van Loo, E. J. (2015). Challenges and prospects for consumer acceptance of cultured meat. *Journal of Integrative Agriculture, 14*(2), 285–294. https://doi.org/10.1016/S2095-3119(14)60884-4

Verhaal, J. C., Khessina, O. M., & Dobrev, S. D. (2015). Oppositional product names, organizational identities, and product appeal. *Organization Science, 26*(5), 1466–1484.

Vermeir, I., & Verbeke, W. (2006). Sustainable food consumption: Exploring the consumer "attitude - behavioral intention" gap. *Journal of Agricultural and Environmental Ethics, 19*(2), 169–194.

Verplanken, B., & Orbell, S. (2003). Reflections on past behavior: A self-report index of habit strength. *Journal of Applied Social Psychology, 33*(6), 1313–1330.

Watson, E. (2019). *Plant-based and cell-cultured 'meat' labeling under attack in 25 states*. Food Navigator. Retrieved from https://www.foodnavigator-usa.com/Article/2019/05/29/Plant-based-and-cell-cultured-meat-labeling-under-attack-in-25-states.

van der Weele, C., & Tramper, J. (2014). Cultured meat: Every village its own factory? *Trends in Biotechnology, 32*(6), 294–296.

WHO. (2015). *Food-based dietary guidelines*. The Netherlands. Retrieved from http://www.fao.org/nutrition/education/food-dietary-guidelines/regions/countries/Netherlands/en.

WHO. (2017). *The burden of foodborne diseases in the WHO European Region*. Retrieved from http://www.euro.who.int/en/health-topics/disease-prevention/food-safety/publications/2017/the-burden-of-foodborne-diseases-in-the-who-european-region-2017.

WHO. (2018). *Noncommunicable diseases - key facts*. Retrieved from https://www.who.int/news-room/fact-sheets/detail/noncommunicable-diseases.

Wilks, M., & Phillips, C. J. C. (2017). Attitudes to in vitro meat: A survey of potential consumers in the United States. *PLoS One, 12*(2), e0171904. https://doi.org/10.1371/journal.pone.0171904

Williams, L. A. (2019). *Top ten trends for 2020* [Webinar]. Retrieved from https://www.foodingredientsfirst.com/Webinars/top-ten-trends-2020.html.

Xiong, H., Dalhaus, T., Wang, P., & Huang, J. (2020). Blockchain technology for agriculture: Applications and rationale. *Frontiers in Blockchain, 3*(7). https://doi.org/10.3389/fbloc.2020.00007

Yamoah, F. A., & Acquaye, A. (2019). Unravelling the attitude-behaviour gap paradox for sustainable food consumption: Insight from the UK apple market. *Journal of Cleaner Production, 217*, 172–184.

Zhang, C., Bai, J., & Wahl, T. I. (2012). Consumers' willingness to pay for traceable pork, milk, and cooking oil in Nanjing, China. *Food Control, 27*(1), 21–28. https://doi.org/10.1016/j.foodcont.2012.03.001

Zhang, G., Zhao, X., Li, X., Du, G., Zhou, J., & Chen, J. (2020). Challenges and possibilities for bio-manufacturing cultured meat. *Trends in Food Science & Technology, 97*, 443–450. https://doi.org/10.1016/j.tifs.2020.01.026

Index

Note: 'Page numbers followed by "*f*" indicate figures and "*t*" indicate tables.'

T

U

V

W

Kolibri 360 GmbH
Ferdinand-Jühlke-Straße 7
99095 Erfurt, Deutschland
produktsicherheit@kolibri360.de